MATHEMATICAL & COMPUTATIONAL BIOLOGY

MATHEMATICAL & COMPUTATIONAL BIOLOGY

Editor

Clímaco Cuevas

scitus academics

Mathematical & Computational Biology

Edited by **Clímaco Cuevas**

This Book is Printed on Acid-Free Paper. ♾

ISBN: 9781681179629
Library of Congress Control Number: 2017949101

616, Corporate Way, Suite 2, 4766,
Valley Cottage, NY 10989

www.scitusacademics.com

Notice

Printed in United States of America.

Preface

Mathematical and Computational Biology, an emerging research area in mathematics, is under rapid growth at worldwide. Mathematics plays a key role in many disciplines of science, primarily as a mathematical modeling tool, including biology. Unlike physics and chemistry, biology is not usually a science associated with mathematics. But because there are quantifiable aspects of life science, mathematics plays a critical role in better understanding the natural world. Basic measurements of living things are often a necessary part of understanding how they change over time is crucial to better understanding living things. Even though physics and biology are very different sciences, mathematicians and biologists developed the "mathematical biology" or "biomathematics" as a recent discipline for the mathematical representation in biology to the theoretical and practical applications in biological, biomedical and biotechnology researches. Similarly, computational biology can play a special role in studying emerging infectious diseases, where a fast characterization of the diseases is often urgently needed before they are widespread.

This volume covers mathematical approaches to gain biological understanding or explain biological phenomena provides biological insight as a result of mathematical analysis or identify and open up challenging new types of mathematical problems that derive from biological knowledge (in the form of data, or theory, or simulation results). By making connections through the application of computational methods among disparate areas of biology, the book also provides substantial new insight into living systems at all scales, from the nano to the macro, and across multiple disciplines, from molecular science, neuroscience and physiology to ecology and population biology. Features works of exceptional significance that further our understanding of living systems at all scales—from molecules and cells, to patient populations and ecosystems—through the application of computational methods. This will be of valuable tool for students, including life and computational scientists, who can take the important findings presented here to the next level of discovery.

Table of Contents

CHAPTER 1

An Effective Numerical Calculation Method for Multi-Time-Scale ematical Models in Systems Biology

Yohei Motomura[1], Hiroyuki Hamada[1,2], Masahiro Okamoto[1,2]

[1]*Graduate School of Systems Life Sciences, Kyushu University, Fukuoka, Japan*

[2]*Synthetic Systems Biology Research Center, Kyushu University, Fukuoka, Japan*

ABSTRACT

The improvements of high-throughput experimental devices such as microarray and mass spectrometry have allowed an effective acquisition of biological comprehensive data which include genome, transcriptome, proteome, and metabolome (multi- layered omics data). In Systems Biology, we try to elucidate various dynamical characteristics of biological functions with applying the omics data to detailed ematical model based on the central dogma. However, such ematical models possess multi-time-scale properties which are often accompanied by time-scale differences seen among biological layers. The differences cause time stiff problem, and have a grave influence on numerical calculation stability. In the present conventional method, the time stiff problem remained because the calculation of all layers was implemented by adaptive time step sizes of the smallest time-scale layer to ensure stability and maintain calculation accuracy. In this paper, we designed and developed an effective numerical calculation method to improve the time stiff problem. This method consisted of ahead, backward, and cumulative algorithms. Both ahead and cumulative algorithms enhanced calculation efficiency of numerical calculations via adjustments of step sizes of each layer, and reduced the number of numerical calculations required for multi-time-scale models with the time stiff problem. Backward algorithm ensured calculation accuracy in the multi-time-scale models. In case studies which were focused on three layers system with 60 times difference in time-scale order in between layers, a proposed method had almost the same calculation accuracy compared with the conventional method in spite of a reduction of the total amount of the number of numerical calculations. Accordingly, the proposed method is useful in a numerical analysis of multi-time-scale models with time stiff problem.

Keywords: Finite Difference Method, Stiff Equation, Multi-Time-Scale, Systems Biology, ematical Analysis

1. INTRODUCTION

Recent improvements in high-throughput biotechnologies such as microarray [1] and mass spectrometry [2] have led to various omics data showing gene expression, protein synthesis, metabolome flux, and cell-cell interactions [3] [4] [5] . The ensuing accumulation of omics data has contributed significantly to ematical models that indicate dynamic characteristics of biological systems, including interactions between genes, proteins, cells, and tissues [6] [7] (Figure 1). Systems biology approaches such as ematical modeling of multiple layers have revealed complex relationships among biological phenomena of varying spatiotemporal scales, and have elucidated mechanisms with high order functions in biological systems [8] [9] [10] [11] [12] . In particular, multi-time-scale models have been applied to analyses of intracellular signal transduction systems such as cell cycle control, cell fate determination, and immune system mechanisms [13] [14] [15] [16] . Moreover, ematical analyses of varying (layer) gene (seconds), metabolism (minutes), and cell (hours) transition rates in biological systems define differences between biological systems and offer important discoveries of disease mechanisms. However, efficient techniques for numerical calculations remain elusive in practical applications of multi-time-scale ematical models.

Multi-time-scale models comprise multiple layers that differ in rates of state change. During conventional numerical calculations of multi-time-scale models, time step sizes that are suitable for the smallest time-scale layer have been adopted for all layers to ensure stability and maintain calculation accuracy. Thus, dynamic behaviors of entire layers are numerically analyzed using excessively reduced time step sizes, leading to

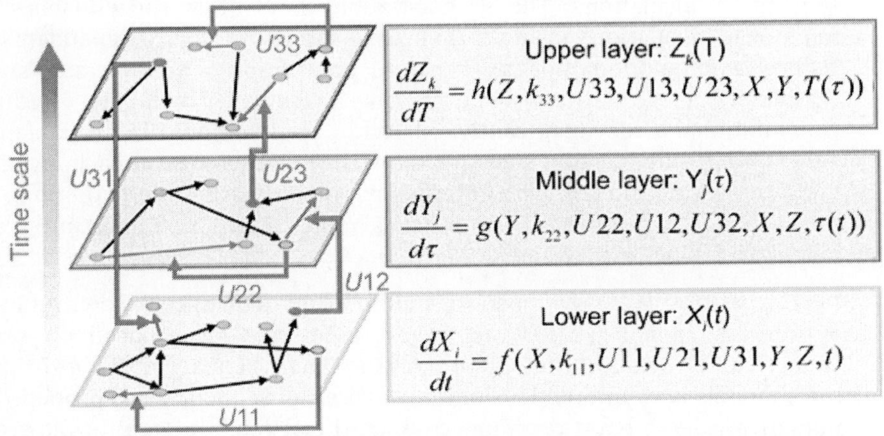

Figure 1. Overview of the multi-time-scale model; This model has three layers and has reactions across layers. The parameters X,Y,Z indicate concentrations; U_{xy} indicates the control variables from layer x to layer y; k_{nn} indicates the matrix of rate constants in layer n.

significant increases in computational demands (time stiff problem) [17] [18] [19] . Numerous implicit methods such as the Radau method [19] and Gear method [20] have been proposed as candidate solutions to the time stiff problem. These methods generate numerical solutions based on calculation sensitivities and stabilities of components in 1 layer. Furthermore, the numerical solutions of these methods are calculated using non-linear simultaneous equations with n unknowns based on n components in the model. In calculating multi-time-scale model using the implicit method, larger time step sizes than those for the smallest time-scale layer can be applied to numerical calculations of all layers because the calculation stability of the implicit method is very high. However, because multi-time-scale models comprise large numbers of components, non-linear simultaneous equations that are calculated using implicit methods become very large. Specifically, although implicit methods suppress increases in computational loads due to excessive reductions in adaptive step sizes, significant increases in volumes of numerical calculations for non-linear simultaneous equations cause failure to eliminate the time stiff problem. Parallel computing with reduced computational cost has been applied to numerical calculations of multi-time-scale models [21] [22] [23] . In contrast, contributions of parallel computing have been limited because analyses of dynamic behaviors of biological systems include numerous sequential calculations. These observations imply that the efficiency of numerical calculations in multi-time-scale models is highly dependent on reduced computational loads. Therefore, application of suitable step sizes to numerical calculations for each time scale layer will likely reduce computational loads significantly. Currently, few methods are available for determining suitable step sizes for numerical calculations of each layer in multi-time-scale models with interactions among layers, and solutions to this problem are essential for practical applications of multi-time-scale models to biological systems.

In this study, we developed a method for dynamically determining appropriate step sizes for the largest time-scale layer based on state changes of the smallest time-scale layer in numerical analysis of multi-time-scale models with interactions among layers. Subsequently, we proposed a numerical method for reducing computational loads of multi-time-scale models (proposed method) and verified the effectiveness of the proposed method using the follow steps:

1) Construction of multi-time-scale model (benchmark model) with interactions among layers that are universally observed in biological systems;

2) Numerical calculation of benchmark models using the conventional method (Control);

3) Numerical calculation of a benchmark model using the proposed method;

4) Comparison of computational loads for proposed and conventional methods;

5) Comparison of numerical solutions for proposed and conventional methods;

6) Discussion of the validity of the proposed method.

Using these procedures, we demonstrated the utility of the proposed method for improving computational efficiency without increasing computational costs of multi- time-scale models with interactions among layers. By reducing computational loads, the proposed method enhances the feasibility of ematical analyses and accommodates greater scales of ematical models, representing a significant contribution to systems biology methods.

2. MATERIAL AND METHODS

2.1. Benchmark Models with Multi-Time-Scales

To design and develop a method that is suitable for multi-time-scale models, we constructed 2 benchmark models (model A and model B) with the time stiff problem (Figure 2) and evaluated the calculation performance of the proposed method. The time stiff problem occurred due to differences in time-scales of each layer by interactions among layers. Thus, these benchmark models satisfied the following conditions: 1) Models included interactions across layers; 2) Models had different time-scales of each layer. Models A and B comprised lower, middle, and upper layers with time scales of seconds, minutes, and hours, respectively. Model A contained inhibition effects such as suppressed expression of anabolic enzymes by metabolic products [24] and negative control of gene expression by the lac repressor protein [25] , and these inhibition effects from upper to lower layers induced the time stiff problem with differences in time- scales of each layer caused by the largest time-scale layer (Figure 2(a)). Model B contained activation effects such as the transcriptional control by RNA polymerase [26] and the control of metabolic flux by enzymes [27] , and these activation effects from lower to upper layers induced the time stiff problem with differences in time-scales of each layer caused by the smallest time-scale layer (Figure 2(b)). Furthermore, these effects of activation and inhibition were expressed using the Hill equation [28] , which empirically explains cooperative effects of oxygen binding to hemoglobin. Equations (1)-(9) show mass balance equations of model A as follows:

$$\frac{\mathrm{d}[X_0]}{\mathrm{d}t} = 0.0 \tag{1}$$

$$\frac{\mathrm{d}[X_1]}{\mathrm{d}t} = k_0 \cdot [X_0] \cdot \frac{B_x}{\left(1 + \left(\frac{Y_5}{K_x}\right)^{n_x}\right)} - k_1 \cdot [X_1] \tag{2}$$

$$\frac{\mathrm{d}[X_i]}{\mathrm{d}t} = k_{i-1} \cdot [X_{i-1}] - k_i \cdot [X_i], (i = 2,3,4,5) \tag{3}$$

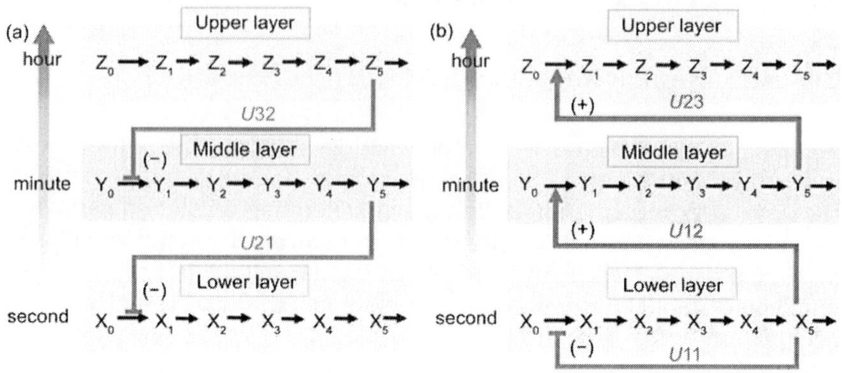

Figure 2. Case studies of multi-time-scale models; We verified the utility of the proposed method using two case studies. Both models have three layers with 60 time differences in time-scale order and were constructed using the Hill equation.

$$\frac{d[Y_0]}{dt}=0.0 \tag{4}$$

$$\frac{d[Y_1]}{dt}=l_0\cdot[Y_0]\cdot\frac{B_y}{\left(1+\left(\frac{[Z_5]}{K_y}\right)^{n_y}\right)}-l_1\cdot[Y_1] \tag{5}$$

$$\frac{d[Y_j]}{dt}=l_{j-1}\cdot[Y_{j-1}]-l_j\cdot[Y_j],\ (j=2,3,4,5) \tag{6}$$

$$\frac{d[Z_0]}{dt}=0.0 \tag{7}$$

$$\frac{d[Z_1]}{dt}=m_0\cdot[Z_0]\cdot B_z-m_1\cdot[Z_1] \tag{8}$$

$$\frac{d[Z_k]}{dt}=m_{k-1}\cdot[Z_{k-1}]-m_k\cdot[Z_k],(k=2,3,4,5) \tag{9}$$

Here, Equations (1)-(3), (4)-(6), and (7)-(9) show magnitudes of change in lower, middle, and upper layers of model A, respectively. Table 1 shows kinetic parameters of Equations (1)-(9), and Equations (10)-(18) show mass balance equations of model B as follows:

Table 1.Constants and parameters in model A.

Parameter	Value	Description
$[X_0],[Y_0],[Z_0]$	5.0	Constant initial concentration (mM)
$[X_i](i=1,2,\cdots,5)$	0.0	Initial concentration (mM)
$[Y_j](j=1,2,\cdots,5)$	0.0	Initial concentration (mM)
$[Z_k](k=1,2,\cdots,5)$	0.0	Initial concentration (mM)
$k_i\ (i=0,1,\cdots,5)$	1.0	Static rate constant (second scale)
$l_j\ (j=0,1,\cdots,5)$	1.0/60	Static rate constant (minute scale)
$m_k\ (k=0,1,\cdots,5)$	1.0/3600	Static rate constant (hour scale)
K_x,K_y	2.5	Repression coefficient
B_x,B_y,B_z	5.0	Maximal expression
n_x,n_y	4.0	Hill coefficient

$$\frac{\mathrm{d}[X_0]}{\mathrm{d}t}=0.0 \tag{10}$$

$$\frac{\mathrm{d}[X_1]}{\mathrm{d}t}=k_0\cdot[X_0]\cdot\frac{B_x}{\left(1+\left(\frac{[X_5]}{K_x}\right)^{n_x}\right)}-k_1\cdot[X_1] \tag{11}$$

$$\frac{\mathrm{d}[X_i]}{\mathrm{d}t}=k_{i-1}\cdot[X_{i-1}]-k_i\cdot[X_i],(i=2,3,4,5) \tag{12}$$

$$\frac{\mathrm{d}[Y_0]}{\mathrm{d}t}=0.0 \tag{13}$$

$$\frac{\mathrm{d}[Y_1]}{\mathrm{d}t}=l_0\cdot[Y_0]\cdot\frac{B_y\cdot[X_5]^{n_y}}{\left(K_y^{n_y}+[X_5]^{n_y}\right)}-l_1\cdot[Y_1] \tag{14}$$

$$\frac{\mathrm{d}[Y_j]}{\mathrm{d}t} = l_{j-1} \cdot [Y_{j-1}] - l_j \cdot [Y_j], (j = 2,3,4,5) \tag{15}$$

$$\frac{\mathrm{d}[Z_0]}{\mathrm{d}t} = 0.0 \tag{16}$$

$$\frac{\mathrm{d}[Z_1]}{\mathrm{d}t} = m_0 \cdot [Z_0] \cdot \frac{B_z \cdot [Y_5]^{n_z}}{\left(K_z^{n_z} + [Y_5]^{n_z}\right)} - m_1 \cdot [Z_1] \tag{17}$$

$$\frac{\mathrm{d}[Z_k]}{\mathrm{d}t} = m_{k-1} \cdot [Z_{k-1}] - m_k \cdot [Z_k] \ (k = 2,3,4,5) \tag{18}$$

Here, Equations (10)-(12), (13)-(15), and (16)-(18) show magnitudes of change in lower, middle, and upper layers of model B. Table 2 shows kinetic parameters of Equations (10)-(18). Normally, the time stiff problem occurs due to differences in time- scales of each component by dynamically changing reaction rates (k_i, l_j, m_k) in model A and B. In this paper, reaction rates were constant in entire time, to focus on the time difference between layers.

2.2. Numerical Solutions of the Benchmark Model Based on the Conventional Method

We obtained numerical solutions of benchmark models (Model A, Model B) shown in Figure 2 using the explicit 4 stage 4th-order Runge-Kutta method [29] as the conventional method. The simulation time was set at 36000 s because the dynamic behavior of the upper layer reached steady state at this time. The adaptive time step size $\mathrm{d}t$ of the conventional method was set to 1.00E−03 s. Figures 3(a)-(c) and Figures 3(d)-(f) show time-dependent changes of component concentrations in models A and B, respectively. Moreover, Figure 3(a) and Figure 3(d), Figure 3(b) and Figure 3(e), and Figures 3(c) and Figure 3(f) show time-dependent changes in component concentrations for simulation times of 0 - 36,000 s (0 - 10 h), 0 - 600 s (0 - 10 min), and 0 - 10 s, respectively. Thus, dynamic behaviors of dependent variables of lower, middle, and upper layers were observed in time-scales of seconds, minutes, and hours, respectively. To verify calculation performance of the proposed method, we defined the results of these calculations as controls.

Table 2. Constants and parameters in model B.

Parameter	Value	Description
$[X_0],[Y_0],[Z_0]$	5.0	Constant initial concentration (mM)
$[X_i](i=1,2,\cdots,5)$	0.0	Initial concentration (mM)
$[Y_j](j=1,2,\cdots,5)$	0.0	Initial concentration (mM)
$[Z_k](k=1,2,\cdots,5)$	0.0	Initial concentration (mM)
$k_i(i=0,1,\cdots,5)$	1.0	Static rate constant (second scale)
$l_j(j=0,1,\cdots,5)$	1.0/60	Static rate constant (minute scale)
$m_k(k=0,1,\cdots,5)$	1.0/3600	Static rate constant (hour scale)
K_x	0.1	Repression coefficient
K_y	0.5	Activation coefficient
K_z	1.5	Activation coefficient
B_x,B_y,B_z	5.0	Maximal expression
n_x,n_y	4.0	Hill coefficient

2.3. Design and Development of Proposed Method

In numerical calculations of the conventional method in multi-time-scale models, the adaptive time step size of the smallest time-scale layer (lower layer) was adopted in calculations for all layers. Hence, numbers of calculation steps of middle and upper layers were equal to that of the lower layer in the conventional method. Here we defined the process of calculating the concentration $S(t+\Delta t)$ of component S at the time $t+\Delta t$ from the concentration $S(t)$ of component S at time t as 1 unit of the calculation (1 step), and Δt was the adaptive time step size. Adaptive time step sizes of middle and upper layers in the conventional method were generally much smaller than suitable for these layers. Use of optimal step sizes for middle and upper layers in numerical calculations reduced the numbers of calculation steps for middle and upper layers. Therefore, we developed effective numerical calculation algorithms (proposed method) that optimally controlled adaptive time step sizes for each layer based on variations of differential component values.

The proposed method comprised ahead, backward, and cumulative algorithms. Ahead and cumulative algorithms reduced the numbers of calculation steps, whereas the backward algorithm ensured calculation accuracy. Initially, the ahead algorithm adopted the arbitrary conventional method for the calculation of the smallest time-scale layer (lower layer) and implemented the iterative numerical integration in the interval

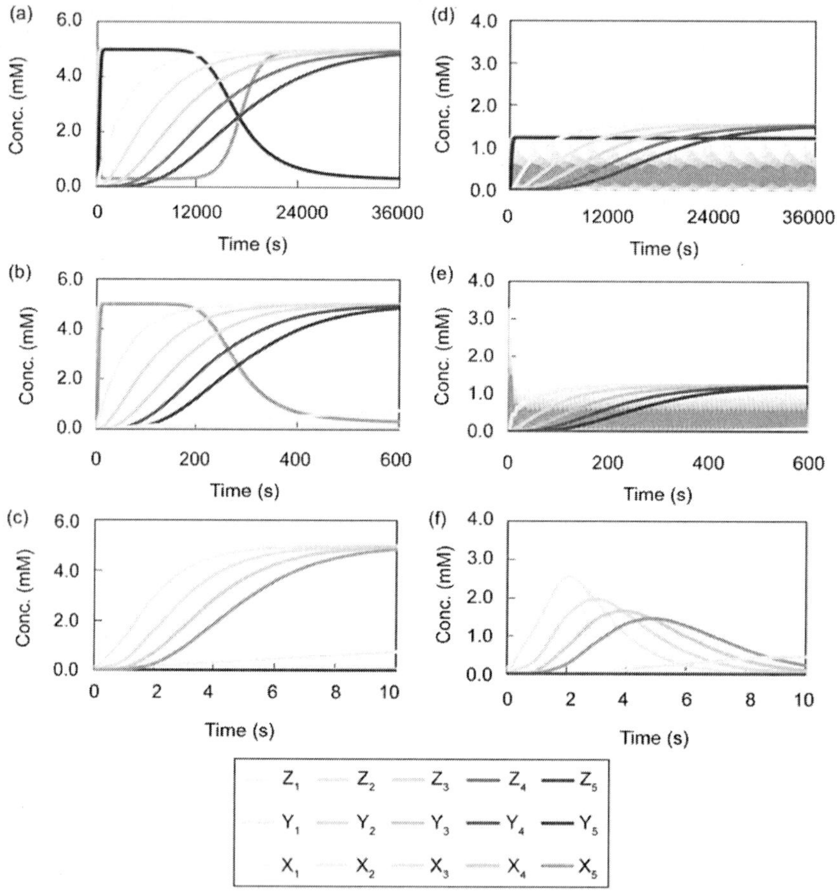

Figure 3. Time course of numerical solutions by the conventional method; The calculation step size dt of the conventional method was set to 1.00E−03 s. (a), (b), and (c) show results of the models A, (d), and (e), and (f) shows the result of model B. Simulation times of (a) and (d) were set to 36,000 s (10 h), those of (b) and (e) were set to 600 s (10 min), and those of (c) and (f) were set to 10 s.

of the arbitrary number of the N_x step, which was set to a predetermined calculation interval. Secondly, the backward algorithm was used to define a predetermined calculation interval in which the differential value was less than a certain value according to the determined calculation interval. Here, the adaptive time step size of the middle layer and the representative value for the lower layer were set to determine the calculation interval and its average integral value of the numerical solution for the lower layer. Subsequently, the cumulative algorithm was used to calculate the dynamics of the 1st step of the middle layer using these values. The proposed method repeated these 3 algorithms until the numerical solution of the upper layer was achieved. Thus, the adaptive time step sizes of middle and upper layers of the proposed method were much larger than those of the conventional method. Moreover, calculation steps of the proposed method were fewer than those of the conventional method, reflecting differing step sizes of proposed and conventional methods. In addition, adaptive time step sizes of middle and upper layers of the proposed method were controlled by the backward algorithm to maintain calculation accuracy. Accordingly, the computational volume of the proposed method was less than that of the conventional method and the calculation accuracies of proposed and conventional methods were comparable. Numerical calculations of the proposed method in the multi-time-scale models comprising the 3 layers of models A and B are described by Equations (19)-(21), which are mass balance equations and targets of the proposed method as follows:

$$\frac{dZ_k}{dt} = h_k(X,Y,Z,t) \tag{19}$$

$$\frac{dY_j}{dt} = g_j(X,Y,Z,t) \tag{20}$$

$$\frac{dX_i}{dt} = f_i(X,Y,Z,t) \tag{21}$$

Equations (22) and (23) describe initial conditions as follows:

$$(X,Y,Z,t) = (X(t=t_0), Y(t=t_0), Z(t=t_0), t_0) \tag{22}$$

$$t_0 = \tau_0 = T_0 \tag{23}$$

Here, the parameters X,Y,Z indicate concentration of components i,j,k identified in each layer t,τ,T of lower, middle and upper layer.

2.3.1. Ahead Algorithm

Initially, numbers of predetermined calculation steps (N_x, N_y, N_z) were set to 60 (Table 3). In the ahead algorithm, the arbitrary conventional method was adopted (Euler method [29] , Runge-Kutta method [29] , Runge-Kutta-Fehlberg method [30] etc.) with the numerical calculation of the smallest time-scale layer (lower layer), and the iterative numerical integration in the interval of the number of N_x steps (predetermined calculation interval) was implemented. The numerical calculation from the initial state to the N_x step was explicitly calculated based on Equations (24) and (25) as follows:

$$X(t=t_{p+1}) = X(t=t_p) + f\left(X(t=t_p), Y(t=t_0), Z(t=t_0), t_p\right)\cdot \mathrm{d}t_p \quad (p=0,1,2,\cdots,N_x-1) \tag{24}$$

$$t_{p+1} = t_p + \mathrm{d}t_p \quad (p=0,1,2,\cdots,N_x-1) \tag{25}$$

Table 3. Parameters of the proposed method.

Parameter	Value	Description
N_x	60.0	Initial number of predetermined calculation step in lower layer
N_y	60.0	Initial number of predetermined calculation step in middle layer
N_z	60.0	Initial number of predetermined calculation step in upper layer
A_x	5.0	Control variable of number of predetermined calculation step in lower layer
A_y	5.0	Control variable of number of predetermined calculation step in middle layer
A_z	5.0	Control variable of number of predetermined calculation step in upper layer

Here, $\mathrm{d}t_p$ indicates the adaptive time step size per step in the lower layer, equations for middle and upper layers were shown in Equation (36) and (46), respectively.

2.3.2. Backward Algorithm

The backward algorithm was used to explore the predetermined calculation interval in which the differential value was less than a certain value according to the determined calculation interval. Here, the number of calculation steps was set to the number of determined calculation steps G_x. Consequently, the backward algorithm could be used to monitor magnitudes of change in the numerical solution of dependent variables in the predetermined calculation interval. Subsequently, the backward algorithm narrowed the determined calculation interval for large changes in numerical solutions of the dependent variable in the predetermined calculation interval and widened that when changes were small. To measure magnitudes of change of all components in the layer, we compared the final differential value D_{x_2} of each component at the time t_{N_x} with the average integral value $\bar{f}$ of the differential value in the predetermined calculation interval using Equation (26) as follows:

$$E = \left| \bar{f} - D_{x\_2} \right| \tag{26}$$

Here, E is evaluation value of each component in the predetermined calculation interval and this evaluation value shows the magnitude of change in the differential value of each component during the predetermined calculation interval. Using Simpson's numerical integration method [29] to the coordinate points (C, D) of time and the differential value, we determined the average integral value $\bar{f}$ of the differential value in the predetermined calculation interval. Equations (27)-(29) show the coordinate points (C, D) of time and the differential value as follows:

$$\left(C_{x\_0}, D_{x\_0}\right) = \left(t_0, f\left(X\left(t = t_0\right), Y\left(t = t_0\right), Z\left(t = t_0\right), t_0\right)\right) \tag{27}$$

$$\left(C_{x\_1}, D_{x\_1}\right) = \left(t_1, f\left(X\left(t = t_1\right), Y\left(t = t_0\right), Z\left(t = t_0\right), t_1\right)\right) \tag{28}$$

$$\left(C_{x\_2}, D_{x\_2}\right) = \left(t_{N_x}, f\left(X\left(t = t_{N_x}\right), Y\left(t = t_0\right), Z\left(t = t_0\right), t_{N_x}\right)\right) \tag{29}$$

We obtained differential values D_{x_M} at the midpoint of the predetermined calculation interval $C_{x\_M} = \left(C_{x\_2} - C_{x\_0}\right)/2$ using the Lagrange interpolation [29] as shown in Equations (30) and (31) as follows:

$$D_{x\_M} = \sum_{u=0}^{2} L_u\left(C_{x\_M}\right) \cdot D_{x\_u} \tag{30}$$

$$L_u\left(C_{x\_M}\right) = \prod_{v=0}^{2(v \neq u)} \frac{C_{x\_M} - C_{x\_v}}{C_{x\_u} - C_{x\_v}} \tag{31}$$

The average integral value $\bar{f}$ was calculated using the coordinate points (C, D) and the differential value D_{x_2} (Equation (32)) as follows:

$$\begin{aligned} \bar{f} &= \frac{\int_{t=C_{x\_0}}^{t=C_{x\_2}} f(X,Y,Z,t)\mathrm{d}t}{C_{x\_2} - C_{x\_0}} \\ &= \frac{\left(\frac{\left(C_{x\_2} - C_{x\_0}\right)}{6} \cdot \left(D_{x\_0} + 4D_{x\_M} + D_{x\_2}\right)\right)}{C_{x\_2} - C_{x\_0}} \\ &= \frac{D_{x\_0} + 4D_{x\_M} + D_{x\_2}}{6} \end{aligned} \tag{32}$$

Evaluation value E (Equation (26)) was calculated as the average integral value $\bar{f}$ and the differential value D_{x_2}. We determined the dependent variable of the lower layer when evaluation value E of all components in the lower layer were less than or equal to the threshold value (α) (E £ α). However, we updated the coordinate points $\left(C_{x\_2}, D_{x\_2}\right)$ to Equation (33) and calculated the D_{x_M}, $\bar{f}$ and evaluation value E when at least one of E were greater than the threshold value (α) (E > α) as follows:

$$\begin{gathered} \left(C_{x\_2}, D_{x\_2}\right) = \left(t_{N_x - S_x}, f\left(X\left(t = t_{N_x - S_x}\right), Y\left(t = t_0\right), Z\left(t = t_0\right), t_{N_x - S_x}\right)\right) \\ \left(S_x = 1, 2, 3, \cdots, N_x - 2\right) \end{gathered} \tag{33}$$

Thereafter, we sequentially increased the number of discarded calculation steps S_x until E of all components in the lower layer were less than or equal to the threshold value (α) (E £ α) and determined the dependent variable of the lower layer. The number of determined calculation steps G_x was set to $N_x - S_x$ when the all evaluation values E were less than or equal to the threshold value (α) (Equation (34)) as follows:

$$G_x = N_x - S_x \tag{34}$$

If E was more than the threshold value (α) at $S_x = N_x - 2$, we set it to $G_x = 2$. Equation (35) shows the coordinate points (C_{x_2}, D_{x_2}) after the number of determined calculation steps G_x was decided as follows:

$$(C_{x\_2}, D_{x\_2}) = \left(t_{G_x}, f\left(X\left(t = t_{G_x}\right), Y\left(t = t_0\right), Z\left(t = t_0\right), t_{G_x}\right)\right) \tag{35}$$

2.3.3. Cumulative Algorithm

After determining the calculation interval of the lower layer with the backward algorithm, the cumulative algorithm was used to implement the numerical calculation of the 1st step in the middle layer using Equations (36)-(39) as follows:

$$Y(t = \tau_1) = Y(t = \tau_0) + g\left(\bar{X}, Y(t = \tau_0), Z(t = \tau_0), \tau_0\right) \cdot \mathrm{d}\tau_0 \tag{36}$$

$$\begin{aligned}
\bar{X} &= \frac{\int_{t=C_{x_0}}^{t=C_{x_2}} X(t)\mathrm{d}t}{C_{x_2} - C_{x_0}} \\
&= \frac{\left(\frac{(C_{x_2} - C_{x_0})}{6} \cdot \left(X(t = C_{x_0}) + 4X(t = C_{x_M}) + X(t = C_{x_2})\right)\right)}{C_{x_2} - C_{x_0}} \\
&= \frac{X(t = C_{x_0}) + 4X(t = C_{x_M}) + X(t = C_{x_2})}{6}
\end{aligned} \tag{37}$$

$$\mathrm{d}\tau_0 = C_{x\_2} - C_{x\_0} = \sum_{j=0}^{G_x - 1} \mathrm{d}t_j \tag{38}$$

$$\tau_{q+1} = \tau_q + \mathrm{d}\tau_q, \left(q = 0,1,2,\cdots,N_y - 1\right) \tag{39}$$

Here, $\mathrm{d}\tau_0$ was equal to the interval from time C_{x_0} to time C_{x_2} and $\overline{X}$ was the time-averaged concentration of the numerical solution of the lower layer in the interval from time C_{x_0} to time C_{x_2}. In addition, $X\left(t = C_{x\_M}\right)$ was calculated using Lagrange interpolation [29]. We applied the conventional method shown in the ahead algorithm to the numerical calculation of the middle layer.

2.3.4. Integration of Whole Algorithms

Calculations of these algorithms gave numerical solutions of lower and middle layers at time τ_1. Subsequently, the numerical solution of the concentration of the component in the interval from the $(G_x + 1)$ step to the $(G_x + N_x)$ step in the lower layer (the predetermined calculation interval) was calculated using the ahead algorithm with the values $X(t = \tau_1)$ and $Y(t = \tau_1)$ as shown in Equation (40) as follows:

$$X\left(t = t_{p+1} + \tau_1\right) = X\left(t = t_p + \tau_1\right) + f\left(X\left(t = t_p + \tau_1\right), Y\left(t = \tau_1\right), Z\left(t = t_0\right), t_p + \tau_1\right) \cdot \mathrm{d}t_p \quad \left(p = 0,1,2,\cdots,N_x - 1\right) \tag{40}$$

Thereafter, the backward algorithm was used to define the number of determined calculation steps G_x and the number of discarded calculation steps S_x in this predetermined calculation interval, and the cumulative algorithm was used to calculate $Y(t = \tau_2)$ of the middle layer. We repeated the procedures in Equations (26)-(40), and the numerical solution of the component concentration of the middle layer was calculated until the N_y step (the predetermined calculation interval in the middle layer) using the following Equations (41) and (42):

$$Y\left(t=\tau_{q+1}\right) = Y\left(t=\tau_q\right)+g\left(\overline{X'},Y\left(t=\tau_q\right),Z\left(t=\tau_0\right),\tau_q\right)\cdot \mathrm{d}\tau_q \quad \left(q=0,1,2,\cdots,N_y-1\right) \tag{41}$$

$$\overline{X'}=\frac{\int_{t=\tau_q}^{t=\tau_{q+1}} X(t)\mathrm{d}t}{\tau_{q+1}-\tau_q} = \frac{\left(\frac{\left(\tau_{q+1}-\tau_q\right)}{6}\cdot\left(X\left(t=\tau_q\right)+4X\left(t=\left(\frac{\tau_{q+1}-\tau_q}{2}\right)\right)+X\left(t=\tau_{q+1}\right)\right)\right)}{\tau_{q+1}-\tau_q} = \frac{X\left(t=\tau_q\right)+4X\left(t=\left(\frac{\tau_{q+1}-\tau_q}{2}\right)\right)+X\left(t=\tau_{q+1}\right)}{6} \tag{42}$$

Here, $\overline{X'}$ denotes the time-averaged concentration of the numerical solution of the lower layer in the interval from time τ_q to τ_{q+1}. The number of determined calculation steps G_y was then decided and the number of discarded calculation steps S_y in middle layer was generated using the backward algorithm. Equations (43)-(45) show the coordinate points of the differential value and the time of the middle layer that is necessary for the corresponding calculation of E (Equation (26)) as follows:

$$\left(C_{y\_0},D_{y\_0}\right)=\left(\tau_0,g\left(X\left(t=\tau_0\right),Y\left(t=\tau_0\right),Z\left(t=\tau_0\right),\tau_0\right)\right) \tag{43}$$

$$\left(C_{y\_1},D_{y\_1}\right)=\left(\tau_1,g\left(X\left(t=\tau_1\right),Y\left(t=\tau_1\right),Z\left(t=\tau_0\right),\tau_1\right)\right) \tag{44}$$

$$\left(C_{y\_2},D_{y\_2}\right)=\left(\tau_{G_y},g\left(X\left(t=\tau_{G_y}\right),Y\left(t=\tau_{G_y}\right),Z\left(t=\tau_0\right),\tau_{G_y}\right)\right) \tag{45}$$

The cumulative algorithm was used to implement the calculation of the 1st step of the upper layer with numerical solutions of lower and middle layers using Equations (46)-(50) as follows:

$$Z(t=T_1)=Z(t=T_0)+h\left(\overline{X''},\overline{Y},Z(t=T_0),T_0\right)\cdot \mathrm{d}T_0 \tag{46}$$

$$\begin{aligned}\overline{X''}&=\frac{\int_{t=C_{y\_0}}^{t=C_{y\_2}}X(t)\,\mathrm{d}t}{C_{y\_2}-C_{y\_0}}\\&=\frac{\left(\frac{\left(C_{y\_2}-C_{y\_0}\right)}{6}\cdot\left(X\left(t=C_{y\_0}\right)+4X\left(t=C_{y\_M}\right)+X(t=C_{y\_2})\right)\right)}{C_{y\_2}-C_{y\_0}}\\&=\frac{X\left(t=C_{y\_0}\right)+4X\left(t=C_{y\_M}\right)+X\left(t=C_{y\_2}\right)}{6}\end{aligned} \tag{47}$$

$$\begin{aligned}\overline{Y}&=\frac{\int_{t=C_{y\_0}}^{t=C_{y\_2}}Y(t)\,\mathrm{d}t}{C_{y\_2}-C_{y\_0}}\\&=\frac{\left(\frac{\left(C_{y\_2}-C_{y\_0}\right)}{6}\cdot\left(Y\left(t=C_{y\_0}\right)+4Y\left(t=C_{y\_M}\right)+Y\left(t=C_{y\_2}\right)\right)\right)}{C_{y\_2}-C_{y\_0}}\\&=\frac{Y\left(t=C_{y\_0}\right)+4Y\left(t=C_{y\_M}\right)+Y\left(t=C_{y\_2}\right)}{6}\end{aligned} \tag{48}$$

$$\mathrm{d}T_0=C_{y\_2}-C_{y\_0}=\sum_{j=0}^{G_y-1}\mathrm{d}\tau_j \tag{49}$$

$$T_{r+1}=T_r+\mathrm{d}T_r\ \ \left(r=0,1,2,\cdots,N_z-1\right) \tag{50}$$

Here, $\mathrm{d}T_0$ was equal to the interval from the time C_{y_0} to the time C_{y_2}, and $\overline{X''}$ was the corresponding time-averaged concentration of the numerical solution of the lower layer. $\overline{Y}$ was the time-averaged concentration of the numerical solution of the middle layer in the same time interval and $X\left(t=C_{y\_M}\right)$ and $Y\left(t=C_{y\_M}\right)$ were calculated using Lagrange interpolation [29] . The numerical calculation of the upper layer shown in Equation (46) applied the same conventional method in the case of the ahead algorithm (Equation (24)). Subsequently, we repeated

Equations (26)-(50) and found the numerical solution from $t = T_0$ to $t = T_{N_z}$ of the dependent variable in the upper layer. Finally, the backward algorithm was used to define the number of determined calculation steps G_z and the number of discarded calculation steps S_z in the upper layer. After completion of upper layer calculations, we calculated iterative numerical integrations in the interval of N_x steps of the lower layer using the values $X\left(t = T_{G_z}\right)$, $Y\left(t = T_{G_z}\right)$ and $Z\left(t = T_{G_z}\right)$ in Equation (51) as follows:

$$X\left(t = t_{p+1} + T_{G_z}\right)$$
$$= X\left(t = t_p + T_{G_z}\right) + f\left(X\left(t = t_p + T_{G_z}\right), Y\left(t = T_{G_z}\right), Z\left(t = T_{G_z}\right), t_p + T_{G_z}\right) \cdot \mathrm{d}t_p$$
$$\left(p = 0, 1, 2, \cdots, N_x - 1\right) \qquad (51)$$

Iterative calculations of component concentrations of all layers were calculated using Equations (26)-(51).

2.3.5. Control of Numbers of Predetermined Calculation Steps

Numbers of predetermined calculation steps $\left(N_x, N_y, N_z\right)$ indicated numbers of calculations with the ahead algorithm. When numbers of predetermined calculation steps $\left(N_x, N_y, N_z\right)$ were excessively large, the backward algorithm discarded the calculation obtained by the ahead algorithm to maintain calculation accuracy. Therefore, the number of predetermined calculation steps $\left(N_x, N_y, N_z\right)$ was closely related to the calculation efficiency of the proposed method, and the number of predetermined calculation steps $\left(N_x, N_y, N_z\right)$ was dynamically changed based on the number of determined calculation steps as shown in Equation (52) as follows:

$$\begin{pmatrix} N_x \\ N_y \\ N_z \end{pmatrix} = \begin{pmatrix} G_x \\ G_y \\ G_z \end{pmatrix} + \begin{pmatrix} A_x \\ A_y \\ A_z \end{pmatrix} \qquad (52)$$

Here, A_x, A_y, A_z were increments of the number of predetermined calculation steps (N_x, N_y, N_z). Accordingly, Equation (52) implicitly controlled the number of predetermined calculation steps of each layer according to the rate of change of the dependent variable.

2.4. Evaluation of Calculation Accuracy

The calculation accuracy of the proposed method was evaluated using Equations (53) and (54). Firstly, we evaluated calculation accuracy according to the consistency of the numerical solution using Equation (53) as follows:

$$V_{ave} = \frac{1}{num} \sum_{a=1}^{num} \frac{Component_{a\_proposed\ method}}{Component_{a\_conventional\ method}} \quad (53)$$

where $Component_{a\_proposed\ method}$ and $Component_{a\_conventional\ method}$ were corresponding to the numerical solution of specific $Component_a\,(a = 1, 2, 3, \cdots, num)$ in the layer of interest using the proposed and conventional method, respectively. The num represented the number of components in the layer of interest. The right side of Equation (53) was the average of proportion of the numerical solution from the proposed method $(Component_{a\_proposed\ method})$ to that of the conventional method $(Component_{a\_conventional\ method})$. Secondly, we evaluated local calculation accuracy according to the standard deviation of the value between the numerical solution of the proposed method and that of the conventional method using Equation (54) as follows:

$$V_{SD} = \sqrt{\frac{1}{num} \sum_{a=1}^{num} \left(\frac{Component_{a\_proposed\ method}}{Component_{a\_conventional\ method}} - V_{ave} \right)^2} \quad (54)$$

Hence, calculation accuracy was evaluated according to V_{ave} and V_{SD}.

3. RESULTS

In the multi-time-scale models (Model A, Model B) shown in Figure 2, we compared numerical solutions of the proposed method with those of the conventional method. In the conventional method, we adopted the explicit 4 stage 4th-order Runge-Kutta method [29] for numerical calculations of all layers. However, the numerical calculation of the lower layer of the proposed method allowed application of the arbitrary method for ordinary differential equations. We also adopted the explicit 4 stage 4th-order Runge- Kutta method [29] for the numerical calculations of the lower layer of the proposed method. Table 1 and Table 2 show kinetic parameters of models A and B, respectively, and the simulation time was set to 36000 s because the dynamic behavior of the upper layer reached steady state at this time. The step size dt of the adaptive time of the explicit 4 stage 4th-order Runge-Kutta method [29] was used in the proposed method and the conventional method, and was set to 1.00E−03 s.

3.1. Case Study 1

As shown in Figure 2(a), model A comprises 18 components and has inhibition effects from upper to middle layers and from middle to lower layers. We compared numerical solutions of the proposed and conventional methods in model A and verified the utility of the proposed method in the multi-time-scale model in terms of calculation efficiency and accuracy. Table 3 shows parameters of initial numbers of predetermined calculation steps (N_x, N_y, N_z) and parameters of control variables for numbers of predetermined calculation steps (A_x, A_y, A_z) of the proposed method. The threshold value (α) of the evaluation value E of the backward algorithm was set to 1.00E−03, 5.00E−04, and 1.00E−04.

3.1.1. Calculation Efficiency of the Proposed Method in Numerical Analyses of Model A

In numerical analyses of model A, the calculation efficiency of the proposed method was compared with that of the conventional method. Figures 4(a)-(c) show time- dependent changes in numbers of accumulated calculation steps for each layer in model A. Here, numbers of accumulated calculation steps for lower, middle, and upper layers were equal to the sum of G_x $(\sum G_x)$, $(\sum G_y)$ G_y and G_z $(\sum G_z)$ within the entire simulation time for model A. We applied the same method to the numerical calculation for the lower layer of the proposed and conventional methods. Hence, numbers of accumulated calculation steps for the lower layer of the proposed method was equal to that of the conventional method (Figure 4(c)). Furthermore, we compared numbers of accumulated calculation steps in middle and upper layers of model A (Figure 4(a) and Figure 4(b)). Accumulated calculation steps in middle and upper layers of the proposed method at 36,000 s (Table 4) were far fewer than those of the conventional method in middle and upper layers of model A. Moreover, the threshold value (α) of the evaluation value E and the number of accumulated calculation steps of the proposed method in middle and upper layers were negatively correlated. Figure 4(d) shows the sum of calculation steps that were discarded by the backward algorithm, which ensured calculation accuracy. In these analyses, the sum of discarded calculation steps was equal to $\sum S_x + \sum S_y + \sum S_z$ within the entire simulation time for model A. Moreover, regardless of the threshold value (α), destruction of the calculation by the backward algorithm occurred during the early stages of the simulation. In addition, destruction of calculations had occurred at 12,000 s when the threshold value (α) was 1.00E−3 or 5.00E−04. Figure 4(e) shows time-dependent changes in the sum of accumulated calculation steps for all layers of model A. The proposed method included the sum of calculation step that were discarded by the backward algorithm as $\sum G_x + \sum G_y + \sum G_z + \sum S_x + \sum S_y + \sum S_z$. Accumulated calculation steps of the pro- posed method for all layers of model A at 36000 s are shown in Table 5 as a proportion of those for the conventional method. Because numbers of accumulated calculation steps of the proposed method in middle and upper layers (decreased numbers of calcu lation steps)

Table 4. Calculation efficiency of the proposed method in middle and upper layers of model A at 36000 s.

Layer	The threshold value (α) of the evaluation value E	Numbers of accumulated calculation steps of the conventional method (A)	Numbers of accumulated calculation steps of the proposed method (B)	Percentage ((B/A) * 100%)
Middle	1.00E−03	3.60E+07	1.13E+04	0.031
	5.00E−04	3.60E+07	1.78E+04	0.049
	1.00E−04	3.60E+07	6.22E+04	0.173
Upper	1.00E−03	3.60E+07	1.86E+02	0.00052
	5.00E−04	3.60E+07	2.97E+02	0.00082
	1.00E−04	3.60E+07	8.02E+02	0.00222

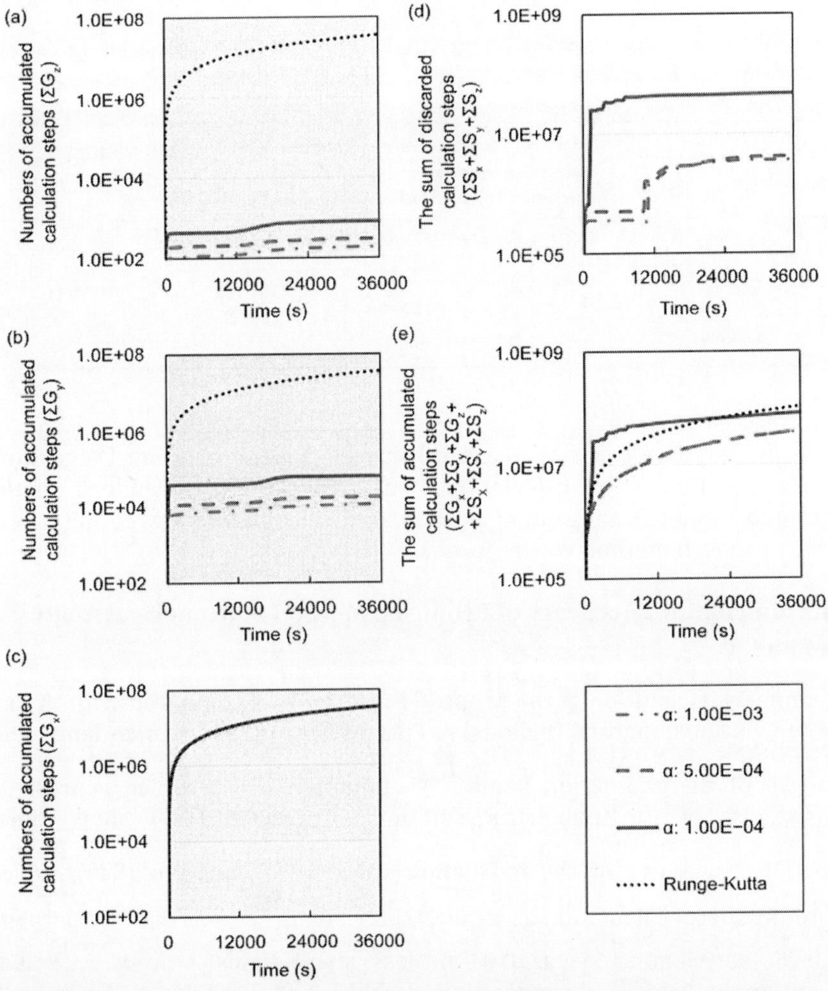

Figure 4. Comparisons of numbers of accumulated calculation steps for the proposed and conventional methods in model A; The calculation step size dt was set at 1.00E−03 s. The threshold value (α) of the evaluation value E of the backward algorithm was set at 1.00E−03, 5.00E−04, and 1.00E−04. (a), (b), and (c) show the sum of numbers of accumulated calculation steps for lower, middle, and upper layers in model A $\left(\sum G_x, \sum G_y, \sum G_z\right)$. (d) shows the sum of calculation steps that were discarded by the backward algorithm in model A $\left(\sum S_x + \sum S_y + \sum S_z\right)$. (e) shows the sum of accumulated calculation steps of all layers in model A $\left(\sum G_x + \sum G_y + \sum G_z + \sum S_x + \sum S_y + \sum S_z\right)$.

Table 5. Calculation efficiency of the proposed method in all layers of model A at 36000 s.

The threshold value(α)of the evaluation value E	Numbers of accumulated calculation steps of the conventional method (A)	Numbers of accumulated calculation steps of the proposed method (B)	Percentage ((B/A)*100%)
1.00E−03	1.08E+08	3.96E+07	36.7
5.00E−04	1.08E+08	4.01E+07	37.1
1.00E−04	1.08E+08	8.24E+07	76.3

were greater than those that were discarded by the backward algorithm (increased numbers of calculation steps), which was dependent on the threshold value (α), the sum of accumulated calculation steps for all layers of the proposed method was reduced.

3.1.2. Calculation Accuracy of Proposed Method in Numerical Analysis of Model A

The numerical solution of the proposed method was compared with that of the conventional method in model A. Figures 5(a)-(c) show time-dependent changes of the evaluation value V_{ave} (Equation (53)), which represents consistency of the numerical solution. Figures 5(d)-(f) show time-dependent changes of the evaluation value V_{SD} (Equation (54)), which represents local calculation accuracy. In any layer, V_{ave} was within the range of 0.95 - 1.05 and V_{SD} was 0.01 or less at all times. Moreover, V_{ave} was asymptotic to 1.0 with decreases in thresholds (α) of the evaluation value E in all layers. Therefore, the calculation accuracy of the proposed method was almost the same as that of the conventional method in numerical analyses of model A.

3.2. Case Study 2

As shown in Figure 2(b), model B comprised 18 components and had activation effects from lower to middle layers and from middle to upper layers. In addition, the effect of negative feedback in the lower layer led to oscillating dynamics in model B. In the pre- sent study, we compared numerical solutions of the proposed and conventional methods in model B and verified the utility of the proposed method in the multi- time-scale model in terms of calculation efficiencies and accuracies. Table 3 shows parameters of initial numbers of predetermined calculation steps (N_x, N_y, N_z) and parameters of control variables for numbers of predetermined calculation steps (A_x, A_y, A_z) of the proposed method. Threshold values (α) of the evaluation value E of the backward algorithm were set at 1.00E−01, 5.00E−02, and 1.00E−02.

3.2.1. Calculation Efficiency of the Proposed Method in Numerical Analyses of Model B

In numerical analysis of model B, the calculation efficiency of the proposed method was compared with that of the conventional method. Figures 6(a)-(c) show time-depen- dent changes in numbers of accumulated calculation steps for each layer of model B.

Here, numbers of accumulated calculation steps of lower, middle, and upper layers were equal to sums of G_x $(\sum G_x)$, $(\sum G_y)$ G_y and G_z $(\sum G_z)$, respectively, within the entire simulation time of model B. Differences between numbers of accumulated calculation steps in the lower layer of the proposed and conventional methods are not demonstrated (Figure 6(c)) as well as in case study 1. However, in further analyses we compared numbers of accumulated calculation steps in middle and upper layers of model B (Figure 6(a) and Figure 6(b)). Table 6 shows numbers of accumulated calculation steps in middle and upper layers of the proposed method at 36000 s in model B relative to those of the conventional method. Numbers of accumulated calculation steps for the proposed method were far fewer than those of conventional method in middle and upper layers of model B. Moreover, threshold values (α) of E from the backward algorithm were negatively correlated with numbers of accumulated calculation steps of the proposed method in middle and upper layers. Figure 6(d) shows the sum of calculation steps that were discarded by the backward algorithm, which was used to ensure calculation accuracy. In these analyses, the sum of discarded calculation steps was equal to $\sum S_x + \sum S_y + \sum S_z$ within the entire simulation time for model B. In case study 2, the calculation was discarded at a constant rate with time. Figure 6(e) shows time-dependent changes of the sum of accumulated calculation steps for all layers in model B. The proposed method included the sum of calculation steps that were discarded by the backward algorithm as $\sum G_x + \sum G_y + \sum G_z + \sum S_x + \sum S_y + \sum S_z$. Table 7 shows the sum of accumulated calculation steps for all layers

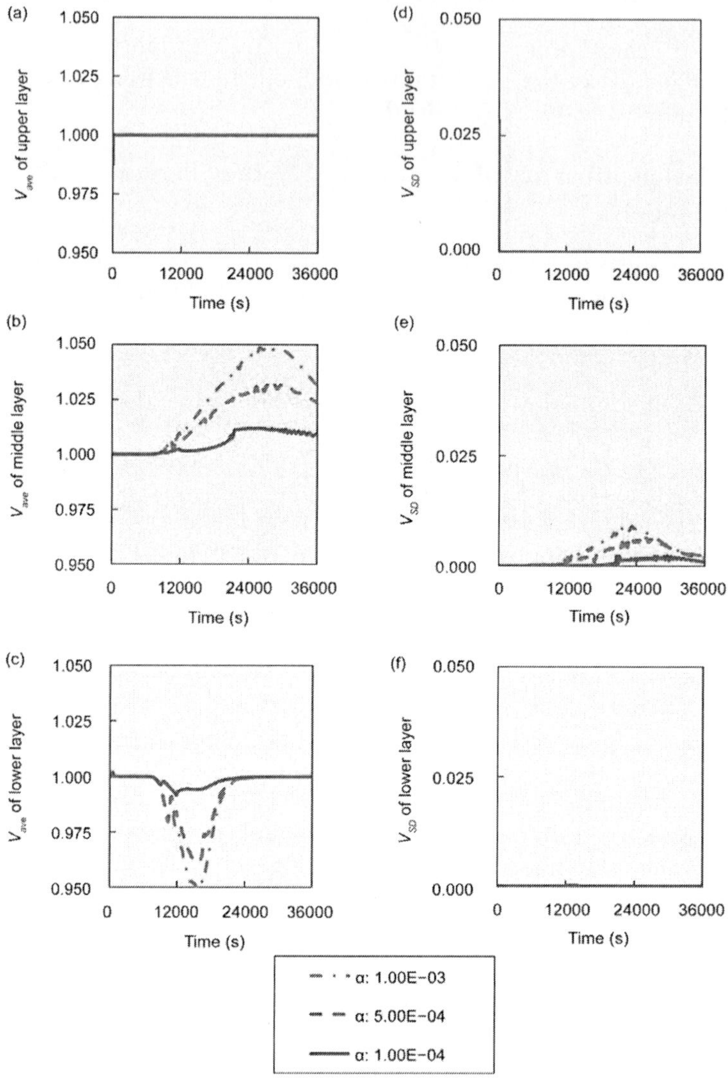

Figure 5. Comparison of calculation accuracies of the proposed and conventional methods in model A; Vertical axes of (a), (b), and (c) show V_{ave} (Equation (53)). Vertical axes of (d), (e), and (f) show V_{SD} (Equation (54)). The threshold value (α) of the evaluation value E of the back- ward algorithm was set at 1.00E−03, 5.00E−04, and 1.00E−04.

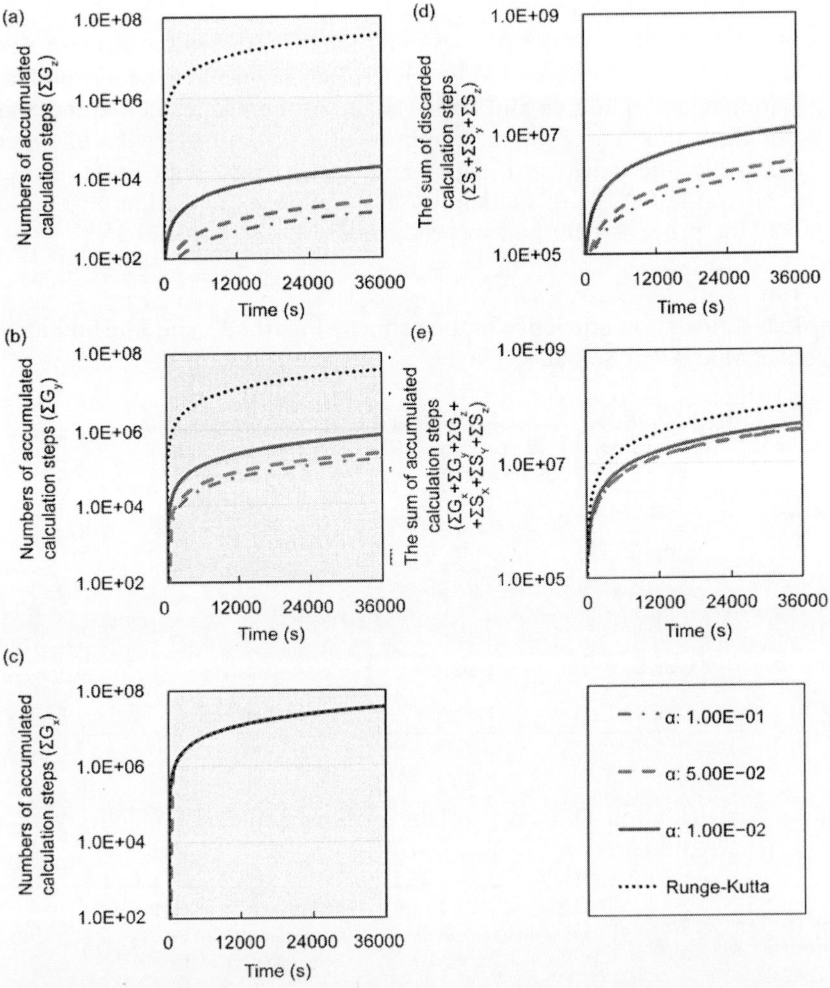

Figure 6. Comparison of numbers of accumulated calculation steps for proposed and conventional methods in model B; The calculation step size dt was set at 1.00E−03. The threshold value (α) of the evaluation value E for the backward algorithm was set at 1.00E−01, 5.00E−02, and 1.00E−02. (a), (b), and (c) show the sum of accumulated calculation steps for lower, middle, and upper layers in model B $\left(\sum G_x, \sum G_y, \sum G_z\right)$. (d) shows the sum of calculation steps that were discarded by the backward algorithm in model B $\left(\sum S_x + \sum S_y + \sum S_z\right)$. (e) shows the sum of accumulated calculation steps for all layers in model B $\left(\sum G_x + \sum G_y + \sum G_z + \sum S_x + \sum S_y + \sum S_z\right)$.

of proposed method at 36000 s in model B as a proportion of that for the conventional method. In case study 2, decreases in numbers of accumulated calculation steps in middle and upper layers of the proposed method were greater than the increases in numbers of calculation steps that were discarded by the backward algorithm, which was dependent on the threshold value (α). Thus, the sum of accumulated calculation steps of all layers of the proposed method was reduced.

Table 6. Calculation efficiency of the proposed method in middle and upper layers of model B at 36,000 s.

Layer	The threshold value (α) of the evaluation value E	Numbers of accumulated calculation steps of the conventional method (A)	Numbers of accumulated calculation steps of the proposed method (B)	Percentage ((B/A) * 100%)
Middle	1.00E−01	3.60E+07	1.67E+05	0.464
	5.00E−02	3.60E+07	2.31E+05	0.642
	1.00E−02	3.60E+07	7.09E+05	1.970
Upper	1.00E−01	3.60E+07	1.39E+03	0.0039
	5.00E−02	3.60E+07	2.82E+03	0.0078
	1.00E−02	3.60E+07	1.88E+04	0.0522

Table 7. Calculation efficiency of the proposed method in all layers of model B at 36,000 s.

The threshold value(α)of the evaluation value E	Numbers of accumulated calculation steps of the conventional method (A)	Numbers of accumulated calculation steps of the proposed method (B)	Percentage ((B/A) * 100%)
1.00E−01	1.08E+08	3.86E+07	35.7
5.00E−02	1.08E+08	3.96E+07	36.7
1.00E−02	1.08E+08	5.00E+07	46.3

3.2.2. Calculation Accuracy of the Proposed Method for Numerical Analysis of Model B

The numerical solution of proposed method was compared with that of the conventional method in model B. Figures 7(a)-(c) show time-dependent changes in the evaluation value V_{ave} (Equation (53)), which represents consistency of the numerical solution, and Figures 7(d)-(f) show time-dependent changes of the evaluation value V_{SD} (Equation (54)), which represents local calculation accuracy. In any layer, V_{ave} was within the range of 0.99 - 1.01, and V_{SD} was 0.01 or less at all times. Therefore, the calculation accuracy of the proposed method was almost the same as that of the conventional method in numerical analyses of model B.

4. DISCUSSION

The advent of high-throughput experimental devices that can accommodate large numbers of samples has allowed simultaneous computation of comprehensive data pertaining to genome, transcriptome, proteome, and metabolome analyses [1] [2] [3] [4] [5] . Consequently, the momentum of theoretical analyses of biological systems that use multi-time-scale models is growing [13] [14] [15] [16] (Figure 1). In theoretical analyses of multi-time-scale models with reactions between layers, the time stiff problem [17] [18] [19] occurs due to differences in time-scales of each layer, leading to significant increases in computation times. In particular, the time stiff problem significantly influences the efficiency of numerical optimizations of system identifications and analyses. Optimization methods are generally used to search for optimum solutions using repeated calculations with varying kinetic parameters for different strategies. For example, system identification by the Real-coded Genetic Algorithm (AREX + JGG) required about 2.0E+06 calculation iterations to estimation parameter values for 112 elements [31] . Calculation times for numerical optimization are generated by multiplying numbers of calculations by the time taken for 1 calculation. Because the time taken for 1 calculation is greatly increased by time stiff problems, calculation times for numerical optimizations also increase linearly. Hence, solutions for the time stiff problem will likely contribute to the efficiency of numerical optimizations. The time stiff problem also occurs in theoretical analyses of natural phenomena, such as the movements of the local clouds and typhoons in simulations of weather conditions [32] and motions and binding of compounds in simulations of chemical reactions [33] . Hence, solutions to the time stiff problem are applicable to varied ematical analyses, including those of biological systems. In the present conventional method, the time stiff problem remained because the calculation of all layers was implemented by adaptive time step sizes of the smallest time-scale layer. Therefore, the present alternative method reduced computation times by controlling adaptive time step sizes for each layer based on variations of differential values of the components.

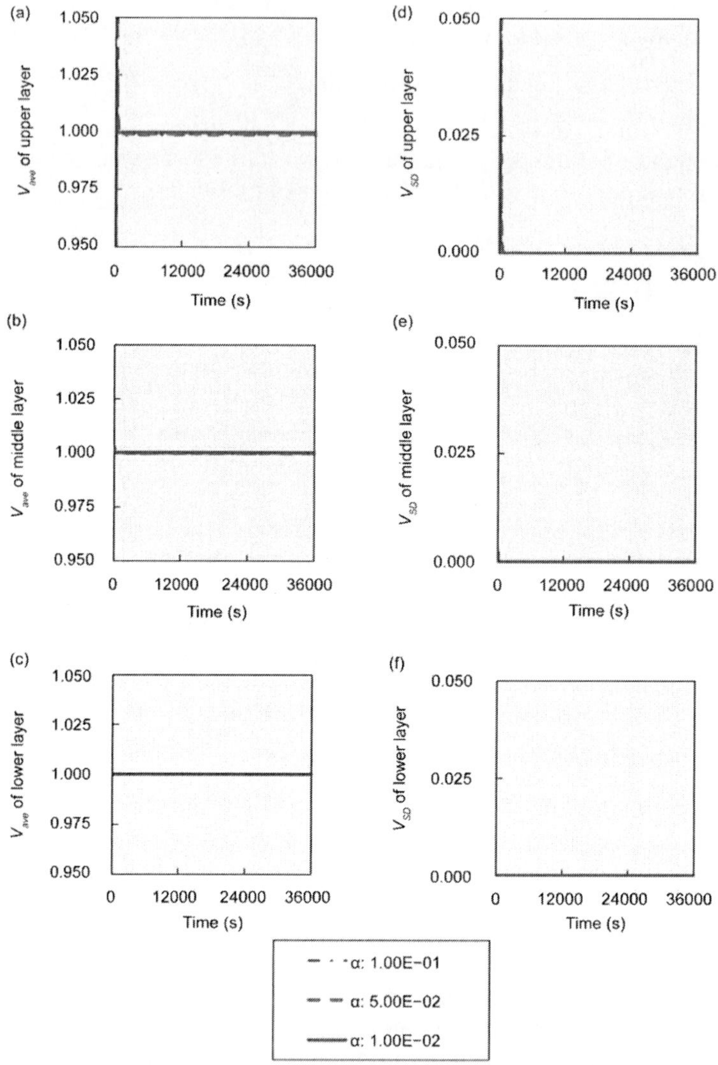

Figure 7. Comparison of calculation accuracies of the proposed and conventional methods in model B; Vertical axes of (a), (b), and (c) show V_{ave} (Equation (53)). Vertical axes of (d), (e), and (f) show V_{SD} (Equation (54)). The threshold value (α) of the evaluation value E of the back- ward algorithm was set at 1.00E−03, 5.00E−04, and 1.00E−04.

The proposed method comprised ahead, backward, and cumulative algorithms. Initially, the ahead algorithm was applied using the conventional method for calculating the smallest time-scale layer (lower layer), and iterative numerical integrations were performed in predetermined calculation intervals. In this study, we used the explicit 4 stage 4th-order Runge-Kutta method [29] to calculate the lower layer and then used the backward algorithm to determine the calculation interval according to the magnitude of change in the differential value during the predetermined calculation interval. Subsequently, the cumulative algorithm calculated the 1st step of the middle layer using the determined calculation interval of the lower layer and the time-averaged concentration in the determined calculation interval of the lower layer. The proposed method identified numerical solutions by repeating the three algorithms for the upper layer. In the conventional method, numbers of calculation steps for middle and upper layers were equal to that of the lower layer so that all layers were calculated according to the adaptive time step size of the lower layer. However, the adaptive time step size of middle and upper layers of the proposed method were much larger than those of the conventional method. In addition, the backward algorithm narrowed the calculation interval for large changes in the numerical solution of the predetermined calculation interval and widened that in the presence of small changes. Accordingly, the proposed method significantly reduced the number of calculation steps for middle and upper layers and maintained calculation accuracy of the backward algorithm. Most current high-speed calculation methods utilize parallel computer resources [21] [22] [23] , which are expedited by dividing the processing of calculations of 1 step between multiple central processing units. In contrast, the proposed method accelerates analyses by reducing numbers of calculation steps. Therefore, the proposed method does not compete with conventional high-speed methods, and can be used in conjunction with various high-speed calculation methods as a calculation module.

In the present study, we created multi-time-scale models as a benchmark (Figure 2) and verified the calculation performance of the proposed method. To this end, we investigated numbers of accumulated calculation steps for each layer of proposed and conventional methods. In case studies 1 and 2, numbers of accumulated calculation steps for the lower layer were comparable in proposed and conventional methods, whereas these were fewer for middle and upper layers of the proposed method than of the conventional method (Figures 4(a)-(c), Figures 6(a)-(c)). Therefore, calculations of middle and upper layers were performed using the proposed method with optimal adaptive time step sizes.

In further analyses, we discarded calculation steps to maintain calculation accuracy in backward algorithm. In case study 1, the sum of discarded calculation steps rapidly increased between the early stages of the simulation and 12,000 s (Figure 4(d)). Moreover, increasing numbers of discarded calculation steps during early stages of the simulation were greatly affected by initial setting values of N_x, N_y, N_z, which are parameters of the proposed method. Accordingly, the ahead algorithm of the proposed method was used to calculate numerical solutions to initial setting values of N_x, N_y, N_z, and the backward algorithm was used to discard calculations and maintain calculation accuracy. Therefore, the present backward algorithm significantly discarded significant numbers of calculations in the early stages of the simulation, because the initial setting values of N_x, N_y, N_z were excessive. Moreover, increasing numbers of discarded calculation steps in the first 12000 s were greatly affected by variations of the dependent variable of model A (Figure 3). These variations reflected decreased inhibition effects of Y_5 based on increases of Z_5 and enhancements of the inhibitory effects of Z_5. To prevent the deterioration of calculation accuracy due to these variations, the backward algorithm was used to adjust the adaptive time step size for each layer to values that corresponded with magnitudes of change in numerical solutions of the dependent variable. Hence, the backward algorithm significantly discarded calculations by 12,000 s in case study 2. The sum of discarded calculation steps also increased linearly with time (Figure 6(d)). Because model B was of the oscillation system, destruction by the backward algorithm occurred in constant cycles. These analyses suggest that the backward algorithm facilitates calculation accuracy.

Numbers of accumulated calculation steps of all layers of proposed and conventional methods were computed for case studies 1 and 2, and decreases by the proposed method for middle and upper layers (Figure 4(a) and Figure 4(b), Figure 6(a) and Figure 6(b)) were greater than increases in those discarded by the calculation steps of the backward algorithm (Figure 4(d), Figure 6(d)). Hence, because reductions in computational volumes by the proposed method were more numerous than those required to maintain calculation accuracy, the proposed method achieved the calculation efficiency of the numerical calculation in case studies 1 and 2 (Figure 4(e), Figure 6(e)).

In comparisons of calculation accuracies of proposed and conventional methods, that of the proposed method was controlled by the threshold (α) of the evaluation value E. The calculation interval that was determined by the backward algorithm was asymptotic to predetermined calculation intervals with the increase of the threshold (α). Time step sizes of middle and upper layers also became larger. Moreover, the numerical solution of the upper layer was reflected in lower and middle layers after 1 step of upper layer calculations. Thus, this reflection time was significantly delayed with excessive step sizes of the upper layer in the presence of high threshold (α) values. This delay caused calculation error in the numerical solution of the proposed method. Hence, threshold (α) values of models that contains reactions from upper to lower layers such as case study 1 need to be smaller than that in the model for case study 2. In this study, threshold (α) values of case study 1 (1.00E−03, 5.00E−04, and 1.00E−04) were set smaller than that of case study 2 (1.00E−01, 5.00E−02, 1.00E−02). Accordingly, at threshold (α) values of 1.00E−03 or 5.00E−04, calculation errors occurred in middle and upper layers for case study 1. However, at threshold (α) < 1.00E−04, calculation errors were avoided. Therefore, in case studies 1 and 2, the calculation accuracy of the conventional method was maintained by the proposed method by setting the optimal threshold (α) value depending on the model (Figure 5, Figure 7).

5. CONCLUSION

In summary, in the present benchmark model, ahead and cumulative algorithms of the proposed method led to calculation efficiency of numerical calculations with adjustments of step sizes of each layer, and reduced the numbers of numerical calculations required for multi-time-scale models with the time stiff problem. Moreover, backward algorithms of the proposed method ensured calculation accuracy in the multi-time- scale model. Accordingly, we suggest that the proposed method is an efficient numerical method for multi-time-scale models.

ACKNOWLEDGEMENTS

This work was supported by Grant-in-Aid for Scientific Research on Innovative Areas 16H01700 and Grant-in-Aid for Challenging Exploratory Research16K12912.

CITE THIS PAPER

Motomura, Y., Hamada, H. and Okamoto, M. (2016) An Effective Numerical Calculation Method for Multi-Time-Scale ematical Models in Systems Biology. Applied ematics, 7, 2241-2268.

REFERENCES

1. Castel, D., Pitaval, A., Debily, M.A. and Gidrol, X. (2006) Cell Microarrays in Drug Discovery. Drug Discovery Today, 11, 13-14.
2. Aebersold, R. and Mann, M. (2003) Mass Spectrometry-Based Proteomics. Nature, 13, 198-207.
3. Bleicher, K.H., Böhm, H.J., Müller, K. and Alanine, A.I. (2003) A Guide to Drug Discovery: Hit and Lead Generation: Beyond High-Throughput Screening. Nature Reviews Drug Discovery, 2, 369-378.
4. Macarron, R., Banks, M.N., Bojanic, D., Burns, D.J., Cirovic, D.A., Garyantes, T., Green, D.V., Hertzberg, R.P., Janzen, W.P., Paslay, J.W., Schopfer, U. and Sittampalam, G.S. (2011) Impact of High-Throughput Screening in Biomedical Research. Nature Reviews Drug Discovery, 10, 188-195.
5. Fischer, E., Zamboni, N. and Sauer, U. (2004) High-Throughput Metabolic Flux Analysis Based on Gas Chromatography-Mass Spectrometry Derived 13C Constraints. Analytical Biochemistry, 325, 308-316.
6. Ge, H., Walhout, A.J. and Vidal, M. (2003) Integrating "Omic" Information: A Bridge Between Genomics and Systems Biology. Trends in Genetics, 19, 551-560.
7. Altaf-Ul-Amin, M., Afendi, F.M., Kiboi, S.K. and Kanaya, S. (2014) Systems Biology in the Context of Big Data and Networks. BioMed Research International, 2014, Article ID: 428570.
8. Hood, L. (2003) Leroy Hood Expounds the Principles, Practice and Future of Systems Biology. Drug Discovery Today, 8, 436-438.
9. Kitano, H. (2002) Systems Biology: A Brief Overview. Science, 295, 1662-1664.
10. Kitano, H. (2002) Computational Systems Biology. Nature, 420, 206-210.
11. Liu, E.T. (2005) Systems Biology, Integrative Biology, Predictive Biology. Cell, 121, 505-506.
12. Papp, B., Notebaart, R.A. and Pál, C. (2011) Systems-Biology Approaches for Predicting Genomic Evolution. Nature Reviews Genetics, 12, 591-602.
13. Yao, Z., Petschnigg, J., Ketteler, R. and Stagljar, I. (2015) Application Guide for Omics Approaches to Cell Signaling. Nature Chemical Biology, 11, 387-397.
14. Bajikar, S.S. and Janes, K.A. (2012) Multiscale Models of Cell Signaling. Annals of Biomedical Engineering, 40, 2319-2327.
15. Helikar, T., Kochi, N., Kowal, B., Dimri, M., Naramura, M., Raja, S.M., Band, V., Band, H. and Rogers, J.A. (2013) A Comprehensive, Multi-Scale Dynamical Model of ErbB Receptor Signal Transduction in Human Mammary Epithelial Cells. PLoS ONE, 8, e61757.
16. Stolarska, M.A., Kim, Y. and Othmer, H.G. (2009) Multi-Scale Models of Cell and Tissue Dynamics. Philosophical Transactions of the Royal Society, 367, 3525-3553.
17. Curtiss, C.F. and Hirschfelder, J.O. (1952) Integration of Stiff Equations. Proceedings of the National Academy of Sciences of the United States of America, 38, 235-243.
18. Sommeijer, B.P. (1993) Parallel-Iterated Runge-Kutta Methods for Stiff Ordinary Differential Equations. Journal of Computational and Applied ematics, 45, 151-168.
19. Hairer, E. and Wanner, G. (1999) Stiff Differential Equations Solved by Radau Methods. Journal of Computational and Applied ematics, 111, 93-111.
20. Gear, C.W. (1971) Numerical Initial Value Problems in Ordinary Differential Equations. Prentice-Hall, Hoboken.
21. Gear, C.W. (1988) Parallel Methods for Ordinary Differential Equations. International Symposium on Vector and Parallel Processors for Scientific Computation, 25, 1-20.
22. Bellen, A., Vermiglio, R. and Zennaro, M. (1990) Parallel ODE-Solvers with Stepsize Control. Journal of Computational and Applied ematics, 31, 277-293.
23. Pope, B.J., Fitch, B.G., Pitman, M.C., Rice, J.J. and Reumann, M. (2011) Performance of Hybrid Programming Models for Multiscale Cardiac Simulations: Preparing for

Petascale Computation. IEEE Transactions on Biomedical Engineering, 58, 2965-2969.

24. Herrero, A., Muro-Pastor, A.M. and Flores, E. (2001) Nitrogen Control in Cyanobacteria. Journal of Bacteriology, 183, 411-425.
25. Lewis, M., Chang, G., Horton, N.C., Kercher, M.A., Pace, H.C., Schumacher, M.A., Brennan, R.G. and Lu, P. (1996) Crystal Structure of the Lactose Operon Repressor and Its Complexes with DNA and Inducer. Science, 271, 1247-1254.
26. Gruber, T.M. and Gross, C.A. (2003) Multiple Sigma Subunits and the Partitioning of Bacterial Transcription Space. Annual Review of Microbiology, 57, 441-466.
27. Suarez, R.K., Staples, J.F., Lighton, J.R.B. and West, T.G. (1997) Relationships between Enzymatic Flux Capacities and Metabolic Flux Rates: Nonequilibrium Reactions in Muscle Glycolysis. Proceedings of the National Academy of Sciences of the United States of America, 94, 7065-7069.
28. Goutelle, S., Maurin, M., Rougier, F., Barbaut, X., Bourguignon, L., Ducher, M. and Maire, P. (2008) The Hill Equation: A Review of Its Capabilities in Pharmacological Modeling. Fundamental & Clinical Pharmacology, 22, 633-648.
29. Press, W.H., Teukolsky, S.A., Vetterling, W.T. and Flannery, B.P. (2007) Numerical Recipes. 3rd Edition, Cambridge University Press, Cambridge.[Citation Time(s):11]
30. Fehlberg, E. (1970) Klassische Runge-Kutta-Formeln vierter und niedrigerer Ordnung mit Schrittweiten-Kontrolle und ihre Anwendung auf Wärmeleitu ngsprobleme. Computing, 6, 61-71.
31. Komori, A., Maki, Y., Nakatsui, M., Ono, I. and Okamoto, M. (2012) Efficient Numerical Optimization Algorithm Based on New Real-Coded Genetic Algorithm, AREX + JGG, and Application to the Inverse Problem in Systems Biology. Applied ematics, 3, 1463-1470.
32. Arakawa, A. and Jung, J.H. (2011) Multiscale Modeling of the Moist-Convective Atmosphere—A Review. Atmospheric Research, 102, 263-285.
33. Duarte, F., Amrein, B.A., Blaha-Nelson, D. and Kamerlin, S.C. (2015) Recent Advances in QM/MM Free Energy Calculations Using Reference Potentials. Biochimica et Biophysica Acta, 1850, 954-965.

CHAPTER 2

Mathematical Modeling of Cardiomyocytes' and Skeletal Muscle Fibers' Membrane: Interaction with External Mechanical Field

Irina V. Ogneva, Nikolay S. Biryukov

Department of Molecular Biomedicine, State Scientific Center of Russian Federation, Institute of Biomedical Problems of the Russian Academy of Sciences, Moscow, Russia

ABSTRACT

We propose a ematic model of muscle cell membrane based on thin-walled elastic rod theory. A deformation occurs in rodents' skeletal and cardiac cells during a period of antiorthostatic suspension. We carried out a quantitative evaluation of the deformation using this model. The calculations showed the deformation in cardiac cells to be greater than in skeletal ones. This data corresponds to experimental results of cell response that appears intense in cardiomyocytes than in skeletal muscle cells. Moreover, the deformation in skeletal and heart muscle cells has a different direction (stretching vs. compression), corresponding to experimental data of different adaptive response generation pathways in cells because of external mechanical condition changes.

Keywords: Mathematical Modeling in Biology; Muscle Cell; Cell Mechanosensitivity; Microgravity

1. INTRODUCTION

Every mechanical system, including living cells, in an external mechanical field is exposed to forces intrinsic to this field. The action of these forces results in mechanical tension that appears in cells. An external influence change (in direction or magnitude) leads to mechanical tension changes in cells and to deformation. The significance of the deformation for the cell depends on its inherent mechanical characteristics and the sensitivity of its mechanosensors.

An external physical signal transformation results in proper cell response generation. At the same time, the clue is the magnitude of the applied force that is able to induce cell response.

All living cells can be divided into two groups: cells that form internal tension only against external influence and cells that can additionally generate force themselves —muscle cells. Muscle cells have a specific structure, a developed cytoskeleton, which occupies most of the cell volume and forms a contractile apparatus. Taking these features into account, one can suppose that a muscle cell mechanosensor is connected to the contractile apparatus, for example with the M-line [1].

However, the muscle cell submembrane cytoskeleton is quite similar to that in non-muscle cells except for certain areas (particularly in areas of M-line and Z-disk projections on the membrane). Hence, cell formation takes place under constant external force action, so one can assume that the first mechanoreception acts to connect with the cell compartment, typical for every living cell. This compartment appears to comprise a membrane and a cortical (submembrane) cytoskeleton. Therefore, the question is what deformations emerge in the muscle fiber membrane after the gravity vector or the fiber contraction rate changes and whether these changes can result in muscle fiber mechanical characteristic change and cell response initiation.

To answer these questions, we need a numerical evaluation of deformations that arise in the sarcolemma after external mechanical conditions change. Such an evaluation requires data about longitudinal and transversal stiffness because muscle cells appear to have a three-dimensional structure [2]. However, this problem turns out to be hard to solve because the contractile apparatus contribution to linear stiffness is several orders greater than the contribution of the sarcolemma. Previously, we succeeded in determining the transversal stiffness of the sarcolemma using atomic force microscopy [3], changing mechanical conditions for both skeletal and cardiac muscles in rodents (**Figure 1**) [4].

An external mechanical condition change was implemented through the common animal antiorthostatic suspension method by tail at an angle of 30° respective to the cage floor (the Ilyin-Novikov method with the MoreyHolton modification is widely used in space physiology to model microgravity effects on a surface [5]). Animal suspension resulted, on the one hand, in a reduction of the external mechanical field on the hind limbs and, on the other hand, in increased mechanical tension in cardiomyocytes. The orientation of the muscle cells (muscle fibers) in the gravity field changed, too.

Nevertheless, the data for only the transversal stiffness of the membrane and cortical cytoskeleton do not afford us the influence of the gravity vector change on the muscle cell membrane and its probable involvement in primary mechanoreception acts. All of the above indicates the necessity of developing a membrane ematical model.

2. EMATICAL MODEL

2.1. Statement of the Problem

Let δ define the thickness of the membrane and cortical cytoskeleton, d is the muscle fiber diameter, and l is its length. Then, we can write the following relations:

$$\frac{\delta}{d} \leq 0.1 \, , \, \frac{d}{l} \leq 0.1$$

This enables us to consider the fiber a long, cylindrical envelope called a thin-walled rod [6].

It is typical for transversal sections of this kind of objects to initially be plane distorted on a surface: W(x, z). W(x, z) is usually called a sectional warping. For closed envelope rods, including muscle fibers, axial uniform warping is also typical, so we can rewrite W(x, z) as W(x). We have $\nabla W(x) = (\nabla\Phi + \underline{x}) \times \underline{k}$ in Saint-Venant's prob-

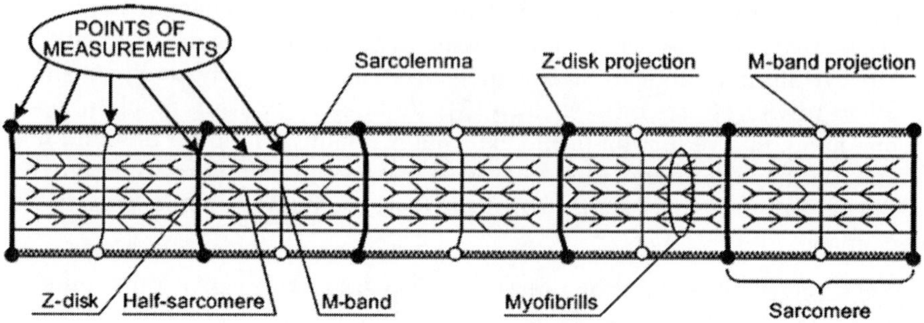

Figure 1. Scheme of muscle fiber with indicated measurement points. Using the AFM surface images, we were able to measure the transverse stiffness of all specific regions of the fiber, such as the contractile apparatus and membrane at different points. For this study, the main parameter is the membrane stiffness between the Z-disk and M-line projections. The structure of this region is universal for all cell types (not just muscle cells).

lem, where Φ is the Prandtl function (underlining indicates that the underlined value is a vector). We can take a sectorial area as W for a thin section. Then, for the part of membrane between the Z-disk and M-line to be considered, we use the thin-walled theory statements. We will consider the cylindrical rod with thin simply connected section F and with volume load f.

For simplicity, we suppose the $z = z_0$ end to be fixed, and we have the surface load $\underline{p}(x)$, which is a net load between the Zdisk and M-line (to be determined later) on the other end $(z = z_1)$. We describe this three-dimensional case using the method of variations [7]:

$$\int_{z_0}^{z_1} \mathrm{d}z \int_F \left(\underline{f}\,\delta \underline{u} - \delta\Pi_3 \right) \mathrm{d}F + \int_F \underline{p}\,\delta \underline{u}\left(\underline{x}, z_1\right) \mathrm{d}F = 0 \quad ,(1)$$

$\underline{u}\left(\underline{x}, z_0\right) = 0$, $\Pi_3 \equiv \mu \left(\underline{\underline{\varepsilon}} \cdot\cdot \underline{\underline{\varepsilon}} + \frac{\nu}{1-2\nu}\left(tr\underline{\underline{\varepsilon}}\right)^2 \right)$ where $\underline{u}$ is the translation vector, Π_3 is the volumetric energy, μ is the Lamé constant, ν is the Poisson ratio, $\underline{\underline{\varepsilon}}$ is the deformation tensor, and $tr\underline{\underline{\varepsilon}}$ is the trace of deformation tensor, the first invariant (double underlining indicates tensors).

To derive equations from the variation principle, we approximate the translation as:

$$\underline{u}(\underline{x},z) \equiv u_z \underline{k} + \underline{u}_\perp = \underline{U}(z) + \underline{\theta}(z) \times \underline{x} + \alpha(z) W(\underline{x}) \underline{k}, \qquad (2)$$

where $\underline{U}(z)$ is the linear translation, $\underline{\theta}(z)$ is the rotation vector, $\alpha(z)$ is the rotation angle per unit length, and $W(\underline{x})$ is the warping function:

$\Delta W = 0$, $\underline{l} \times \underline{k} \cdot \nabla W|_{\partial F} = \underline{l} \cdot \underline{x}$,

$\underline{l}$ is the basis vector of the Lagrangian coordinate.

Let us suppose that there is a lack of transversal shifts; then:

$$\underline{U}'_\perp = \underline{\theta}_\perp \times \underline{k}, \quad \alpha(z) = \theta'_z. \qquad (3)$$

Taking into account Equation (3), approximation (2) becomes:

$$\underline{u}(\underline{x},z) = (\underline{U}_z - \underline{U}'_\perp \underline{x} + \theta'_z W)\underline{k} + (\underline{U}_\perp + \theta_z \underline{k} \times \underline{x}), \qquad (4)$$

That is:

$$u_z = U_z - \underline{U}'_\perp \underline{x} + \theta'_z W, \; \underline{u}_\perp = \underline{U}_\perp + \theta_z \underline{k} \times \underline{x}. \qquad (5)$$

From the classical theory of elasticity, we know that:

$$\underline{\underline{\varepsilon}} = \nabla \underline{u}^S, \qquad (6)$$

where S is the symmetrization symbol.

Using (4), Equation (6) can be rewritten:

$$\underline{\underline{\varepsilon}} = U'_z \underline{k}\underline{k} - \underline{U}''_\perp \underline{x}\underline{k}\underline{k} + \theta''_z W \underline{k}\underline{k} + \theta'_z \nabla W \underline{k}\underline{k} + \theta'_z (\underline{k} \times \underline{x}) \underline{k}$$
$$= \varepsilon_z \underline{k}\underline{k} + \theta'_z (\nabla \Phi \times \underline{k}\underline{k})^s \qquad ,(7)$$

where $\nabla W + \underline{k} \times \underline{x} = \nabla \Phi \times \underline{k}$ and $\varepsilon_z = U'_z - \underline{U}''_\perp \cdot \underline{x} + \theta''_z W$.

Then:

$$\underline{\underline{\varepsilon}} \cdot\cdot \underline{\underline{\varepsilon}} = \varepsilon_z^2 + \theta_z'^2 \frac{|\nabla\Phi|^2}{2}, \ \left(tr\underline{\underline{\varepsilon}}\right)^2 = \varepsilon_z^2 \tag{8}$$

Substituting (8) into (1), we obtain a new volumetrical energy density:

$$\Pi_3 = \mu\left(\varepsilon_z^2 + \theta_z' \frac{|\nabla\Phi|^2}{2} + \frac{\nu}{1-2\nu}\varepsilon_z^2\right)$$

$$= \mu\frac{1-\nu}{1-2\nu}\varepsilon_z^2 + \frac{\mu}{2}\theta_z'^2|\nabla\Phi|^2$$

Since $\mu = \dfrac{E}{2(1+\nu)}$:

$$\Pi_3 = \frac{1}{2}\left(\hat{E}\varepsilon_z^2 + \mu\theta_z'^2|\nabla\Phi|^2\right), \tag{9}$$

where $\hat{E} = E\dfrac{1-\nu}{(1+\nu)(1-2\nu)}$, and E is Young's modulus.

Taking into account (7), (9) becomes:

$$\Pi_3 = \frac{\hat{E}}{2}\left(U_z' - \underline{U}_\perp''\underline{x} + \theta_z''W\right)^2 + \frac{\mu}{2}\theta_z'^2|\nabla\Phi|^2$$

$$= \frac{\hat{E}}{2}\left[U_z'^2 + \left(\underline{U}_\perp''\underline{x}\right)\left(\underline{U}_\perp''\underline{x}\right) + \theta_z''^2W^2 + 2U_z'\theta_z''W\right.$$

$$\left. -2U_z'\underline{U}_\perp''\underline{x} - 2\underline{U}_\perp''\underline{x}\theta_z''W\right] + \frac{\mu}{2}\theta_z'^2|\nabla\Phi|^2 \tag{10}$$

Integrating (10) in the section:

$$\Pi_1 \equiv \int_F \Pi_3 \mathrm{d}F$$

$$= \frac{\hat{E}}{2}\left[U_z'^2 \int_F \mathrm{d}F + \underline{U}_\perp'' \left(\int_F \underline{x}\underline{x} \mathrm{d}F \right) \underline{U}_\perp'' + \theta_z''^2 \int_F W^2 \mathrm{d}F \right.$$

$$\left. + 2U_z'\theta_z'' \int_F W \mathrm{d}F - 2U_z'\underline{U}_\perp'' \int_F \underline{x} \mathrm{d}F - 2\underline{U}_\perp''\theta_z'' \int_F \underline{x} W \mathrm{d}F \right]$$

$$+ \frac{\mu}{2}\theta_z'^2 \int_F |\nabla\Phi|^2 \mathrm{d}F$$

$$= \frac{\hat{E}}{2}\left(FU_z'^2 + \underline{U}_\perp'' \underline{\underline{J}} \underline{U}_\perp'' + \theta_z''^2 I_0 - 2\theta_z'' \underline{U}_\perp'' \underline{\underline{J}} \underline{\eta} \right) + \frac{\mu}{2} C \theta_z'^2 , \qquad (11)$$

where $\underline{\underline{J}} = \int_F \underline{x}\underline{x} \mathrm{d}F$, $I = \int_F W^2 \mathrm{d}F$, $\int_F W \mathrm{d}F = 0$, $\underline{\underline{J}}\underline{\eta} = \int_F W \underline{x} \mathrm{d}F$, $C = \int_F |\nabla\Phi|^2 \mathrm{d}F = 2\int_F \Phi \mathrm{d}F$.

Let us now determine the work of volume loads:

$$\int_F \underline{f}\,\delta\underline{u} \mathrm{d}F = \int_F \underline{f}\left(\delta\underline{U} + \underline{k} \times \underline{x}\delta\theta_z - \underline{x}\underline{k}\delta\underline{U}_\perp' + W\underline{k}\delta\theta_z' \right) \mathrm{d}F$$

$$= \delta U_z \int_F f_z \mathrm{d}F + \delta\underline{U}_\perp \int_F \underline{f}_\perp \mathrm{d}F + \underline{k}\delta\theta_z \int_F \underline{x} \times \underline{f} \mathrm{d}F$$

$$- \underline{k} \times \delta\underline{U}_\perp' \int_F \underline{f} \times \underline{x} \mathrm{d}F + \delta\theta_z' \int_F \underline{f}\,\underline{k} W \mathrm{d}F$$

$$= q_z \delta U_z + \underline{q}_\perp \delta\underline{U}_\perp + \underline{m}\left(\underline{k} \times \delta\underline{U}_\perp' + \underline{k}\delta\theta_z \right) + b\delta\theta_z' , \qquad (12)$$

where $q_z \equiv \int_F f_z \mathrm{d}F$, $\underline{q}_\perp \equiv \int_F \underline{f}_\perp \mathrm{d}F$, $\underline{m} \equiv \int_F \underline{x} \times \underline{f} \mathrm{d}F$, $b \equiv \int_F f_z W \mathrm{d}F$ is the distributed bimoment per unit length.

Similarly, a work of the volume load on the end:

$$\int_F \underline{p}\delta\underline{u}\mathrm{d}F = Q_{1z}\delta U_z + \underline{Q}_{1\perp}\delta\underline{U}_\perp$$

$$+\underline{M}_1\left(\underline{k}\times\delta\underline{U}'_\perp + \underline{k}\times\delta\theta_z\right) + B_1\delta\theta'_z, \qquad (13)$$

where $Q_{1z} = \int_F p_z \mathrm{d}F$, $\underline{Q}_{1\perp} = \int_F \underline{p}_\perp \mathrm{d}F$, $\underline{M}_1 = \int_F \underline{x}\times\underline{p}\mathrm{d}F$, $B_1 = \int_F \underline{p}_z W \mathrm{d}F$ is the bimoment on the end.

Taking into account Equations (12) and (13), the variational equation becomes:

$$\int_{z_0}^{z_1}\left[q_z\delta U_z + \underline{q}_\perp\delta\underline{U}_\perp + \underline{m}\left(\underline{k}\times\delta\underline{U}'_\perp + \underline{k}\delta\theta'_z\right) + b\delta\theta'_z\right]\mathrm{d}z$$

$$-\int_{z_0}^{z_1}\left[\frac{\partial\Pi_1}{\partial U'_z}\delta U'_z + \frac{\partial\Pi_1}{\partial\underline{U}''_\perp}\delta\underline{U}''_\perp + \frac{\partial\Pi_1}{\partial\theta'_z}\delta\theta'_z + \frac{\partial\Pi_1}{\partial\theta''_z}\delta\theta''_z\right]\mathrm{d}z$$

$$+Q_{1z}\delta U_z + \underline{Q}_{1\perp}\delta\underline{U}_\perp + \underline{M}_1\left(\underline{k}\times\delta\underline{U}'_\perp + \underline{k}\delta\theta_z\right) + B_1\delta\theta'_z = 0 \qquad .(14)$$

From (14), we obtain differential equations and limits:

$$Q_z = \frac{\partial\Pi_1}{\partial U'_z} = \hat{E}FU'_z \quad Q'_z + q_z = 0,$$

$$z_1: Q_z = Q_{1z}, \quad \underline{Q}'_\perp + \underline{q}_\perp = 0, \quad \underline{M}'_\perp + \underline{k}\times\underline{Q}_\perp + \underline{m}_\perp = 0,$$

$$\underline{M}_\perp = \underline{k}\times\frac{\partial\Pi_1}{\partial\underline{U}''_\perp} = \hat{E}\underline{k}\times\underline{\underline{J}}\left(\underline{U}''_\perp - \underline{\eta}\theta''_z\right) \qquad ,(15)$$

$$z_1: \underline{Q}_\perp = \underline{Q}_{1\perp}, \quad \underline{M}_\perp = \underline{M}_{1\perp}$$

$$M'_z + m_z = 0, \quad M_z = \mu C\theta'_z - B' - b,$$

$$B = \frac{\partial\Pi_1}{\partial\theta''_z} = \hat{E}\left(I\theta''_z - \underline{U}''_\perp\underline{\underline{J}}\underline{\eta}\right),$$

$$z_1: M_z = M_{1z}, \quad B = B_1, \quad \underline{u}\left(\underline{x}, z=0\right) = 0$$

The set of equations derived in (15) gives us a chance to completely describe muscle fiber membrane behavior as a long cylindrical envelope and to find the potential energy Π_3 in the case described.

Let us assume that there is no external moment influence and that section warping and transversal shift contribution are negligible in comparison with the longitudinal component. This assumption is justified because of the specific muscle cell structure. Then, solving (15), we can find U_z:

$$U_z = \frac{1}{\hat{E}F}\left[-\iint_F \int f_z \mathrm{d}F\mathrm{d}z + \iint_F \int f_z \mathrm{d}F\mathrm{d}z \Big|_{z=0} + z\left(\iint_F f_z \mathrm{d}F\mathrm{d}z + \int_F p_z \mathrm{d}F \right)\Bigg|_{z=z_1} \right], \qquad (16)$$

where f_z is the volume load, p_z is the surface load, F is the section area, $\hat{E}$ is the adduced Young's modulus, and ν is the Poisson ratio.

2.2. External Mechanical Loading

An external mechanical field acts on the whole organism and launches a number of processes, leading to nervous activation change in skeletal muscles, a liquid shift in the cranial direction, and, as a result, to a volume load change in the heart.

These processes results in the muscle fiber membrane becoming subjected to the following forces: f_{ca} by the contractile apparatus as a result of nervous activation, p_{hy}, hydrostatic pressure (only for cardiomyocytes), and f_{gr}, the gravity. Nervous activation by intracellular signal mechanisms launch results in mechanical tension that arises in a muscle fiber because of myosin head and actin filament interaction, which is transmitted into the sarcolemma by the cortical cytoskeleton. Let us suppose that this interaction is uniform distributed over the length of the contractile apparatus. Therefore, it can be represented as a periodical function with the period of T, which equals the distance between two successive myosin heads.

Then $f_{ca} = f_0 n \sin\left(\frac{2\pi}{T} z\right)$, where f_0 is the force generated by the single bridge, approximately 3 - 5 pN, n represents a number of bridges, which can be determined as the ratio of the fiber length (l) to the distance between two successive bridges, approximately 43 nm, per fiber volume. Gravity also acts on a muscle cell depending on the cell orientation. The specific volumetric force in this case may be represented as $f_{gr} = \rho g \cos\varphi \cdot \frac{z}{z_1}$, where

φ is the angle between the gravity vector and the fiber longitudinal axis direction, g is the free-fall acceleration, and ρ is the liquid density. The surface load on the end is a net load of gravity and hydrostatic pressure; $p_z = \rho d g \sin\varphi + p_{hy}$, where d is the fiber diameter.

Then, the external forces become:

$$f_z = f_{ca} + f_{gr} = f_0 \frac{4}{\pi d^2 T} \sin\left(\frac{2\pi}{T} z\right) + \rho g \cos\varphi \cdot \frac{z}{z_1},$$

$$p_z = \rho d g \sin\varphi + p_{hy}. \quad (17)$$

3. NUMERICAL EXAMPLES

As an experimental model, we use a rodent's antiorthostatic suspension (**Figure 2**).

We have determined in previous experiments Young's modulus of the sarcolemma of different skeletal muscles [4] and a rat's left ventricle [8,9]. Moreover, the introduction of the nifedipin system into rats resulted in increased sarcolemma transversal stiffness in skeletal muscles [10]. We have also determined the muscle fiber diameter. Based both on these data and ratio (16), we can find characteristic longitudinal deformations of M. soleus fibers and cardiomyocytes (**Table 1**).

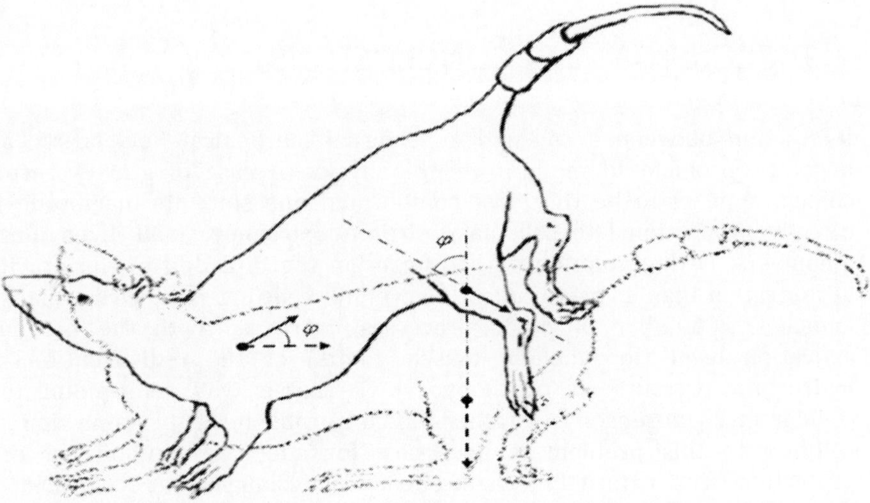

Figure 2. The Ilyin-Novikov method with the Morey-Holton modification of a rodent's antiorthostatic suspension [5]. The dotted arrows indicate the direction of muscle cells' dominant axis (z axis) under normal conditions, and the solid arrows represent the suspension case.

Table 1. Longitudinal absolute and relative deformations of different muscle cells after antiorthostatic suspension at an angle of 30°.

Muscle Object		M. soleus	Heart
Rats	U_z, μm	4.29	9.5
	U_z/l	2.15×10^{-4}	6.31×10^{-2}
Rats + nifedipin (10 mg per day per 1 kg of animal weight)	U_z, μm	1.98	-
	U_z/l	0.99×10^{-4}	

Results obtained for the following initial data: $\delta = 30$ nm, $\rho = 10^3$ kg/m^3, $g = 9.8$ m/s^2, $\varphi = 30°$, $f_0 = 4 \times 10^{-12}$ N, $l = 2 \times 10^{-2}$ m (for m. soleus fibers), and $l = 1.5 \times 10^{-4}$ m, $d = 2 \times 10^{-5}$ m, $p_{hy} = 50$ Pa (for cardiomyocytes). $T = 43 \times 10^{-9}$ m, $z = 2.4 \times 10^{-6}$ m, $z_1 = 1.8 \times 10^{-6}$ m.

4. DISCUSSION AND CONCLUSIONS

Interaction between a cell and an external mechanical field is still an unsolved problem in modern cell biophysics. A case of gravity vector change appears to be the most complicated one since its magnitude is directly proportional to cell mass, so it is extremely small. Even more ambiguous is the situation with changes in the direction of the gravity vector rather than its magnitude. However, for cells in tissue, such a change launches a number of other processes connected with the nervous activation level (for skeletal muscle fibers) or the redistribution of hydrostatic pressure (in cardiomyocytes). That is why determining the cellular mechanosensor is a rather difficult problem. At the same time, a solution to this problem is necessary for adequate methods of cell protection (from external mechanical condition changes that occur during, for example, a space flight) to be developed.

We have previously supposed that the submembrane cytoskeleton, the most universal compartment for different types of cells, is a mechanosensor [4,9,11,12]. However, the question regarding the significance of deformations arising in the cortical cytoskeleton still cannot be experimentally proved.

Therefore, in this work, we decided to theoretically evaluate deformations that appear due to the effects described above. To obtain a proper estimation, we developed a muscle fiber sarcolemma model as a thin-walled rod. A series of assumptions enabled us to obtain an analytical solution, which makes it possible to estimate the deformation to the lower limit. The results showed that the absolute deformations in skeletal muscles and in the heart are comparable, but the relative deformation of the myocardium is dramatically greater. Nifedipin introduction changes the mechanical properties of muscle cells such that the deformations there appear to be minimal.

The numerical estimation results highly correspond with experimental data about the cell response formation speed: changes in cardiomyocytes occur intense than in soleus fibers, and this fact remains unexplained [4,8]. We should note the different signs of deformations in the heart and in the soleus: one case involved stretching and the one compression, which could mean the launching of different cell response signal pathways.

Irina V. Ogneva, Nikolay S. Biryukov Thus, according to the results of the ematical model developed and numerical experiments indicate that the external mechanical load change could lead to cortical cytoskeleton deformation. The deformation of cardiomyocytes is greater than that of soleus muscle fibers. Furthermore, the deformations have different signs in the heart and in the soleus: one involves describes stretching and the other compression. One can suppose that the different signs of deformation result in different cell responses, probably due to various proteins dissociating from the cortical cytoskeleton in each case. In addition, the deformation magnitude may influence the speed of the cell response; this fact corresponds to experimental data. In total, the results obtained from ematical modeling explain a series of experimental data, particularly various start-up times of the cell response in the heart and soleus muscles, as well as different adaptive protein patterns in these cells as a result of external mechanical condition changes and suggests a new approach to the search for cellular mechanosensors.

REFERENCES

1. A. A. Shabarchin and A. K. Tsaturyan, "Proposed Role of the M-Band Sarcomere Mechanics and Mechano-Sensing: A Model Study," Biomechanics and Modeling in Mechanobiology, Vol. 9, No. 2, 2010, pp. 163-175.
2. J. M. Ervasti, "Costameres: The Achille's Heel of Herculean Muscle," The Journal of Biological Chemistry, Vol. 278, No. 16, 2003, pp. 13591-13594.
3. I. V. Ogneva, D. V. Lebedev and B. S. Shenkman, "Transversal Stiffness and Young's Modulus of Single Fibers from Rat Soleus Muscle Probed by Atomic Force Microscopy," Biophysical Journal, Vol. 98, No. 3, 2010, pp. 418-424.
4. I. V. Ogneva, "Transversal Stiffness of Fibers and Desmin Content in Leg Muscles of Rats under Gravitational Unloading of Various Durations," Journal of Applied Physiology, Vol. 109, No. 6, 2010, pp. 1702-1709.
5. E. Morey-Holton, R. K. Globus, A. Kaplansky and G. Durnova, "The Hindlimb Unloading Rat Model: Literature Overview, Technique Update and Comparison with Space Flight Data," Advances in Space Biology and Medicine, Vol. 10, 2005, pp. 7-40.
6. V. Z. Vlasov, "Thin-Walled Elastic Rods," Fizmatgiz, USSR, Moscow, 1959, 574 pages.
7. V. V. Eliseev, "Mechanics of Elastic Bodies," Izdatelstvo SPbSTU, Saint-Peterburg, 1999, 336 pages.
8. I. V. Ogneva and I. B. Ushakov, "The Transversal Stiffness of Skeletal Muscle Fibers and Cardiomyocytes in Control and after Simulated Microgravity," In: C. L. Frewin, Ed., Atomic Force Microscopy Investigations into Biology: From Cell to Protein, InTech, Chroatia, 2012, 354 pages.
9. I. V. Ogneva, T. M. Mirzoev, N. S. Biryukov, O. M. Veselova and I. M. Larina, "Structure and Functional Characteristics of Rat's Left Ventricle Cardiomyocytes under Antiorthostatic Suspension of Various Duration and Subsequent Reloading," Journal of Biomedicine and Biotechnology, Vol. 2012, 2012, Article ID: 659869, 11 pages.
10. I. V. Ogneva and E. G. Altaeva, "Effects of Nifedipine on the Mechanical Properties of Sarcolemma and Modulation of Calcium Accumulation Dynamics in Fibers of the Rat Soleus Muscle under Short-Term Hypogravity Conditions," Biofizika, Vol. 55, No. 5, 2010, pp. 918-924.
11. I. V. Ogneva, "Transversal Stiffness and Beta-Actin and Alpha-Actinin-4 Content of the M. soleus Fibers in the Conditions of a 3-Day Reloading after 14-Day Gravitational Unloading," Journal of Biomedicine and Biotechnology, Vol. 2011, 2011, Article ID: 393405, 7 pages.
12. I. V. Ogneva, "Cell Mechanosensitivity: Mechanical Properties and Interaction with Gravitational Field," BioMed Research International, Vol. 2013, 2013, Article ID: 598 461, 17 pages.

CHAPTER 3

Topology in Biology: Singularities and Surgery Transformations in Metazoan Development and Evolution

Valeria V. Isaeva[1,2], Nickolay V. Kasyanov[3], Eugene V. Presnov[4]

[1]*A.N. Severtsov Institute of Ecology and Evolution of the Russian Academy of Science, Moscow, Russia*

[2]*A.V. Zhirmunsky Institute of Marine Biology, Far East Branch of Russian Academy of Sciences, Vladivostok, Russia*

[3]*Institute of Theory and History of Architecture and Town Planning, Moscow, Russia*

[4]*The Volcani Center, Gilat, Israel*

ABSTRACT

The review presents a topological description and interpretation (analysis) of some events in metazoan development and evolution through the use of well-known ematical concepts and theorems (using topological approach). It is the topological language that can provide strict and adequate description of various phenomena in developmental and evolutionary transformations. Topological singularities inevitably arising and transforming during early development destroy the preexisting pattern of symmetry. The symmetry breaking of preexisting spatial pattern plays a critical role in biological morphogenesis in development and evolution. Some events of early development are interpreted in terms of symmetry breakdown and related to well-known ematical theorems. A topological inevitability of some developmental events through the use of classical topological concepts is discussed. The topological approach makes it possible to consider the succession of spherical surgeries, which change the topological genus of an animal body surface. We model the biological shape as a set of smooth, closed, oriented surfaces—membrane or epithelial layers. Membrane and epithelial surfaces are boundary layers, interfaces between a living structure and its environment, ensuring metabolism. Toroid forms as well as fractal structures in metazoans can be considered as functionally optimized biological design and attractors in biological morphogenesis. The epithelial surface is an interface between the internal medium of an organism and the outside environmental medium; topological and fractal transformations during metazoan evolution and development increase this interface, ensuring better adaptation of organism to the environment. Fractal structures as well as toroid forms can be considered as a functionally optimized design in Metazoa. Topological methodology reveals a certain set of topological rules constraining and directing biological morphogenesis during evolution and development.

Keywords: Topological Surgery, Singularity, Symmetry Breakdown, Evolutionary Biology, Embryology

1. INTRODUCTION

The living organisms inhabit and develop in the real physical space and are organized according to the properties of this space. There are general principles in physics, geometry and topology certainly is applied to many situations in biology [1] . Spatial organization of living systems has a dynamical character and includes such discontinuities as topological surgeries and symmetry breaking. Topological language and models can provide strict and adequate description and modeling of biological objects and events since topology operates with the most general properties of spaces as ematical subjects. Topological modifications of metazoan body patterns in development and evolution are discrete, qualitative steps in biological morphogenesis. Discrete character of biological morphogenesis involves qualitative topological discontinuities, topological catastrophes [2] -[4] .

Topological analysis of DNA structure and functions showed the power of the topological methodology see [5] [6] . Following R. Thom [2] -[4] we applied topological concepts and some theorems to describe and interpret developmental and evolutionary form dynamics. In our publications we introduce the topological language into descriptions of biological development at different levels of organization [7] -[15] .

The topological approach to description of biological forms and morphogenetic processes has become a more common practice covering the fields of evolutionary and developmental biology, animal behavior and self-or- ganization, neurology, metagenomic biology and biomedicine [9] [10] [14] -[19] . Topological studies by molecular biologists became the field of biochemical topology e.g. [5] [6] . Accordingly, the topological analysis of metazoan transformation can be called topological biology or biological topology [14] [15] .

The topological methodology makes it possible to consider as a whole the succession of shape transformations during metazoan development and evolution. Topological approach reveals also a certain set of topological rules constraining and directing biological morphogenesis during evolution and development.

2. TOPOLOGICAL SINGULARITIES AND SYMMETRY TRANSFORMATIONS IN EARLY DEVELOPMENT

Because an animal body is formed from the spherical egg cell, symmetry breakdown is one of the fundamental processes of development. Symmetry breaking and symmetry propagation are fundamental processes in biological morphogenesis, in metazoan evolution and development [1] [20].

In most metazoans, the animal-vegetal (anterior-posterior) axis of the egg and future embryo is established during oogenesis and usually depends on the position of oocyte with respect to its cellular and non-cellular surrounding. The first symmetry breakdown is associated with the emergence of morphological and functional gradients along animal-vegetal axis of an oocyte. It was shown that spatial anisotropy in distribution of gene products in ooplasm determines the polarity of egg and embryo specifying polar axes of future organism [21] . Cytoskeleton rearrangements break cell symmetry [22] . Actin and tubulin were shown to polymerize forming cytoskeletal actin filaments and microtubules that define the shape of a cell, its polarity and migration, intracellular transport, and the mitotic spindle forming during division. Both the actin and microtubule cytoskeletal elements work together to convert this initial asymmetry into a global cell polarity, for example, in polarization of Drosophila oocytes [20] [22] . A polarized cellular architecture is fundamental to formation of embryonic axes [20] . Egg cytoskeleton functions as a global morphogenetic determinant, which directs and maintains the anisotropy of molecular information in the ooplasm so determining axis polarity of an egg and a future organism [14] . Different organisms use different mechanisms to establish polarity but all rely mainly on the actin and microtubule cytoskeletons. Radially symmetrical arrays of microtubules in animal cells are often called asters, and the presence of such aster breaks the cell symmetry [22] . In egg cells, the aster as visualized topological singularity localizes near animal egg pole.

The axial polarity of an egg is manifested in transcellular ion flows, which generates the extracellular electric field as signals and effectors of the egg polarity [23] (Figure 1(a)).

Egg cell polarization as the emergence of the topological singularity (or singularities) of the vector field on a sphere is inevitable according to the Poincaré-Hopf theorem: "For any smooth vector field given on two-dimensional sphere S^2, a singular point of the field exists" [9] -[12] . The vector field on the sphere has at least one singularity, i.e. the point where the value of the field becomes zero (Figure 2(a)). This singular point may be either animal or vegetal pole, or both setting the animal-vegetal differences. This singularity breaks the spherical symmetry of the egg, polarizing the egg sphere [9] [10] [12] [14] .

In some animals, at least in chordates, the second symmetry breaking takes place following gamete contact and fusion; this symmetry breaking process involves the asymmetrical flow of cortical actin networks [22] . Sperm penetration in an egg triggers the whole cascade of phenomena, resulting in the wave of calcium release, the electric current of egg activation (Figure 1(b)) and a rearrangement of cortical structure [24] [25] . In chordates, besides preexisting animal egg pole, a new singularity emerges following penetration of a spermium and its spermaster that determines dorso-ventral axis of the egg and future animal see [15] [22] [26] . So the local order of sperm penetration is connected to the integral order of the egg, zygote and developing organism. This symmetry breaking process involves the ooplasmic segregation as a flow of cortical actin networks and displacement of cortical egg components [14] [26] . Ooplasmic segregation creates the spatial-temporal vector field of the cortical displacement, and a singularity of this field is the fixed point of ooplasmic segregation. We consider ooplasmic segregation as a mapping a ball into itself [9] [10] [12] . According to Brouwer's theorem on fixed point, any continuous mapping of the ball into itself has a fixed point. During ooplasmic segregation, subcellular components of an egg are moving with respect to a fixed, stationary point in the course of continuous deformation. The local order of ooplasmic segregation is connected to integral order of the zygote and developing organism [9] [12] . Egg symmetry transformations provide the most important developmental information determining axial pattern of future animal.

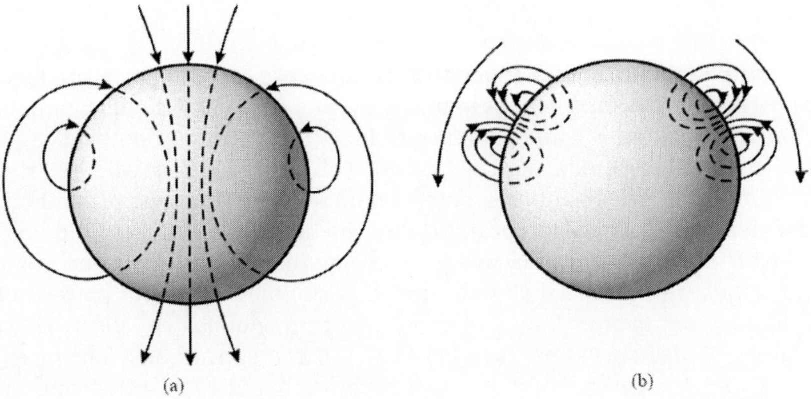

Figure 1. Electric currents in eggs of the frog Xenopus laevis: (a) The electric field in an oocyte (after [23] , modified); (b) The wave of the activation current following sperm penetration (after [27] modified).

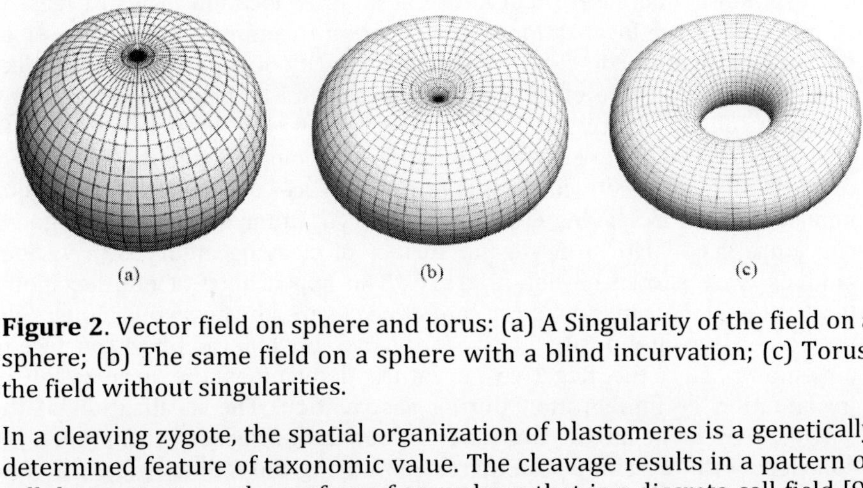

Figure 2. Vector field on sphere and torus: (a) A Singularity of the field on a sphere; (b) The same field on a sphere with a blind incurvation; (c) Torus, the field without singularities.

In a cleaving zygote, the spatial organization of blastomeres is a genetically determined feature of taxonomic value. The cleavage results in a pattern of cellular contacts on the surface of an embryo that is a discrete cell field [9] [10] [12] [14] . Each cell of the field has a number of its edges corresponding to the number of neighboring cells. There are only five homogenous discrete fields on the sphere corresponding to five regular polyhedrons, Platonic bodies. During synchronous cleavage divisions, only the first four blastomeres can create a homogeneous field on the embryo surface (a cleaving embryo is considered as topological polyhedron, not geometrical one [9] [10] . After third division (8 blastomeres) the field pattern of cellular contacts on the embryo surface becomes topologically inhomogeneous inevitably that means the symmetry breaking of the discrete cell field. In hexagonal cell network, presence of cells with another number of sides (for example, with 5 or 7 sides) can be considered as a topological dislocation of homogeneous graph cells [28] (Figure 3(a), Figure 3(b)). This means that the symmetry of the field is breaking, and this symmetry breaking is a consequence of Euler's theorem on polyhedra [10] [12] . The number of adjacent cells for any cell of the discrete cell field is the local order of the field, and the general number of vertices, edges and faces of the given graph is the integral order of the field. So a discrete field on the spherical surface of a cleaving embryo inevitably has topological singularities, and the local singularities determine the integral pattern of a developing organism. Gauss-Bonnet theorem defines a discrete cell curvature, Gauss curvature (see [9] [10]). If the sum of adjacent faces is more than 6, the curvature of the face becomes negative. It was proposed a causal relation between cell negative curvature, which leads to incurvation (invagination) and following morphogenetic cell movement during gastrulation [9] [11] [12] [16] . According to the Gauss-Bonnet theorem, an increase of curvature at some location has to be counterbalanced by a corresponding decrease of curvature at another location [16] . At least in some animals, the localization of invagination (or immigration, ingression) during gastrulation coincides with the singularity of negative curvature i.e., with the highest value of the cell field on vegetal hemisphere of embryos, whereas cells of positive curvature are present near animal pole [9] [10] [12] . Figure 3(c), Figure 3(d) demonstrates topological inhomogeneities of discrete fields of cell contacts on the surface of animal and vegetal hemispheres of a cleaving crustacean embryo. Other examples of negative curvature singularities on vegetal surface of cleaving embryos in various animals were shown earlier [9] [11] . The appearance of non-hexagonal faces is inevitable since a sphere cannot be covered by hexagons alone. Cells with a negative and positive curvature inevitably emerge on the surface of an embryo, and the negative curvature determines the localization of invagination or immigration during gastrulation. The localization of the singularities of negative curvature on vegetal surface of these embryos coincides with the site of cell ingression during the following gastrulation process, whereas cells of positive curvature are present on the animal pole surface. Thus, gastrulation may be related to an inevitable emergence of discrete curvature that directs development in three-dimensional space.

Gastrulation is the first spherical surgery in development, a topological catastrophe. The completion of gastrulation is topological transition from the sphere with singularity (singularities) to the torus, which is a topologically stable object with homogeneous discrete field, without singularities [9] [10] [12] (Figure 2). So a postgastrulation embryo or larva with through intestine tube is a topologically stable object. According to L. Wolpert, "the most important time in your life is not birth, marriage, or death, but gastrulation" (cited after [16] [27] [29]). Gastrulation is a way of topological stabilization in embryogenesis; consequently, gastrulation is indeed a very important event in our life.

Gastrulation is an extensive rearrangement of cells in an embryo. In embryos, cell migration pathways during gastrulation and other morphogenetic processes [27] visualize the morphogenetic vector fields with singular points. Thus, the breaking of spatial symmetry of preexisting pattern is inevitable and plays a critical role in early development.

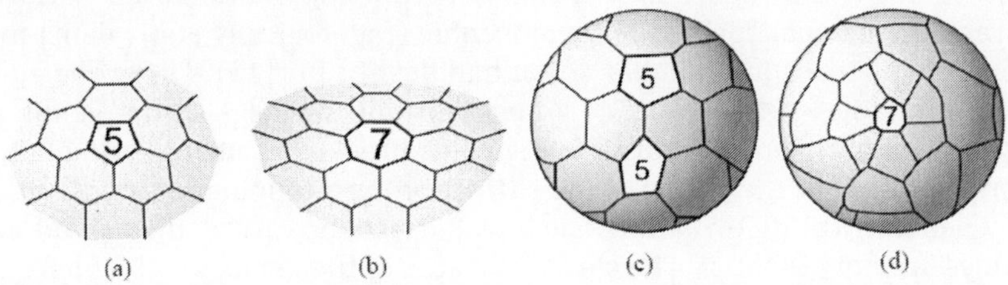

Figure 3. Topological inhomogeneities of cell discrete fields fields: (a) The cell of positive curvature among hexagonal cells; (b) The cell of negative curvature; (c) Two cells of positive curvature on animal hemisphere of a crustacean embryo; (d) The cell of negative curvature on vegetal hemisphere of the same embryo (after [10] , modified).

3. TOPOLOGICAL SURGERIES IN METAZOAN EVOLUTION AND DEVELOPMENT

In metazoan evolution, appearance of the epithelial tissue was one of the most significant innovations [30] [31] . During metazoan development, epithelia determine the increasing complexity of the body changing their local curvature [16] . Such epithelial morphogenetic movements as the folding, spreading, budding, cavitation, and delamination produce reorganization of whole body plan [30] . Epithelial cells form cohesive cell sheets; the structural and functional connectivity of an epithelial layer during epithelial morphogenesis is ensured by the system of specialized intercellular contacts that integrate cells into a united morphological and functional entity [30] [31] .

To translate traditional morphological descriptions of a metazoan morphogenesis into topological language the external shape of an organism is modeled as a set of smooth, closed, orientable surfaces formed by epithelial layers neglecting the thickness of the cellular layers [7] [8] [12] [13] [15] [16] . The establishment of apical-basal polarity of epithelial cells generates two surfaces of an epithelium with different components and functions [32] [33] . Thus, epithelial cell layers are characterized by connectivity, closeness of an intact surface and apical-bas- al polarity, anisotropy.

Metazoan morphogenesis may be represented as topological modification(s) of epithelial surfaces of spherical or toroidal shape. According to the theorem of elementary topology, any closed surface in three-dimensional space is homeomorphic (topologically equivalent) to the sphere with a certain number (p) of handles [8] [9] [11] -[14] . The sphere with p handles sets a class of homeomorphic surfaces of the genus p. If there are no topological surgeries, such as breaking (cutting) and gluing of epithelial sheets, the genus of the surface (p) is a topological invariant, and any geometrical deformations such as surface curvature, linear and angle values are not essential. The closed surfaces of the genus p = 0 (sphere), p = 1 (torus), p = 2 (double torus, "pretzel") and so on give a topological classification. The differences between these surfaces have a topological, qualitative nature [3] .

Topological handles in biological objects are usually realized as channels (or canals, holes), for example, the digestive tube. We consider only epithelized through channels as topological handles. Repeated addition of "handles" or creation of "holes", "tunnels" has led to formation of double, triple, and higher order toroids [16] . An adult metazoan organism topologically is a sphere with a set of handles. Metazoan morphogenesis may be represented as topological modification(s) of closed epithelial surfaces.

Topological transitions in metazoan development occur due to topological surgeries realized by "cutting" and "gluing" of epithelial sheets [7] [8] , as "self-wounding and healing" [16] . Local topological surgeries in epithelial morphogenesis involve complex processes at the cellular level: sheet disintegration followed by cell adhesion and cytoskeletal cooperation giving rise to a newly formed cell sheet (or sheets).

We can see two kinds of topological surgeries of epithelial surfaces in biological morphogenesis. Spherical surgeries change the genus of the surface. Another kind of topological surgeries changes the connectivity of epithelial layers by the separation of additional closed epithelial surfaces from preexisting ones or by the fusion of preexisting layers. The developmental modifications of connectivity do not change the topological pattern of the outer surface of animal body, so only the spherical surgeries changing the genus of the surface are considered here.

Topological patterns in metazoan evolution, from sponges to chordates, were analyzed using the genus of the surface as a topological invariant. We considered here (Figure 4) only basic topological archetypes of spatial organization for each taxon, neglecting variations and modifications inside the taxon. We used the interpretation of metazoan phylogeny, which consider three major clades of Bilateria: Deuterostomia, Lophotrochozoa, and Ecdysozoa (see, e.g., [34]).

In evolution of Metazoa we can find several topological transitions [12] [13] . In solitary cnidarians and such lower Bilateria as Acoelomorpha with a blind gut the kind of the surface is p = 0. The first important topological surgery in evolution of Bilateria (Triploblastica) is the appearance of a through intestinal tube instead of a blind gut. Among Bilateria, animals of many phyla have the stable topological pattern with the through digestive tube without any other through channels, and their external surface is topologically homeomorphic to the torus, p = 1 (Figure 4, Figure 5(a)). The appearance of a through gut is a topological innovation of great evolutionary importance resulting in the differentiation of digestive system and better utilization of nutrients from the external medium. Topologically the digestive tube is outer surface of an organism as well as ectodermal surface of the organism; both ecto- and entodermal epithelia form united continuous surface, the outer surface of an organism.

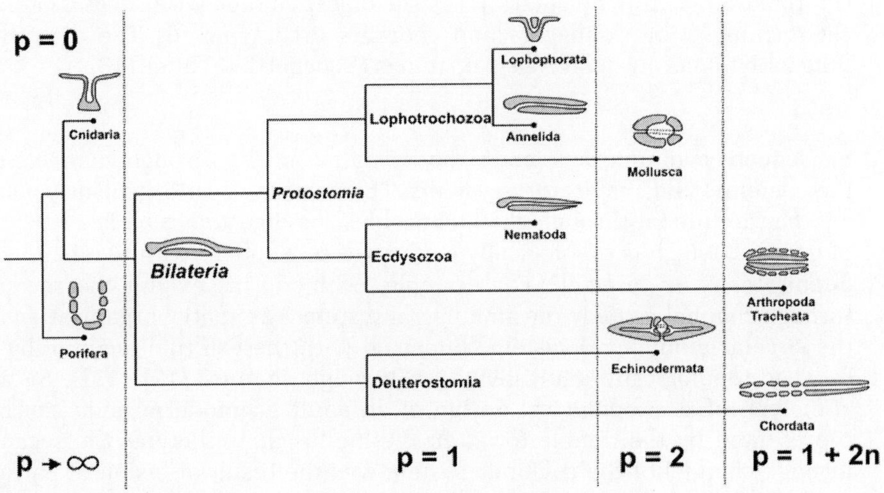

Figure 4. The scheme of main topological body patterns in Metazoa using genus of surface (p).

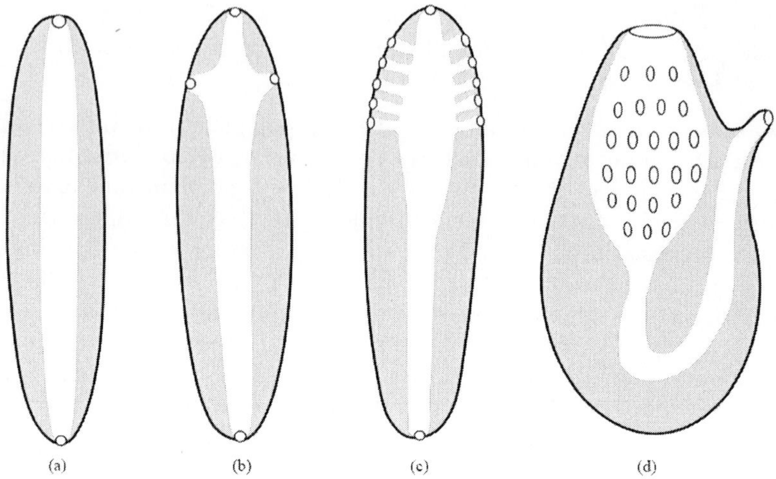

Figure 5. Topological patterns in some hemichordates and lower chordates: (a) A simple digective tube with oral and anal openinhs in most invertebrates; (b) The digestive canal with two additional lateral openings (Pterobranchia, Appendicularia); (c) The digestive tube with several lateral slit openings (Enteropneusta and chordate archetype; (d) The digestive tube with numerous pharyngeal openings (Urochordata) after [13] .

Subsequent evolution has led to the formation of the through channels of the coelomic and respiratory systems. The surface of an organism having another additional through channel besides the digestive tube is a surface of genus 2, which is topologically equivalent to a sphere with two handles, double torus or "pretzel". For example, echinoderms exhibit the second through channel, namely the ambulacral system. The spatial organization of the circular ambulacral system connected with the external medium by a canal is topologically equivalent to a through channel [12] [13] . So an archetype of the topological design of an adult echinoderm body can be represented by the double torus. In the molluscan archetype, the second through channel of the coelomic system was the result of fusion of paired coeloms via coelomoducts. Acquisition of the surface of the genus $p = 2$ may be considered as a side branch in the evolution of both Deuterostomia (in Echinodermata) and Protostomia (in some Lophotrochzoa).

The next evolutionary level of topological design in animals was attained through the development of the respiratory system. The system of paired tracheal tubules with lateral connections (of ectodermal origin) in Tracheata (higher terrestrial arthropods) or paired gill slits (of entodermal origin) in a chordate archetype evolved (Figure 5(c)). Accordingly, a parallelism of topological and functional innovations in the evolution of deuterostomes and of some protostomes (among Ecdysozoa) is obvious as a transition to the surface of the genus p = 1 + 2n, by adding an even number of channels (handles) of the respiratory system. The creation of through respiratory system is the topological modification of the surface, which results in better utilization of oxygen from outside medium flowing through the organism and more intensive metabolism. Through channels of the respiratory system are filled with external medium and may be considered as an outer surface of the animal body.

There are various evolutionary modifications of the archetypical body plan in many phyla due to additional topological surgeries. Colonial cnidarians are "multi-oral" organisms, i.e., multiple tori. In Spongia and in some Cnidaria, an increase of genus p up to very high values and the shaping of topologically complicated fractal systems are evident. In some scyphomedusae and creeping ctenophores branching and anastomosing canals of gastro-vascular system result in the increasing of genus p up to indefinite high values. In some flat worms (Platyhelminthes) complicated topological and fractal patterns have arisen on the base of the branched gut and excretory system [13] [14] .

Hemichordata have a through gut and several pairs of gill slits (p = 1 + 2n) similar to chordate archetype; in some Appendicularia, the digestive canal has two additional lateral openings (Figure 5(b), Figure 5(c)). Adult mammals and some other vertebrates (Chordata, Vertebrata) also exhibit a triple-torus pattern because their nostrils are connected to the gastrointestinal tract [16] ; gill slits in vertebrates are transient structures disappearing during embryogenesis. In birds, the genus of the body surface attains very high values because of numerous inner anastomoses in their bronchial system, which forms the bronchiolar network [13] . In ascidians (Chordata: Urochordata), the body surface is a high-order torus due to the presence of numerous gill openings (Figure 5(d)).

Thus, topological transformations of the body surface during metazoan evolution resulted in better distribution of flows to and from the external medium, used as the source of nutrients and oxygen and the sink of excreta, so ensuring greater metabolic intensity and better adaptation of organism to the environment.

An analysis of topological organization of the closed epithelial layer that covers the outer body surface as a continuous envelope is applicable to the topological organization of the body surface in metazoan development [12] -[14] [16] . During embryonic development, the surface of most metazoan animals undergoes sequential spherical surgeries, which change the topological genus of the surface, and developmental morphogenesis may be represented by series of topological surgeries of the closed body surface.

Local topological surgeries inevitably lead to global topological modifications of biological forms, as the genus of the surface is a global property.

No modification is observed in topology of the body surface during embryonic development of those animals which in adult state have the zero genus of the surface as some solitary cnidarians and Acoelomorpha [12] [13] [16] .

The surfaces of an egg, blastula as well as early gastrula in metazoan animals are surfaces of the genus 0, homeomorphic to a sphere (Figure 6(a)). In most Metazoa, the development of an organism beginning from the surface of the genus 0 leads to an adult animal which genus of the surface differs from zero. A transition from blind archenteron to through digestive tube is realized during gastrulation by the formation of another opening besides the blastopore (a primary mouth); the last becomes the definitive oral opening in Protostomia, or anal opening in Deuterostomia. The epithelial surface of an embryo after gastrulation or a larva possessing a through intestine tube (having both oral and anal openings) is a surface of genus 1, topologically equivalent to a sphere with one handle or to a torus Figure 6(b)). In animals having in the adult state the genus of the body surface $p = 1$ (for instance, in round and annelid worms), only the transition $p = 0 \rightarrow p = 1$ occurs during development.

The surface of an organism having another additional through channel besides the digestive tube is the surface of genus 2, topologically equivalent to a sphere with two handles, double torus or "pretzel". For example, sea urchins and other echinoderms exhibit a second through channel, formed by the ambulacral system (Figure 6(c)). In sea urchins, the spatial organization of the circular ambulacral system connected with the external medium by a canal is topologically equivalent to a through channel (see [12] [13]). There are two topological trans- formations: $p = 0 \rightarrow p = 1 \rightarrow p = 2$ in development of sea urchins and other echinoderms (as well as in molluscan development).

During embryonic development of the Tracheata (among Arthropoda) and some representatives of Hemichordata, the following transformations of the genus of the body surface take place: $p = 0 \rightarrow p = 1 \rightarrow p = 1 + 2n$ [12] [13] .

The detailed analysis of topological form dynamics in ontogenesis may provide additional possibilities for metazoan description and classification. For example, we have considered the different dynamics of numerical characteristics of four types of channels (gut, ambulacral system, hydropore, and pores through the madrepore plate) during embryogenesis, metamorphosis and in adult animals in different classes of Echinodermata [12] .

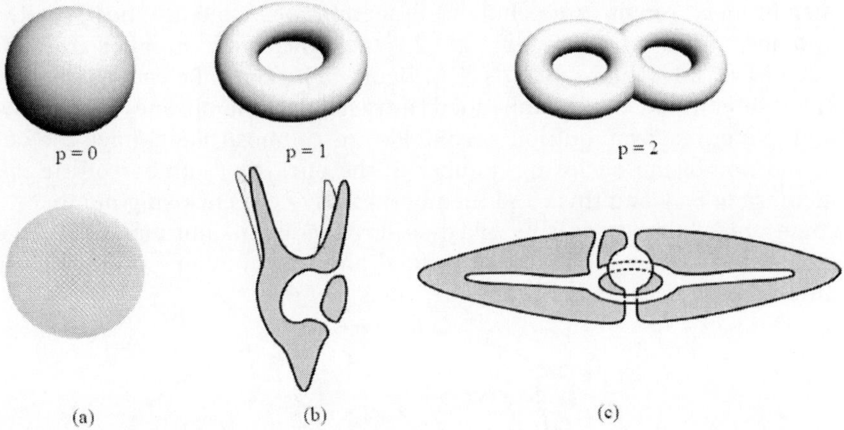

Figure 6. Topological patterns in sea urchin development: (a) An egg; (b) A larva (pluteus); (c) Adult animal.

Topological surgeries of another kind modify connectivity of germ layers and result in the separation of a set of additional closed epithelial surfaces from preexisting ones, for example, enterocoelic formation of mesoderm in Deuterostomia, neurulation and eye cap formation in Chordata and so on [7] [8] . The topological surgeries of connectivity do not change the genus of the outer surface. The spatial organization of evolutionarily advanced animals may be represented topologically as an outer epithelial envelope of a certain genus p embracing a number of inner closed epithelial surfaces embedded inside the outer envelope.

At cell level, topological approach was used to describe cell membrane system and membrane dynamics [12] [14] [16] [35] . Cell membrane systems as well as epithelia can be described as closed and orientable surfaces. Cell membrane systems undergo topological surgeries of the connectivity with perpetual separations of additional closed membrane surfaces from pre-existing ones and, inversely, fusions of membrane surfaces. These topological surgeries modifying the spatial organization of cell membrane system are the dynamic expression of cell functions.

The spatial organization of a eukaryotic cell may be represented topologically as a number of inner closed membrane surfaces of cell organelles embedded inside the outer cell membrane. Following Y. Tashiro [32] we introduce the enclosing (embedding) number to numerically evaluate membrane surfaces sequentially enclosed one into another (as Russian dolls). The enclosing number for outer cell membrane (embedded inside an external space) is 1, the enclosing number for most intracellular membranes (outer mitochondrial membranes, membranes of Golgi systems, nuclear envelope) is 2; the enclosing number of inner mitochondrial membranes is 3 (Figure 7). The nuclear envelope in eukaryotic cells is an embedded intracellular membrane system with multiple pores, or a multiple toroid. Plastids in most plants have the outer membrane of the enclosing number 2, the enclosing number of the inner membrane is 3, and thylacoid membranes have the enclosing number 4. In some unicellular organisms, plastids have more membranes [36] ; these membranes are sequentially embedded one into another, so the enclosing number of thylacoid membranes becomes 5 or 6.

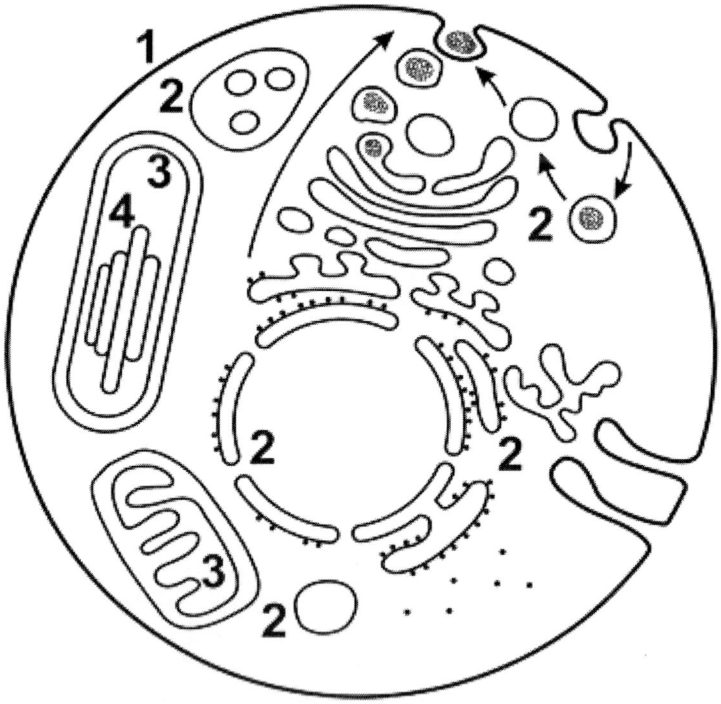

Figure 7. Topological model of membrane systems of a eukaryotic cell; membrane enclosing numbers are indicated.

4. CONCLUSIONS

Topological singularities inevitably emerging and transforming in biological morphogenesis disrupt preexisting symmetry pattern. The symmetry of a metazoan body, including the scale symmetry of fractal structures, the translational symmetry of metamerism and other variants of morphofunctional iterations is an efficient means of modular morphogenesis. Evolutionary developmental biology includes the concept of modularity, according to which ontogenesis consists of relatively independent processes, dissociable modules; structural elements of the developing organism may also be presented as discrete dissociable units. Dissociation of developmental modules and their recombination leads to evolutionary transformations with duplications and subsequent divergence, insertions, deletions and substitutions (see [37] [38]). This kind of evolutionary transformations was called "evolutionary cut and paste" [37] . Evolutionary cutting and pasting are the same as topological breaking (cutting) and gluing. Essentially, "cut and paste" approach is topological one.

The architecture of gene regulatory networks is modular (see [38] [39]). In evolution, duplications of genes, gene clusters, modules of gene networks and whole genomes were revealed. The co-linearity phenomenon as a mapping of the linear arrangement of Hox-genes in their cluster (or clusters) in the chromosome (chromosomes) into the spatial and temporal sequence of gene expression along a body axis also can be interpreted in terms of topology as genotype-phenotype mapping [15] . Hox-genes are known as genes-architects determining the body plan in Bilateria [27] . The deletion of a Hox-gene or a disturbance of the gene order in the Hox-cluster can cause large-scale macro-evolutionary transformation in body morphology (see, e.g., [40] [41]). Thus, topological rearrangement by cutting (breaking) and pasting (glueing) can be found at all levels of biological organization.

Symmetry breaking during oogenesis and early development is a fundamental event establishing the main axial coordinates of a future organism, while in later development the scale of symmetry breaking is decreasing. Topological rehandlings are regular processes in ontogeny and in phylogeny and involve the emergence of discontinuities [1] . Topological singularities inevitably emerge and transform in biological morphogenesis. The inevitability of egg cell polarization as the emergence of the topological singularity of the vector field on spheres follows from the Poincaré-Hopf theorem. The presence of a fixed point on the cortical sphere of an egg cell is inevitable for ooplasmic segregation according to Brouwer's theorem. Zygote division creates pattern of contacts on the surface of an embryo, i.e., a discrete morphogenetic field with inevitable topological singularities.

According to the Gauss-Bonnet theorem, cells with a negative or positive curvature inevitably emerge on the surface of an embryo, and the negative curvature can determine the location of invagination during gastrulation. Gastrulation is a topological transition from the sphere with heterogeneity and singularity of the field to the torus with a homogeneous field.

Subsequent metazoan morphogenesis may be represented as a succession of spherical surgeries of closed orientable epithelial surfaces. The local topological surgery leads to the global topological modification of biological forms. The epithelial surface is an interface between the internal medium of an organism and the outside environmental medium; topological and fractal transformations during metazoan evolution and development increases this interface, ensuring metabolic efficiency and better adaptation of organism to the environment.

The course and outcome of biological evolution is strongly influenced by constraints; evolution and development result in discontinuities and directionality of morphological transformations [42] . The inevitable topological dependence and constraints are integrated, fitted into genetically determined processes of biological morphogenesis, which cannot be independent of physical and topological organization of our space [14] [15] . Topological methodology reveals a certain set of topological rules that constrain and direct biological morphogenesis [9] [14] .

Valeria V. Isaeva, Nickolay V. Kasyanov, Eugene V. Presnov (43,144)">43] . Just as branching structures, toroid forms can be considered as functionally optimized biological design and topological attractors in biological morphogenesis.

REFERENCES

1. Bouligand, Y. (1996) Morphological Singulatities and Macroevolution. Memorie della Societa Italiana di Scienze Naturali e de Museo Civico di Storia Naturale di Milano, 27, 89-94.
2. Thom, R. (1969) Topological Models in Biology. Topology, 8, 313-335.
3. Thom, R. (1996) Qualitative and Quantitative in Evolutionary Theory with Some Thoughts on Aristotelian Biology. Memorie della Societa Italiana di Scienze Naturali e de Museo Civico di Storia Naturale di Milano, 27, 115-117.
4. Thom, R. (1997) Structural Stability and Morphogenesis: An Outline of a General Theory of Models. Addison-Wesley Publishing Co., New York.
5. Wasserman, S.A. and Cozzarelli, N.R. (1986) Biochemical Topology: Applications to DNA Recombination and Replication. Science, 232, 951-960.
6. Monastyrsky, M.I. (2007) Topology in Molecular Biology. Springer-Verlag, Berlin.
7. Maresin, V.M. and Presnov, E.V. (1985) Topological Approach to Embryogenesis. Journal of Theoretical Biology, 114, 387-398.

8. Presnov, E.V., Malyghin, S.N. and Isaeva, V.V. (1988) Topological and Thermodynamic Structure of Morphogenesis. In: Lamprecht, I. and Zotin, A.I., Eds., Thermodynamics and Pattern Formation in Biology, Walter de Gruyter, Berlin, 337-370.
9. Presnov, E., Isaeva, V. and Kasyanov N. (2010) Topological Determination of Early Morphogenesis in Metazoa. Th- eory in Bioscience, 129, 259-270.
10. Presnov, E., Isaeva, V. and Kasyanov, N. (2014) Topological Invariance in Biological Development. Axioes, 24, 117-135.
11. Presnov, E.V. and Isaeva, V.V. (1991) Local and Global Aspects of Biological Morphogenesis. Speculations in Science and Technology, 14, 68-75.
12. Presnov, E.V. and Isaeva, V.V. (1996) Topological Classification: Onto- and Phylogenesis. Memorie della Societa Italiana di Scienze Naturali e de Museo Civico di Storia Naturale di Milano, 27, 89-94.
13. Isaeva, V., Presnov, E. and Chernyshev, A. (2006) Topological Patterns in Metazoan Evolution and Development. Bulletin of ematical Biology, 68, 2053-2067. (s):9]
14. Isaeva, V.V., Kasyanov, N.V. and Presnov, E.V. (2008) Analysis Situs of Spatial-Temporal Architecture in Biological Morphogenesis. In: Kelly, J.T., Ed., Progress in ematical Biology Research, Nova Science Publishers, New York, 141-189
15. Isaeva, V.V., Kasyanov, N.V. and Presnov, E.V. (2012) Topological Singularities and Symmetry Breaking in Development. Biosystems, 109, 280-298.
16. Jockush, H. and Dress, A. (2003) From Sphere to Torus: A Topological View of the Metazoan Body Plan. Bulletin of ematical Biology, 65, 57-65.
17. Greenblum, S., Turnbaugh, P.J. and Borenstein, E. (2012) Metagenomic Systems Biology of the Human Gut Microbiome Reveals Topological Shifts Associated with Obesity and Inflammatory Bowel Disease. Proceedings of the National Academy of Sciences of the United States of America, 109, 594-599.
18. Dabaghian, Y., Memoli, F., Frank, L. and Carlsson, G. (2012) A Topological Paradigm for Hippocampal Spatial Map Formation Using Persistent Homology. PLoS Computational Biology, 8.
19. Niizato, T., Murakami, H. and Gunji, Y.P. (2014) Emergence of the Scale-Invariant Proportion in a Flock from the Metric-Topological Interaction. Biosystems, 119, 62-68.
20. Li, R. and Bowerman, B. (2010) Symmetry Breaking in Biology. Cold Spring Harbor Perspectives in Biology, 2, a003475.
21. Nüsslein-Volhard, C. (1991) Determination of the Embryonic Axes of Drosophila. Development. Supplement, 1, 1-10.
22. Mullins, R.D. (2010) Cytoskeletal Mechanisms for Breaking Cell Symmetry. Cold Spring Harbor Perspectives in Biology, 2, a003392.[Citation Time(s):5]
23. Nuccitelli, R. (1984) The Involvement of Transcellular Ion Currents and Electric Fields in Pattern Formation. In: Malacinski, G.M. and Bryant, S.V., Eds., Pattern Formation: A Primer in Developmental Biology, MacMillan, London, 23-46.
24. Vacquer, V.D. (1981) Dynamic Changes of the Egg Cortex. Developmental Biology, 84, 1-26.
25. Kirschner, M.W. and Gerhart, J.C. (2005) The Plausibility of Life. Yale University Press, Yale, New Haven.
26. Gilbert, S.F. (2000) Developmental Biology. Sinauer Associates, Inc., Sunderland.
27. Kline, D. and Nuccitelli, R. (1985) The Wave of Activation Current in the Xenopus Egg. Developmental Biology, 111, 471-487.
28. Rivier, N., Miri, M.F. and Oguey, C. (2005) Plasticity and Topological Defects in Cellular Structures: Extra Matter, Folds and Crab Moulting. Colloids and Surfaces A, 263, 39-45.
29. Vogel, G. (2000) Tracking the Movements that Shape an Embryo. Science, 288, 86-87.
30. Kolega, J. (1986) The Cellular Basis of Epithelial Morphogenesis. In: Browder, W., Ed., Developmental Biology, Vol. 2, Plenum Press, New York, 103-143.
31. Zallen, J.A. (2007) Planar Polarity and Tissue Morphogenesis. Cell, 129, 1051-1063.

32. Nelson, C.M. (2009) Geometric Control of Tissue Morphogenesis. Biochimica et Biophysica Acta-Molecular and Cell Research, 1793, 903-910.
33. Vladar, E.K., Antic, D. and Axelrod, J.D. (2009) Planar Cell Polarity Signaling: The Developing Cell's Compass. Cold Spring Harbor Perspectives in Biology, 1, a002964.
34. Halanych, K.M. (2004) The New View of Animal Phylogeny. Annual Review of Ecology, Evolution and Systematics, 35, 229-256.
35. Tashiro, Y. (1983) Subcellular Compartments and Protein Topogenesis. Cell Structure and Function, 8, 91-107.
36. Whatley, J.M. and Whatley, E.R. (1984) Evolutionary Aspects of the Eukaryotic Cell and Its Organelles. In: Lindskens, H.F. and Heslop-Harrrison, J., Eds., Cellular Interactions, Springer Verlag, Berlin, 18-58.
37. Shubin, N.H. (1998) Vertebrate Palaeontology: Evolutionary Cut and Paste. Nature, 394, 12-13.
38. Raff, R.A. and Sly, B.J. (2000) Modularity and Dissociation in the Evolution of Gene Expression Territories in Development. Evolution and Development, 2, 102-113.
39. Peter, I.S. and Davidson, E.H. (2011) Evolution of Gene Regulatory Networks Controlling Body Plan Development. Cell, 144, 970-985.
40. Minelli, A. (2003) The Development of Animal Form: Ontogeny, Morphology, and Evolution. Cambridge University Press, Cambridge.
41. Deutsch, J.S. and Mouchel-Vielh, E. (2003) Hox Genes and the Crustacean Body Plan. BioEssays, 25, 878-887.
42. Sará, M. (1999) New Perspectives on the Role of Constraints in Evolution. Rivista Biologica/Biological Forum, 92, 29-52.
43. Thomas, R.D.K. and Reif, W.E. (1993) The Skeleton Space: A Finite Set of Organic Designs. Evolution, 47, 341-360.

CHAPTER 4

Computational Approaches for Biomarker Discovery

Malik Yousef[1,2], Naim Najami[1,3], Loai Abedallah[2,4], Waleed Khalifa[1,2]

[1]*The Institute of Applied Research, The Galilee Society, Shefa Amr, Israel*

[2]*Computer Science, The College of Sakhnin, Sakhnin, Israel*

[3]*Department of Biology, The Academic Arab College of Education, Haifa, Israel*

[4]*Department of ematics, University of Haifa, Haifa, Israel*

ABSTRACT

Computational biology plays a significant role in the discovery of new biomarkers, the analyses of disease states and the validation of potential biomarkers. Biomarkers are used to measure the progress of disease or the physiological effects of therapeutic intervention in the treatment of disease. They are also used as early warning signs for various diseases such as cancer and inflammatory diseases. In this review, we outline recent progresses of computational biology application in research on biomarkers discovery. A brief discussion of some necessary preliminaries on machine learning techniques (e.g., clustering and support vector machines—SVM) which are commonly used in many applications to biomarkers discovery is given and followed by a description of biological background on biomarkers. We further examine the integration of computational biology approaches and biomarkers. Finally, we conclude with a discussion of key challenges for computational biology to biomarkers discovery.

Keywords: Computational Biology, Biomarker Discovery, Machine Learning

1. INTRODUCTION

Machine learning is the subfield of artificial intelligence which focuses on methods to construct computer programs that learn from experience with respect to some class of tasks and a performance measure [1] .

Machine learning enables one to generate automatic rules based on observation of the appropriate examples by the learning machine. However, the selection and design of the features that will be considered in order to represent each example for the learning process are very important and influence the classifier performance [2] .

Each instance in any dataset used by the machine learning methods is presented by a sequence of features, where each instance has the same number and types of features. The features can be categorical (i.e. gender), numerical (i.e. weight, size, age), and Boolean (i.e. sick? married?). So, the algorithms of the machine learning were asked to explain the relationships between the features in the data.

There are two major settings of learning schemes in machine learning. One is called unsupervised learning, where no prior information is given to the learner regarding the data or the output. It studies how systems can learn to represent particular input data in a way to find natural partitions (grouping/clustering) of patterns. Clustering is a simple classical method of the unsupervised learning, which partitions the data set into clusters, so that the data in each subset share some common trait according to some defined distance measure.

The main goal of clustering is to reduce the amount of data by grouping similar data items together. Most of the unsupervised learning methods use a measure of similarity between patterns in order to group them into clusters. The simplest of these involves defining a distance between patterns. For patterns whose features are numeric, the distance measure can be ordinary Euclidean distance between two instances, or Manhattan distance or any other similarity function. Figure 1describes a simple example where a given set of sample points is clustered into three clusters around three different centers. It was noted that the clustering algorithm got data where no prior information was given. Moreover, the centers and clusters shapes were unknown, and then the learner studied the data and clustered it into three clusters as shown in the example.

Clustering methods [3] - [8] can be divided into four basic types:

Ÿ Exclusive clustering (e.g., K-means algorithm [9]).

Ÿ Overlapping clustering (e.g., fuzzy C-means algorithm [10] and improved by Bezdek [11]).

Ÿ Hierarchical clustering (e.g., hierarchical clustering algorithm which was defined by Johnson in 1967 [12]).

Ÿ Probabilistic (e.g., expectation-maximization (EM) algorithm [13]).

In the other setting, which is a termed supervised learning, the instances are given with known labels; its main goal is to build a classifier which then makes predictions about future instances to assign their class labels. The dataset will be divided into two partitions: one as a training dataset, and the second as a testing data set. The classifier was built according to the training dataset and its performance was measured by the performance of the classifier over the testing dataset. The example in Figure 2 has two classes (red class and blue class) and the classifier was asked to classify the black square point. In this case, if the classifier classifies the new instances according to the first nearest neighbor, then the black square will be classified as a blue class.

Some of the popular classification algorithms are:

Decision trees are trees that classify instances by sorting them based on feature values. Each node in a decision tree represents a feature in an instance to be classified, and each branch represents a value that the node can assume. Instances are classified starting at the root node and sorted based on their feature values [14] .

Support vector machines (SVMs) are a learning machine developed by Vapnik [15] . The performance of this algorithm, as compared to other algorithms, has proven to be particularly useful for the analysis of various classification problems, and has recently been widely used in the bioinformatics field [16] - [18] . Linear SVMs are usually defined as SVMs with linear kernel. The training data for linear SVMs could be linear non-separable and then soft-margin SVM could be applied. Linear SVM separates the two classes in the training data by producing the optimal separating hyper-plane with a maximal margin between class 1 and class 2 samples (Figure 3). Given a training set of labeled examples $(x_i, y_i)\ i = 1, \cdots, l$ where $x_i \in R^l$ and $y_i \in \{+1, -1\}$, the support vector machines (SVMs) find the separating hyper-plane of the form $w \cdot x + b = 0\ w \in R^l, b \in R$.

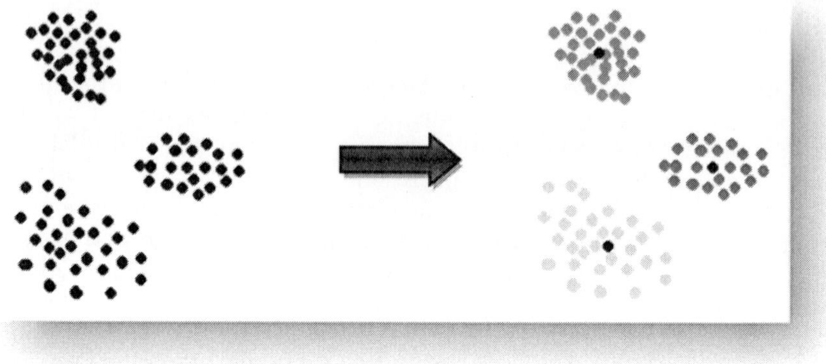

Figure 1. Data samples clustered into three clusters around three different centers.

Here, w is the "normal" of the hyper-plane. The constant b defines the position of the hyper-plane in the space. One could use the following formula as a predictor for a new instance: $f(x) = \text{sign}(w \cdot x + b)$ (for more information see Vapnik [15]).

2. BIOMARKER-BIOLOGICAL BACKGROUND

A biomarker is a gene, protein/peptide or metabolite present in a biological system, used to indicate a physiological or pathological state that can be recognized or monitored [19] - [21] . Biomarker discovery is a challenging process; a good biomarker has to be sensitive, specific and its test highly standardized and reproducible.

Genomic studies provide scientists with methods to quickly analyze genes and their products en masse. DNA microarray technologies permit systematic approaches to biological discovery that has begun to have a profound impact on biological research, pharmacology, and medicine. The ability to obtain quantitative information about the complete transcription profile of cells promises to be an exceptionally powerful means to explore basic biology, diagnose disease, facilitate drug development, tailor therapeutics to specific pathologies, and generate databases with information about living processes [22] .

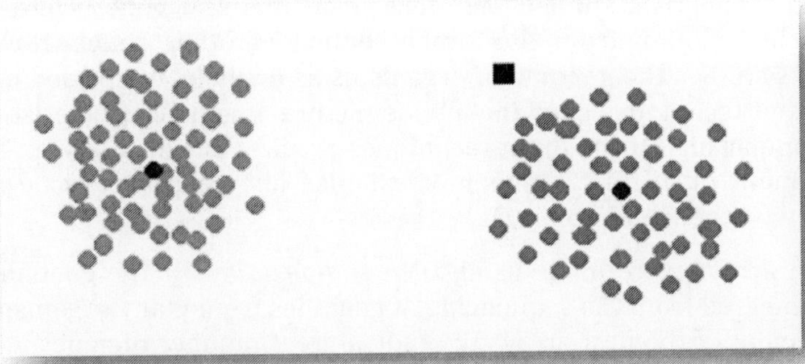

Figure 2. An example of how the classifiers work.

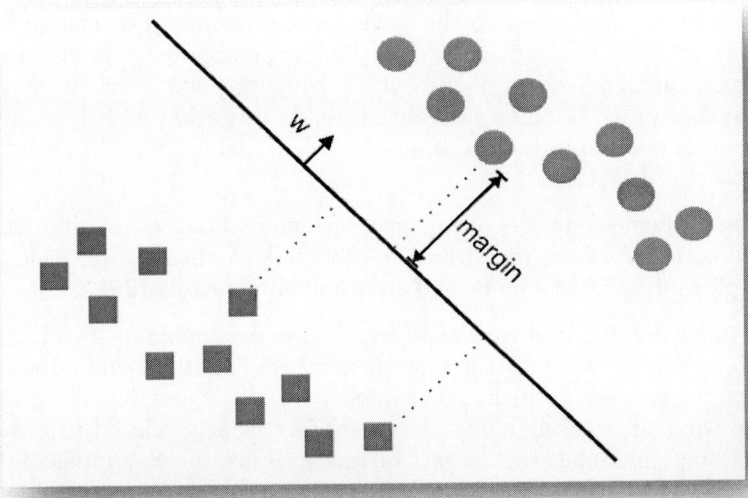

Figure 3. A simple linear support vector.

Gene expression studies bridge the gap between DNA information and trait information by dissecting biochemical pathways into intermediate components between genotype and phenotype. These studies open new avenues for identifying complex disease genes and biomarkers for disease diagnosis and for assessing drug efficacy and toxicity.

For years, scientists studied one gene at a time and genes were indeed studied in isolation from the larger context of other genes. Nowadays, genomics studies the genome of organisms as a whole. It is based on high-throughput techniques allowing a wide picture of gene characteristics. The most popular high throughput techniques are arrays, which are an orderly arrangement of a great number of samples allowing large-scale studies (http://www.gene-chips.com/).

The first arrays were made using DNA samples. This is the genomic area, which emerged from the sequencing of genomes from many organisms. The development of the first arrays to study a great number of genes at a time started many years ago [23] and has widely expanded since then. But now, we can array DNA and RNA probes, proteins, antibodies and even biological samples allowing new types of research. Furthermore, other types of high throughput techniques are currently developing, for instance to study metabolites.

Genomics is thus linked to the development of new biotechnology which covers a broad field of disciplines (biology, computer science, chemistry, physics, and engineering) and which converge and work in synergy to advance rapidly. Genomics involves the identification of organism's genes and understanding how the genes work using new biotechnological approaches [24] .

Recent advances in genomics are bringing about a revolution in our understanding of the molecular mechanisms of phenotypes, including the complex interplay of genetic and environmental factors [25] .

Genomics divided into two basic areas, structural genomics and functional genomics (also called post-ge- nomic area) [26] . In the former, the target of research is DNA which corresponds to the genetic background of organisms. Structural genomics is therefore clearly related to genetics. In functional genomics, the targets of research are the key molecules which give life to the cells: RNA, proteins and also metabolites (which are both biologically active molecules within cells and tissues). Functional genomics allows the detection of genes that are turned on/off at any given time depending on environmental factors.

Today, genomics has induced two new paradigms in biology. The first paradigm is a new approach allows the study of the complex network through which genes and proteins communicate. It implies the combination of expertise from biologists, engineers, chemists, and computer scientists. This multidisciplinary approach allows the development of systems biology. The second paradigm is a direct consequence of more available information derived from genomics studies. The raw data needs to be analyzed and then to be used in the systemic approach indicated above. This led the development of bioinformatics which needs the use of computers to manage biological information.

The practical applications of gene expression analyses are numerous and only beginning to be realized. One particularly powerful application of gene expression analyses is biomarker identification, which can be used for disease risk assessment, early detection, prognosis, prediction response to therapy, and preventative measures is a challenging task for cancer prevention and the improvement of treatment outcomes. Approaches to cancer biomarker discovery include genomic, epigenomic, transcriptomic, and proteomic analyses.

Current efforts in the laboratory focus on the identification of biomarkers in chronic lymphocytic leukemia, lung cancer and colon cancer. Among the biomarkers we consider are plasma microRNAs (miRNAs). miRNAs are a class of small RNAs that function as regulators of gene expression. Alteration of gene expression patterns due to dysregulation of miRNAs is a common theme in tumorigenesis. High concentrations of cell-free miRNAs originating from the primary tumor have been found in the plasma of cancer patients, and several lines of evidence indicate that plasma miRNAs are associated with specific vesicles called exosomes. Plasma miRNAs have emerged as a promising source of cancer biomarkers [27] .

A recent discovery of quantifiable circulating cancer-associated miRNAs, expose the immense potential for their use as novel minimally invasive biomarkers for breast and other cancers [28] .

In this section, structural genomics and the different fields of functional genomics (RNA studies, proteomics, metabolomics) will be first detailed for the reader to better understand what genomics is, before the description of the two new paradigms in biology derived from genomics (systemic approaches, bioinformatics).

3. COMPUTATIONAL APPROACHES FOR BIOMARKER DISCOVERY

DNA microarray technologies permit systematic approaches to biological discovery that has begun to have a profound impact on biological research, pharmacology, and medicine. The ability to obtain quantitative information about the complete transcription profile of cells promises to be an exceptionally powerful means to explore basic biology, diagnose disease, facilitate drug development, tailor therapeutics to specific pathologies, and generate databases with information about living processes [22] .

Gene expression studies bridge the gap between DNA information and trait information by dissecting biochemical pathways into intermediate components between genotype and phenotype. These studies open new avenues for identifying complex disease genes and biomarkers for disease diagnosis and for assessing drug efficacy and toxicity.

The practical applications of gene expression analyses are numerous and only beginning to be realized. One particularly powerful application of gene expression analyses is biomarker identification, which can be used for disease risk assessment, early detection, prognosis, prediction response to therapy, and preventative measures.

4. COMPUTATIONAL BIOMARKER (FEATURES) SELECTION

Classification of samples from gene expression datasets usually involves small numbers of samples and tens of thousands of genes. The problem of selecting those biomarker genes that are important for distinguishing the different sample classes being compared poses a challenging problem in high dimensional data analysis and the potential biomarkers are important in improvement in diagnostics and therapeutics development. A variety of methods to address these types of problems have been implemented [29] - [35] . These methods can be divided into two main categories: those that rely on filtering methods and those that are model-based or so-called wrapper approaches [29] [31] .

W. Pan [35] has reported a comparison of different filtering methods, highlighting similarities and differences between three main methods. The filtering methods, although faster than the wrapper approaches, are not particularly appropriate for establishing rankings among significant genes, as each gene is examined individually and correlations among the genes are not taken into account. Although wrapper methods appear to be more accurate, filtering methods are presently more frequently applied to data analysis than wrapper methods [31] .

Li and Yang [36] compared the performance of support vector machine (SVM) algorithms and ridge regression (RR) for classifying gene expression datasets and also examined the contribution of recursive procedures to the classification accuracy. Their study explicitly shows that the way in which the classifier penalizes redundant features in the recursive process has a strong influence on its success. They concluded that RR performed best in this comparison and further demonstrate the advantages of the wrapper method over filtering methods in these types of studies.

Guyon et al. [37] compared the usefulness of RFE (for SVM) against the "naïve" ranking on a subset of genes. The naïve ranking is just the first iteration of RFE to obtain ranks for each gene. They found that SVM-RFE is superior to SVM without RFE and also to other multivariate linear discriminant methods, such as linear discriminant analysis (LDA) and mean-squared-error (MSE) with recursive feature elimination. See Figure 4(b) for the procedure of SVM-RFE.

Xiong, Fang et al. [38] Propose a general framework to incorporate feature (gene) selection into pattern recognition in the process to identify biomarkers. Using this framework, they develop three feature wrappers that search through the space of feature subsets using the classification error as measure of goodness for a particular feature subset being "wrapped around": linear discriminant analysis, logistic regression, and support vector machines.

Yousef, Jung et al. [39] , describe a new method for gene selection and classification, which is comparable to or better than some methods which are currently applied. Their method (SVM-RCE) combines the K-means algorithm for gene clustering and the machine learning algorithm, SVMs [15] , for classification and gene cluster ranking. The SVM-RCE method differs from related classification methods in that it first groups genes into correlated gene clusters by K-means and then evaluates the contributions of each of those clusters to the classification task by SVM. One can think of this approach as a search for those significant clusters of gene which have the most pronounced effect on enhancing the performance of the classifier. See Figure 4(a) that illustrates the procedure of SVM-RCE. Moreover, Lin-Kai [40] propose an improved method of SVM-RCE called ISVM-RCE. ISVM-RCE eliminates genes within the clusters instead of removing a cluster of genes when the number of clusters is small. Six data sets are used to test the performance of ISVM-RCE and compared their performances with SVM-RCE and linear-discriminant-analysis-based RFE (LDA-RFE). The results show that ISVM-RCE greatly reduces the time cost of SVM-RCE, meanwhile obtains comparable classification performance as SVM-RCE, while LDA-RFE is not stable.

Recently, Grate [41] has described a technique for discovering small sets of genes (3 or less). The technique is based on brute force approach of exhaustive search through all genes, gene pairs and some cases triple of genes. The combination is analyzed with classification method looking for those combinations that form training error-free classifiers.

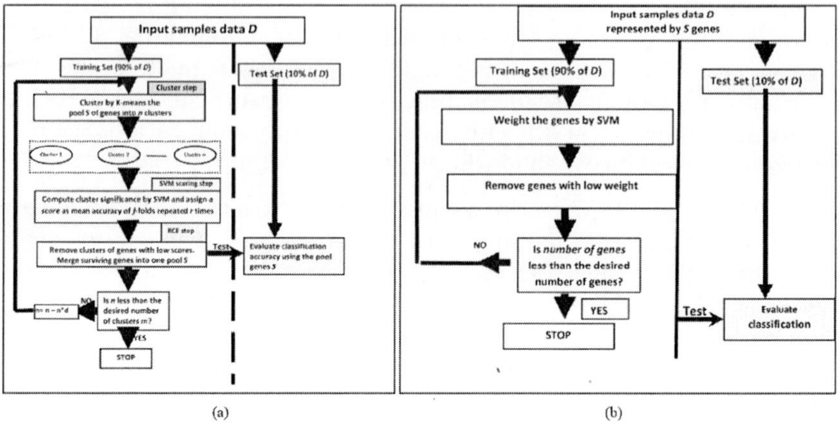

Figure 4. (a) Recursive cluster elimination (RCE) procedure with SVM; (b) Recursive feature elimination (RFE) procedure with SVM.

Robustness of biomarkers is an important issue, as it may greatly influence subsequent biological validations. In addition, a more robust set of markers may strengthen the confidence of an expert in the results of a selection method. Abeel [42] has proposed a general framework for the analysis of the robustness of a biomarker selection algorithm. The framework is based on ensemble feature selection, where multiple feature selections are combined in order to increase the robustness of the final set of selected features. The proposed methodology is evaluated on four microarray datasets showing increases of up to almost 30% in robustness of the selected biomarkers, along with an improvement of about 15% in classification performance. A different approach to deal with inconsistent cancer biomarkers due to bioinformatics artifacts, was proposed by [43] . The approach is based on using multiple data sets from microarrays, mass spectrometry, protein sequences, and other biological knowledge in order to improve the reliability of cancer biomarkers. The study presents a novel Bayesian network (BN) model which integrates and crosses-annotates multiple data sets related to prostate cancer. The computational results show that the method is able to find biologically relevant biomarkers with highest reliability [44] .

Some data is composed from multiple category or classes. For such a data a special methods of biomarker selection is required. [45] has proposed classification method is based on two schemes: error-correcting output coding (ECOC) and pairwise coupling (PWC). The biomarker pattern for distinguishing each disease category from another one is achieved by the development of an extended Markov blanket (EMB) feature selection method.

The study of [46] demonstrates that the machine learning approach can be used to detect a small subset of biomarker genes from high dimensional datasets and generate a classifier for multiple classes. A multiclass support vector machine (MC-SVM) method and an unsupervised K-mean clustering method were applied to independently refine the classifier, producing a smaller subset of 39 and 30 classifier genes, separately, with 11 common genes being potential biomarkers.

Yousef et al. [47] developed a new algorithm called recursive network elimination (RNE) with SVM. The main idea is to integrate network information with recursive feature elimination based on SVM. First, filter one thousand genes selected by t-test from training set so that only genes that map to a gene network database remain. Then to the remaining genes the gene expression network analysis tool (GXNA) [48] is applied to form n clusters of genes that are highly connected in the network. Using these clusters linear SVM is used to classify the samples and a weight is assigned to each cluster based on its significance to the classification. The clusters with less information are removed while retaining the remainder for the next classification step. This process is repeated until an optimal classification result is attained.

5. A COMPARATIVE PERFORMANCE

Pirooznia, M. et al. [49] conducted a study to compare the performance (with cross validation) of different machine learning algorithm such as SVM, RBF Neural Nets, MLP Neural Nets, Bayesian, Decision Tree and Random Forrest methods. Additionally, the efficiency of the feature selection methods including support vector machine recursive feature elimination was compared. The data set consists from eight different binary (two class) microarray datasets. The performance was very well and in average higher than 90%. As expected, this study shows that in most cases the accuracy is improved with feature selection. Additionally, this study reports about stability of the top 50, 100, 200 genes with SVM-RFE. Actually those genes will be serving later as a biomarker for diagnostic diseases.

Yousef et al. [39] compare the performance of SVM-RCE against the popular SVM-RFE method to reported in most cases that SVM-RCE is with better results as in average of 6 datasets is 96% while SVM-RFE with 92% accuracy.

6. CONCLUSIONS

In this review, we proposed many computational approaches which are critical for mining high-dimensional data in order to effectively discover biomarkers. Best data mining approach would to integrate different approaches to arrive at an effective algorithm; however most of the suggested methods ignore existing biological knowledge and treat all the genes equally. Information about gene networks or pathways should be incorporated into a classifier to improve both the predictive performance and interpretability of the resulting biomarker genes as suggested by [47] and [50] .

Moreover, we suggest developing more algorithms that incorporate biological knowledge, extracted from existing database with the procedure of feature selection to have more accurate biological results.

ACKNOWLEDGEMENTS

This study was supported by The College of Sakhnin and The Institute of Applied Research—The Galilee Society. Also this study was supported by a grant from the Ministry of Science, Israel.

COMPETING INTERESTS

Malik Yousef, Naim Najami,, Loai Abedallah,, Waleed Khalifa, No competing interests.

REFERENCES

1. Mitchell, T. (1997) Machine Learning. McGraw-Hill, New York.
2. Malik Yousef, N.N. and Khalifav, W. (2010) A Comparison Study between One-Class and Two-Class Machine Learning for MicroRNA Target Detection. Journal of Biomedical Science and Engineering, 3, 347-252.
3. Jain, A.K. and Dubes, R.C. (1988) Algorithms for Clustering Data. Prentice Hall, Englewood Cliffs.
4. Hartigan, J. (1975) Clustering Algorithms. Wiley, New York.
5. Tryon, R.C. and Bailey, D.E. (1973) Cluster Analysis. McGraw-Hill, New York.
6. Sneath, P.H.A. and Sokal, R.R. (1973) Numerical Taxonomy. Freeman, San Francisco.
7. Anderberg, M.R. (1973) Cluster Analysis for Applications. Academic Press, New York.
8. Jardine, N. and Sibson, R. (1971) ematical Taxonomy. Wiley, London.

9. MacQueen, J.B. (1967) Some Methods for Classification and Analysis of Multivariate Observations. Proceedings of 5-th Berkeley Symposium on ematical Statistics and Probability, 1, 281-297.
10. Dunn, J.C. (1973) A Fuzzy Relative of the ISODATA Process and Its Use in Detecting Compact Well-Separated Clusters. Journal of Cybernetics, 3, 32-57.
11. Bezdek, J.C. (1981) Pattern Recognition with Fuzzy Objective Function Algoritms. Plenum Press, New York.
12. Johnson, S.C. (1967) Hierarchical Clustering Schemes. Psychometrika, 32, 241-254.
13. Dempster, A.P., Laird, N.M. and Rubin, D.B. (1977) Maximum Likelihood from Incomplete Data via the EM Algorithm. Journal of the Royal Statistical Society, 39, 1-38.
14. Yuan, Y. and Shaw, M.J. (1995) Induction of Fuzzy Decision Trees. Fuzzy Sets and Systems, 69, 125-139.
15. Vapnik, V. (1995) The Nature of Statistical Learning Theory. Springer, New York.
16. Donaldson, I., Martin, J., De Bruijn, B., Wolting, C., Lay, V., Tuekam, B., Zhang, S.D., Baskin, B., Bader, G., Michalickova, K., Pawson, T. and Hogue, C.W.V. (2003) PreBIND and Textomy—Mining the Biomedical Literature for Protein-Protein Interactions Using a Support Vector Machine. BMC Bioinformatics, 4, 11.
17. Pavlidis, P., Weston, J., Cai, J. and Grundy, W.N. (2001) Gene Functional Classification from Heterogeneous Data. Proceedings of the 5th Annual International Conference on Computational Biology, Montreal, 22-25 April 2001, 249- 255.
18. Haussler, D. (1999) Convolution Kernels on Discrete Structuresed. Technical Report UCSCCRL-99-10, Baskin School of Engineering, University of California, Santa Cruz.
19. Novak, K. (2006) Taking out the Trash. Nature Reviews Cancer, 6, 92.
20. Novak, K. (2006) Marked Aggression. Nature Reviews Cancer, 6, 96.
21. Goymer, P. (2006) Different Treatment. Nature Reviews Cancer, 6, 94-95.
22. Young, R.A. (2000) Biomedical Discovery with DNA Arrays. Cell, 102, 9-15.
23. Hergenhahn, M., Muhlemann, K., Hollstein, M. and Kenzelmann, M. (2003) DNA Microarrays: Perspectives for Hypothesis-Driven Transcriptome Research and for Clinical Applications. Current Genomics, 4, 543-555.
24. ESRC (Economic and Social Research Council) (2002) Genomics Scenario Project 2. Overview and Forecasts of the Applications of Genomics. http://www.cric.ac.uk /cric/projects/genomics/overview.pdf
25. Collins, F.S., Green, E.D., Guttmacher, A.E. and Guyer, M.S. (2003) A Vision for the Future of Genomics Research. Nature, 422, 835-847.
26. Eggen, A. (2003) Basics and Tools of Genomics. Outlook on Agriculture, 32, 215-217.
27. Jeffrey, S.S. (2008) Cancer Biomarker Profiling with microRNAs. Nature Biotechnology, 26, 400-401.
28. Heneghan, H.M., Miller, N., Lowery, A.J., Sweeney, K.J. and Kerin, M.J. (2010) MicroRNAs as Novel Biomarkers for Breast Cancer. Journal of Oncology, 2010, Article ID: 950201.
29. Wang, Y., Tetko, I.V., Hall, M.A., Frank, E., Facius, A., Mayer, K.F.X. and Mewes, H.W. (2005) Gene Selection from Microarray Data for Cancer Classification—A Machine Learning Approach. Computational Biology and Chemistry, 29, 37-46.
30. Li, T., Zhang, C.L. and Ogihara, M. (2004) A Comparative Study of Feature Selection and Multiclass Classification Methods for Tissue Classification Based on Gene Expression. Bioinformatics, 20, 2429-2437.
31. Inza, I., Larrañaga, P., Blanco, R. and Cerrolaza, A.J. (2004) Filter versus Wrapper Gene Selection Approaches in DNA Microarray Domains. Artificial Intelligence in Medicine, 31, 91-103.
32. Zhang, X.G., Lu, X., Shi, Q., Xu, X.Q., Leung, H.C.E., Harris, L.N., et al. (2006) Recursive SVM Feature Selection and Sample Classification for Mass-Spectrometry and Microarray Data. BMC Bioinformatics, 7, 197.

33. Duan, K.B., Rajapakse, J.C., Wang, H.Y. and Azuaje, F. (2005) Multiple SVM-RFE for Gene Selection in Cancer Classification with Expression Data. IEEE Transactions on NanoBioscience, 4, 228-234.
34. Yang, X.W., Lin, D.Y., Hao, Z.F., Liang, Y.C., Liu, G.R. and Han, X. (2003) A Fast SVM Training Algorithm Based on the Set Segmentation and k-Means Clustering. Progress in Natural Science, 13, 750-755.
35. Pan, W. (2002) A Comparative Review of Statistical Methods for Discovering Differentially Expressed Genes in Replicated Microarray Experiments. Bioinformatics, 18, 546-554.
36. Li, F. and Yang, Y.M. (2005) Analysis of Recursive Gene Selection Approaches from Microarray Data. Bioinformatics, 21, 3741-3747.
37. Guyon, I., Weston, J., Barnhill, S. and Vapnik, V. (2002) Gene Selection for Cancer Classification Using Support Vector Machines. Machine Learning, 46, 389-422.
38. Xiong, M., Fang, X. and Zhao, J. (2001) Biomarker Identification by Feature Wrappers. Genome Research, 11, 1878- 1887.
39. Yousef, M., Jung, S., Showe, L.C. and Showe, M.K. (2007) Recursive Cluster Elimination (RCE) for Classification and Feature Selection from Gene Expression Data. BMC Bioinformatics, 8, 144.
40. Luo, L.K., Huang, D.F., Ye, L.J., Zhou, Q.F., Shao, G.F. and Peng, H. (2011) Improving the Computational Efficiency of Recursive Cluster Elimination for Gene Selection. IEEE/ACM Transactions on Computational Biology and Bioinformatics, 8, 122-129.
41. Grate, L. (2005) Many Accurate Small-Discriminatory Feature Subsets Exist in Microarray Transcript Data: Biomarker Discovery. BMC Bioinformatics, 6, 97.
42. Abeel, T., Helleputte, T., Van de Peer, Y., Dupont, P. and Saeys, Y. (2009) Robust Biomarker Identification for Cancer Diagnosis with Ensemble Feature Selection Methods. Bioinformatics, 26, 392-398.
43. Deng, X., Geng, H. and Ali, H.H. (2007) Cross-Platform Analysis of Cancer Biomarkers: A Bayesian Network Approach to Incorporating Mass Spectrometry and Microarray Data. Cancer Informatics, 3, 183-202.
44. Huang, H.C., Jupiter, D. and VanBuren, V. (2010) Classification of Genes and Putative Biomarker Identification Using Distribution Metrics on Expression Profiles. PLoS ONE, 5, e9056.
45. Oh, J.H., Kim, Y.B., Gurnani, P., Rosenblatt, K.P. and Gao, J.X. (2008) Biomarker Selection and Sample Prediction for Multi-Category Disease on MALDI-TOF Data. Bioinformatics, 24, 1812-1818.
46. Li, Y., Wang, N., Perkins, E.J., Zhang, C.Y. and Gong, P. (2010) Identification and Optimization of Classifier Genes from Multi-Class Earthworm Microarray Dataset. PLoS ONE, 5, e13715.
47. Yousef, M., Ketany, M., Manevitz, L., Showe, L.C. and Showe, M.K. (2009) Classification and Biomarker Identifica- tion Using Gene Network Modules and Support Vector Machines. BMC Bioinformatics, 10, 337.
48. Nacu, S., Critchley-Thorne, R., Lee, P. and Holmes, S. (2007) Gene Expression Network Analysis and Applications to Immunology. Bioinformatics, 23, 850-858. http://dx.doi.org/10.1093/bioinformatics/btm019
49. Pirooznia, M., Yang, J.Y., Yang, M.Q. and Deng, Y.P. (2008) A Comparative Study of Different Machine Learning Methods on Microarray Gene Expression Data. BMC Genomics, 9, S13.
50. Tai, F. and Pan, W. (2007) Incorporating Prior Knowledge of Predictors into Penalized Classifiers with Multiple Penalty Terms. Bioinformatics, 23, 1775-1782.

CHAPTER 5

A Systems Biological Perspective of Cellular Stress-Directed Programmed Cell Death

Orsolya Kapuy*, Beáta Lizák, Ibolya Stiller, Gábor Bánhegyi

Department of Medical Chemistry, Molecular Biology and Pathobiochemistry, Semmelweis University, Budapest, Hungary

ABSTRACT

Each eukaryotic cell of multicellular organisms must be able to maintain its integrity by sensing both external and internal stimuli. The primary goal of the generated response mechanism is to drive back the system to the former or to a new homeostatic state. Moreover, the response has to provide an accurate survival-or-death decision to avoid any "misunderstanding" and its unwanted consequences. New data revealed that a systems-level crosstalk of molecular networks has an essential role in achieving the correct characteristic of the response. Although many molecular components of these processes already have been revealed, several elements and regulatory connections of crosstalk are still missing. These "gaps" of the complex control networks make hardly impossible to present comprehensive models. Therefore we approach the questions from a systems biology aspect by combining the experimental results with the special technique of ematical modelling. In this short report we discuss some novel and preliminary data gained by this approach on the crosstalk between life and death decisions under cellular stress, to get a systems biological view of these networks.

Keywords: Systems Biology; Cellular Stress; Autophagy; Apoptosis; Crosstalk; ematical Modelling

1. INTRODUCTION

For delicate cellular balance sustained by various signal transduction pathways and control elements, the most important criteria are that adequate answers must be formulated if and only if stimuli exist [1] . To understand this process, let's imagine a regulating system for a jet plane fully wired via a lots of switches and arms, while each task (e.g. landing, lightening) is directed by a well-defined control module. The regulatory modules of cellular network contain a variety of enzyme proteins which can influence the activity of other proteins in positive or negative direction. The dynamic features of the response mechanism can be described by signal-response curves [1] . Regulatory networks implementing decision making process can be mainly characterized with two different phenomena. In one case a network is designed for broad dynamic ranges and proportionate responses [2] . This corresponds to that feature of a jet plane, when e.g. the pilot can choose the height of the plane above mean sea level continuously in a broad range during flight. According to the wiring of the regulatory elements linear, hyperbolic or sigmoid signal-response curves can be generated [1] [3] (Figure 1(a), upper panel). In the other case, the cellular decision making occurs in a binary way with two discrete stable states separated by an unstable one [3] (Figure 1(a), lower panel). Using our metaphor of jet plane, this can be symbolized for e.g. with the two separated states of the wheels (IN or OUT). These discontinuous switch-like characteristic results in an abrupt change as signal level reaches a critical threshold value in either direction (see grey dotted arrows on Figure 1(a), lower panel). The key regulatory components of the cellular signalling networks are generally wired with positive feedback loops (either mutual activation or antagonism) to achieve all-or-none decision [4] -[6] . Depending on the feature of the control network, the switch can be reversible or irreversible. Reversible response mechanism assumes that the system is able to switch off sensing the elimination of the signal, while one-way switch guarantees that the response element remain in its active state even if the stimuli get disappeared [1] (see black dotted arrow on Figure 1(a), lower panel).

2. THE CHARACTERISTIC FEATURES OF CELLULAR LIFE-OR-DEATH DECISION

Usually the cellular decision making process between life and death can be described with an all-or-none characteristic. The suicide network of apoptosis, as a type of cell death mechanisms has been extensively revealed; nowadays plenty of articles are focusing on the cellular self-eating driven by autophagy, too. Since the huge numbers of the regulatory components and the complexity of their connections make hardly impossible to understand the characteristic behaviour of these mechanisms [7] , therefore a novel approach called systems biology gets widely used which involves both experimental techniques and ematical modelling [8] [9] .

2.1. Committing Suicide by Apoptosis

Apoptosis removes aberrant or damaged cells, but it also has an important role in eliminating cells during embryonic development and immune system maturation [10] . Apoptotic cell death can be easily recognized by cellular shrinkage, mitochondrial permeabilization and DNA fragmentation [11] . The programmed cell death driven by apoptosis has two characteristic rules: 1) cell death response must not be triggered by week signals and 2) once apoptosis is triggered cells should commit to finish the process (i.e. point-of-no-return) [12] . It is very important that the cell must not hesitate on the borderline of two well-separated states (i.e. ON and OFF state of apoptosis); otherwise not only a wrong decision, but also a not finished process might be fatal as well [1] [13] . This is based on principles similar to the take-off decision points of jet planes. The plane is not allowed to start taking-off before reaching the crucial velocity of the plane on the runway. In addition, the pilot cannot break the taking off either after induction, otherwise the plane would crash. Therefore the two states of the plane (grounded and took off) are separated with a discontinuous step. While the pilot is reminded by certain signs of instruments, in cells there are the control modules, which meet the obligation for apoptosis following the realization of the level of cellular stress.

2.2. Autophagy-Dependent Self-Cannibalism

Another type of programmed cell death is autophagy. Autophagy is an evolutionary conserved cellular digestive process whereby cytosolic contents are sequestered in vesicles and delivered to the lysosome for degradation [14] [15] . The basic level of autophagy is essential for housekeeping while increased level of autophagic process plays a key role in cellular differentiation, growth control, cell defence, tissue remodelling and different type of cellular stresses (e.g. hypoxia) [14] [16] . Down-regulation of the key sensor of both growth factors and nutrient availability (so called mTOR pathway) results in induction of autophagy, too [17] . Therefore autophagy is not really a self-killing mechanism, rather a cytoprotective process of cells found in eukaryotes from yeast to mammalians [16] [18] [19] .

3. THE IMPORTANCE OF CROSSTALK BETWEEN AUTOPHAGY AND APOPTOSIS

The control network of both autophagy and apoptosis is complex. In addition, an increasing number of novel results confirm that these two mechanisms are not independent from each other; rather they are connected at various regulatory levels generating crosstalk between them [20] -[22] . While autophagy usually results in survival, apoptosis induces cell death, therefore their crosstalk seems to have an essential role in achieving a well-balanced cellular response reacting to various stress signals (e.g. endoplasmic reticulum (ER) stress, nutrient deprivation) [22] .

Although an increasing number of experimental data analyses the key components of this crosstalk, relatively few data are focusing on exploring the dynamical features of the crosstalk of the system. Nevertheless it is really an interesting question how the regulatory network can choose between autophagy-dependent survival and apoptosis-induced cell death in respect with stress signal. It has already been proven that autophagy precedes apoptosis during cellular stress [23] , but how the system switches between them is still a mystery.

3.1. The Signal-Response Profile of Both Autophagy and Apoptosis under Cellular Stress

Based on experimental data our systems biological analysis has shown recently the crucial feedback loops of the regulatory network participating in decision-making process between the life and death [13] . The different components of the regulatory networks were sorted into three groups, called crosstalk element, autophagy inducer and apoptosis inducer and they were connected with crucial feedback loops (Figure 1(b)). By definition autophagy inducers (such as Beclin1, Ambra1, Atg3, Atg4, Atg5) are able to promote the self-eating process, while apoptosis inducers (e.g. Caspase-3,

Caspase-6, Caspase-8, Calpain) enhance apoptotic cell death. Experimental data are suggested that cFLIP, vFLIP, Bcl2 and Bcl_{XL} have a regulatory effect on both autophagy and apoptosis inducers [22] . Therefore these molecules are considered as crosstalk elements. Our analysis suggests that the crosstalk element is able to act negatively on both autophagy and apoptosis resulting in a direct link be tween the two mechanisms [13] . However, autophagy and apoptosis inducers are antagonists generating a mutual exclusion. This opposition in itself provides a peculiar regulatory motif, so-called double negative feedback in the system. In addition, there is a positive feedback between apoptosis inducer and inactive autophagy inducer. The apoptosis inducer-dependent inactive form of autophagy directly promotes apoptotic cell death (Figure 1(b)).

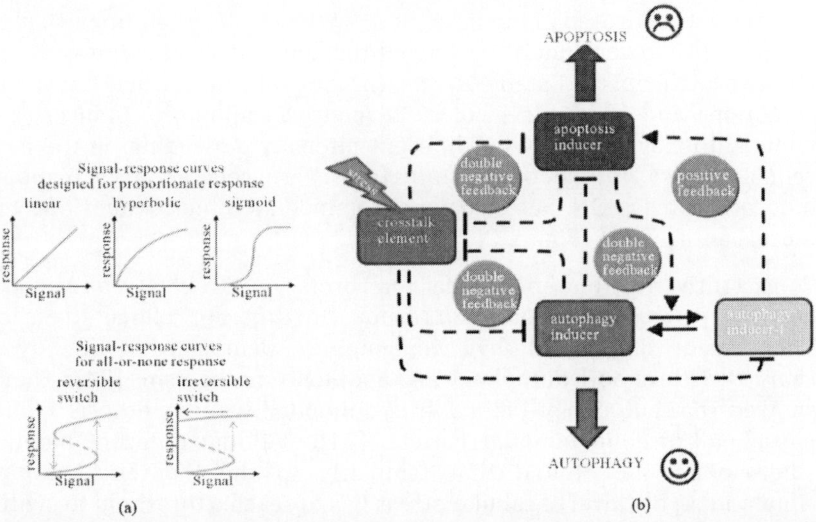

Figure 1. Studying cellular decision making process. (a) The possible signal-response curves designed for proportionate response (upper panel) and for all-or-none response (lower panel). Solid lines denote stable states, while dashed line presents the unstable ones. (b) The wiring diagram and the key network motifs of autophagy-apoptosis crosstalk. The active forms of the elements are grouped in dark-coloured boxes, while light-coloured box denotes the apoptosis inducer-cleaved form. The experimentally proved regulatory connections between molecules are presented by dashed lines, while solid arrows show biochemical reactions. Blocked end lines denote inhitition.

Plotting the signal-response curve of both autophagy and apoptosis inducers, it could be seen that autophagy inducer follows a sigmoid activation profile, while apoptosis inducer has a switch-like characteristic according to the level of cellular stress (Figure 2(a), lower panel). Under normal physiological conditions autophagy has a basic level, which is high enough to block the apoptosis inducer. With respect to stress autophagy inducer quickly gets activated through down-regulation of crosstalk element, meanwhile apoptosis inducer remains fully inactive. Although crosstalk element has almost been disappeared promoting apoptosis inducer the sigmoid increase of autophagy inducer is able to down-regulate apoptotic cell death (Figure 2(a)). However reaching a given threshold value of the signal (i.e. when cellular stress becomes excessive) apoptosis inducer gets activated immediately and switches off the autophagic process (Figure 2(a), lower panel). The signal-response curve of apoptosis inducer has two distinguished stable states. In one state (at normal physiological conditions and at low level of cellular stress) apoptosis inducer is OFF, while autophagy works with different intensity depending on the level of cellular stress. In the other state (at severe cellular stress) apoptosis inducer switches ON, autophagy gets immediately inactivated due to the mutual antagonism (Figure 2(a)).

Consistently with our theoretical interpretation, Xu et al. confirmed that autophagy preceded apoptosis in response to different cellular stress levels. They used single cell live imaging to demonstrate the dynamic characteristic of both autophagy and apoptosis mechanisms [24] . They also showed that autophagy induction is unimodal, while apoptosis induction has an all-or-none bimodal pattern [24] . Adding different amount of staurosporin they found that (similarly to the theoretical computer simulations) the level of cellular stress has to reach a threshold to switch on the apoptotic process. Otherwise autophagy wins against apoptosis and the self-killing pathway will be only transiently activated, which would not be sufficient to the fatal cellular decision. Unsurprisingly, inhibition of autophagy by mutating any key components of the survival process results in premature activation of apoptosis in response to cellular stress [24] .

3.2. The Dynamic Properties of Crosstalk between Autophagy and Apoptosis

The mutual antagonism between autophagy and apoptosis ensures a precise decision between life and death depending on the level of cellular stress, but one question has not been answered yet; namely, what is the mechanism to control that first autophagy is activated under cellular stress then the cell later allows the suicide process?

To our knowledge the crosstalk element has a crucial role to render the correct order of autophagy dependent survival and apoptotic cell death under cellular stress. According to the effect of crosstalk element on programmed cell death mechanisms there are two possibilities. In the first case the crosstalk element (e.g. Bcl2) inhibits both autophagy and apoptosis inducers [13] . When cellular stress is imposed the activity of autophagy inducer would become fully active earlier than the apoptosis inducer, while the level of crosstalk element gets decreased by cellular stress (Figure 2(a), lower panel). Although the crosstalk element was trying to switch on the apoptosis, the autophagy inducer hindered the cell death process. If the stress situation sustained the apoptosis inducer would be eventually active. In line with this step the autophagy inducer immediately switched off (see Figure 2(a), lower panel). Even if the crosstalk element is completely diminished afterwards, the survival mechanisms could not function any more, i.e. the whole energy of the cell is devoted to execute suicide.

In an alternative model we claim that the crosstalk element is the so-called unfolded protein response (UPR) induced by ER stress. Cellular stress acts positively on crosstalk element resulting in its hyperbolic activation (Figure 2(b)). ER stress via UPR activates both autophagy and apoptosis. Autophagy came first, while apoptosis induction requires higher concentration of cellular stress to down-regulate the survival process via a double negative feedback loop (Figure 2(b)). The systems-level analysis shows that the crosstalk element has an important role in ensuring the correct order of the two mechanisms during cellular stress, while the feedback loops between autophagy and apoptosis inducers determine the dynamic profiles of the control system [Kapuy, unpublished results].

Although the underlying mechanism is very similar, different types of cellular stressors are able to trigger various components of the upstream regulators. For example, in ER stress mTOR pathways have a key role to promote apoptosis via down-regulating autophagy [25] -[27] . A double negative feedback loop was experimentally confirmed between autophagy and mTOR, while apoptosis activates mTOR supposing a positive linkage between them [25] . According to the experimental data it was suggested with theoretical techniques that both the positive effect on apoptosis inducer and the negative effect on autophagy inducer by mTOR is essential to expand the window of survival when mTOR is inhibited by rapamycin pre-treatment [Kapuy, submitted results]. These extra feedback loops might have a role in securing the robustness of the answer mechanism.

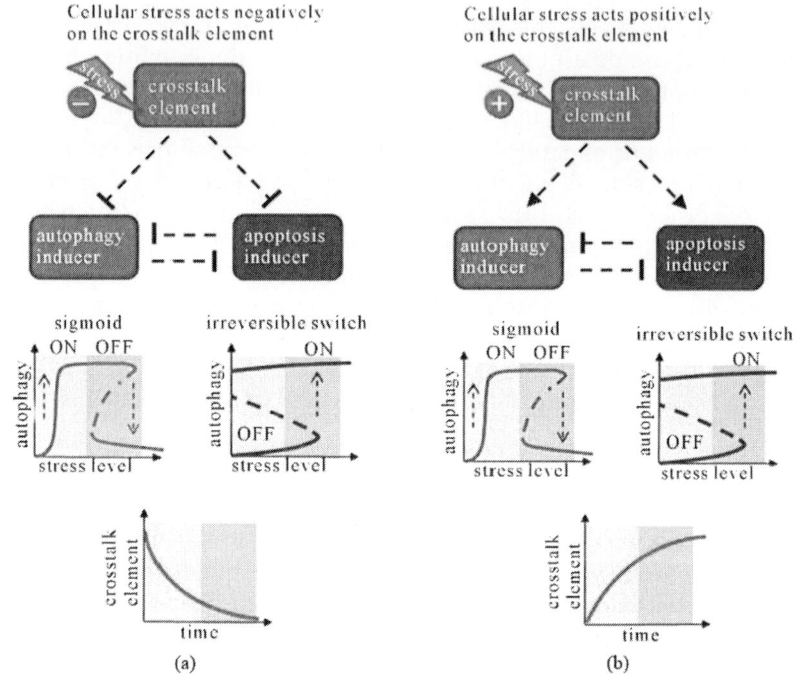

Figure 2. The dynamic characteristic of life-or-death decision under celullar stress. The simplified wiring diagram of autophagy-apoptosis crosstalk (upper panel), the signal-response curve of both autophagy and apoptosis and the time course of crosstalk element (lower panel) are presented, where (a) crosstalk element inhibits both autophagy and apoptosis inducers; (b) crosstalk element activates both autophagy and apoptosis inducers. Upper panel: the autophagy inducer, the apoptosis inducer and the crosstalk element are depicted by isolated green, blue and red boxes, respectively. Dashed lines show how the elements can influence each other. Lower panel: solid lines denote stable states, while dashed line denotes the unstable state. Light grey background refers to tolerable stress, while dark grey background shows excessive level of cellular stress.

4. CONCLUSIONS

Recent experimental data [24] together with our analysis [13] showed that the systems-level double negative and positive feedbacks guarantee the precisely executed life-or-death decision under cellular stress. A crosstalk between cytoprotective autophagy and self-killing apoptosis ensures that the sigmoid induction of autophagy always precedes the switch-like activation of cell death even at excessive level of cellular stress. This process is controlled by various crosstalk elements (e.g. Bcl2). Our preliminary results suggest that crosstalk elements can act on both autophagy and apoptosis inducers positively and negatively, but the strength of these regulatory connections is crucial to render a transient activation for autophagy, even under fatal cellular stress. Compartmentation of the crosstalk element can also help to maintain the proper order of autophagy and apoptotic cell death [21] , while extra feedback loops (e.g. mTOR-UPR crosstalk) might have a role in securing the robustness of the answer mechanism. However, the exploration of these systems-level feedbacks will be a challenge for future research.

The number of elements participating in the crosstalk between autophagic and apoptotic pathways is continuously growing. Despite the experimental efforts, the network of survival-or-death decision-making is hardly understood. These circumstances necessitate the use of alternative, non-experimental approaches. Systems biology offers the possibility to develop models with a predictive value. These kinds of models have been successfully applied for the description of different parts of the autophagic and apoptotic pathways. However, a comprehensive model for the whole network of crosstalking is painstakingly missing due to the very high number of nodes and edges in the complete system. Thus, a royal road to understand the process of decision-making can be the development of comprehensive but less detailed models. Since autophagy-apoptosis crosstalk induced by cellular stress is tightly connected to a variety of diseases, such as diabetes, various liver and neurodegenerative diseases (e.g. Parkinson's or Huntington's diseases), the long-term significance of these works implies medical purposes.

ACKNOWLEDGEMENTS

Orsolya Kapuy, Beáta Lizák, Ibolya Stiller, Gábor Bánhegyi (OTKA-PD 104110). OK is thankful to Bolyai Research Scholarship of the Hungarian Academy of Sciences. We thank Dr. Enuo He for critically reading the manuscript.

REFERENCES

1. Tyson, J.J., Chen, K.C. and Novak, B. (2003) Sniffers, Buzzers, Toggles and Blinkers: Dynamics of Regulatory and Signaling Pathways in the Cell. Current Opinion in Cell Biology, 15, 221-231.

2. Biggar, S.R. and Crabtree, G.R. (2001) Cell Signaling Can Direct Either Binary or Graded Transcriptional Responses. The EMBO Journal, 20, 3167-3176.
3. Tyson, J.J. and Novak, B. (2010) Functional Motifs in Biochemical Reaction Networks. Annual Review of Physical Chemistry, 61, 219-240.
4. Tyson, J.J., Chen, K. and Novak, B. (2001) Network Dynamics and Cell Physiology. Nature Reviews Molecular Cell Biology, 2, 908- 916.
5. Tyson, J.J., Csikasz-Nagy, A. and Novak, B. (2002) The Dynamics of Cell Cycle Regulation. BioEssays, 24, 1095- 1109.
6. Novak, B., et al. (2010) Regulated Protein Kinases and Phosphatases in Cell Cycle Decisions. Current Opinion in Cell Biology, 22, 801-808.
7. Tyson, J.J., et al. (2011) Dynamic Modelling of Oestrogen Signalling and Cell Fate in Breast Cancer Cells. Nature Reviews Cancer, 11, 523-532.
8. Arkin, A.P. and Schaffer, D.V. (2011) Network News: Innovations in 21st Century Systems Biology. Cell, 144, 844- 849.
9. Nurse, P. and Hayles, J. (2011) The Cell in an Era of Systems Biology. Cell, 144, 850-854.
10. Taylor, R.C., Cullen, S.P. and Martin, S.J. (2008) Apoptosis: Controlled Demolition at the Cellular Level. Nature Reviews Molecular Cell Biology, 9, 231-241.
11. Elmore, S. (2007) Apoptosis: A Review of Programmed Cell Death. Toxicologic Pathology, 35, 495-516.
12. Zhang, T., Brazhnik, P. and Tyson, J.J. (2009) Computational Analysis of Dynamical Responses to the Intrinsic Pathway of Programmed Cell Death. Biophysical Journal, 97, 415-434.
13. Kapuy, O., et al. (2013) A Cellular Stress-Directed Bistable Switch Controls the Crosstalk between Autophagy and Apoptosis. Molecular BioSystems, 9, 296-306.
14. Levine, B. and Kroemer, G. (2008) Autophagy in the Pathogenesis of Disease. Cell, 132, 27-42.
15. Wirawan, E., et al. (2012) Autophagy: For Better or for Worse. Cell Research, 22, 43-61.
16. Ravikumar, B., et al. (2010) Regulation of Mammalian Autophagy in Physiology and Pathophysiology. Physiological Reviews, 90, 1383-1435.
17. Hay, N. and N. Sonenberg (2004) Upstream and Downstream of mTOR. Genes & Development, 18, 1926-1945.
18. Wang, S.Y., et al. (2011) Core Signaling Pathways of Survival/Death in Autophagy-Related Cancer Networks. The International Journal of Biochemistry & Cell Biology, 43, 1263-1266.
19. Levine, B. and Klionsky, D.J. (2004) Development by Self-Digestion: Molecular Mechanisms and Biological Functions of Autophagy. Developmental Cell, 6, 463-477.
20. Maiuri, M.C., et al. (2007) Self-Eating and Self-Killing: Crosstalk between Autophagy and Apoptosis. Nat Rev Mol Cell Biol, 8, 741-752.
21. Eisenberg-Lerner, A., et al. (2009) Life and Death Partners: Apoptosis, Autophagy and the Cross-Talk between Them. Cell Death Differ, 16, 966-975.
22. Gordy, C. and He, Y.W. (2012) The Crosstalk between Autophagy and Apoptosis: Where Does This lead? Protein Cell, 3, 17-27.
23. Ogata, M., et al. (2006) Autophagy Is Activated for Cell Survival after Endoplasmic Reticulum Stress. Molecular and Cellular Biology, 26, 9220-9231.
24. Xu, Y., Yuan, J. and Lipinski, M.M. (2013) Live Imaging and Single-Cell Analysis Reveal Differential Dynamics of Autophagy and Apoptosis. Autophagy, 9, 1418-1430.
25. Appenzeller-Herzog, C. and Hall, M.N. (2012) Bidirectional Crosstalk between Endoplasmic Reticulum Stress and mTOR Signaling. Trends in Cell Biology, 22, 274-282.http://dx.doi.org/10.1016/j.tcb.2012.02.006
26. Walter, P. and Ron, D. (2011) The Unfolded Protein Response: From Stress Pathway to Homeostatic Regulation. Science, 334, 1081-1086.

CHAPTER 6

Eigenvalues of Jacobian Matrices Report on Steps of Metabolic Reprogramming in a Complex Plant-Environment Interaction

Thomas Nägele*, Wolfram Weckwerth

Department of Molecular Systems Biology, University of Vienna, Vienna, Austria

ABSTRACT

ematical modeling of biochemical systems aims at improving the knowledge about complex regulatory networks. The experimental high-throughput measurement of levels of biochemical components, like metabolites and proteins, has become an integral part for characterization of biological systems. Yet, strategies of ematical modeling to functionally integrate resulting data sets is still challenging. In plant biology, regulatory strategies that determine the metabolic output of metabolism as a response to changes in environmental conditions are hardly traceable by intuition. ematical modeling has been shown to be a promising approach to address such problems of plant-environment interaction promoting the comprehensive understanding of plant biochemistry and physiology. In this context, we recently published an inversely calculated solution for first-order partial derivatives, i.e. the Jacobian matrix, from experimental high-throughput data of a plant biochemical model system. Here, we present a bioematical strategy, comprising 1) the inverse calculation of a biochemical Jacobian; 2) the characterization of the associated eigenvalues and 3) the interpretation of the results with respect to biochemical regulation. Deriving the real parts of eigenvalues provides information about the stability of solutions of inverse calculations. We found that shifts of the eigenvalue real part distributions occur together with metabolic shifts induced by short-term and long-term exposure to low temperature. This indicates the suitability of ematical Jacobian characterization for recognizing perturbations in the metabolic homeostasis of plant metabolism. Together with our previously published results on inverse Jacobian calculation this represents a comprehensive strategy of ematical modeling for the analysis of complex biochemical systems and plant-environment interactions from the molecular to the ecosystems level.

Keywords: Inverse Problems; Jacobian Matrix; Eigenvalue; Plant Systems Biology; Metabolomics

1. INTRODUCTION

The molecular organization of plant metabolism is of high complexity. This complexity arises from a sophisticated level of subcellular organization and of highly interlaced regulatory circuits interconnecting different levels of biochemical organisation. ematical modeling is a powerful way to improve the current understanding of plant cells on a molecular level [1]. It is of particular interest to understand how plant metabolism is affected by changes in environmental conditions due to the central position of plants in our ecosystems [2]. A prominent example of how ematical modeling can promote the understanding of plant-environment interaction is given by the modeling approach of photosynthesis which was developed decades ago [3,4] and is central to current model approaches on global net primary production (NPP) on earth [5,6]. Focusing the question how ematical modeling can be applied to functionally integrate highthroughput experiments on the metabolome of plant cells, we recently published a strategy of inversely calculating the Jacobian matrix of a biochemical network from experimental high-throughput data [7,8]. When a system of ordinary differential equations (ODEs) is applied to describe a biochemcial system of interest, the Jacobian matrix represents the first-order partial derivative of functions f_i of metabolite concentrations M_i with respect to metabolite concentrations (Equation (1)):

$$\text{Jacobian matrix} : J = \begin{pmatrix} \frac{\partial f_1}{\partial M_1} & \cdots & \frac{\partial f_1}{\partial M_n} \\ \vdots & \ddots & \vdots \\ \frac{\partial f_n}{\partial M_1} & \cdots & \frac{\partial f_n}{\partial M_n} \end{pmatrix} \quad (1)$$

The metabolite functions f_i are defined by the system of ODEs describing time-dependent changes of metabolite concentrations (Equation (2)):

$$\frac{\mathrm{d}c_i(t)}{\mathrm{d}t} = f_i(M_1, \cdots, M_n) = Nv(\boldsymbol{M}(t), \boldsymbol{p}(t)) \quad (2)$$

Here, M(t) is the n-dimensional vector of metabolite concentrations, v is an r-dimensional vector of metabolic fluxes and N describes the m x r stoichiometrix matrix of the biochemical reaction network. If the full stoichiometry of a large biochemical network cannot be resolved experimentally, a metabolic interaction matrix, N_I, can be derived by interconnecting the experimentally accessible components. The term $v(\boldsymbol{M}(t), \boldsymbol{p}(t))$ indicates that the metabolic flux depends on time-variant metabolite concentrations, i.e. the substrates, products and effectors of the enzymatic reaction, and also on time-variant parameters, e.g. enzyme kinetic parameters or thermodynamic constraints. Hence, to determine the Jacobian matrix of a metabolic system the stoichiometry as well as the timevariant parameters or fluxes must be known. While the method of metabolic reconstruction allows the computational assisted construction of genome-scale stoichiometric matrices [9], the experimental analysis of kinetic parameters is highly laborious and becomes almost impossible for huge reaction networks with several thousand metabolites and reactions [10]. To overcome this limitation and to derive the Jacobian matrix directly from experimental data an inverse problem can be formulated [8, 10,11] (Equation (3)):

$$JC + CJ^{\mathrm{T}} = -2D \quad (3)$$

Here, J represents the Jacobian matrix, C is the covariance matrix derived from the experimental data and D is the so-called fluctuation matrix integrating metabolite fluctuations which can be modelled by a Langevintype equation (Equation (4)):

$$\dot{\Xi}_k = \sum_l J_{kl} \Xi_l + \sqrt{2D_k}\, \psi_k(t) \quad (4)$$

Time-dependent changes in the matrix of metabolite concentrations Ξ are directly linked to the Gaussian noise function $\psi(t)$. Stationary solutions of the Langevin-type equation are linked to the covariance matrix by a corresponding Fokker-Planck equation and finally result in (Equation (3)) [11,12]. Applying this inverse approach to experimental metabolomics data of leaf material from the genetic model plant Arabidopsis thaliana, we could unravel regulatory instances in plant primary metabolism contributing to the establishment of a new metabolic homeostasis induced by changes in light intensity and temperature [7]. Based on these findings, we now present an approach for characterization of the solution set of the inverse problem by deriving the eigenvalues of the Jacobian matrix. The real part of an eigenvalue provides information about the stability properties of the solution set [13]. We show that these stability properties can report on shifts in plant metabolism which are due to a perturbation of environmental conditions.

2. MODEL GENERATION AND INVERSE CALCULATION

Model construction and experimental analysis of metabolite content was described previously [7]. Here, we provide the model structure in a hierarchical layout (**Figure 1**) designed with the open source software package CellDesigner™ (V4.3; http://celldesigner.org). The model file is provided on request in Systems Biology Markup Language (SBML). We used this model structure to derive the metabolic interaction matrix, N_I, allowing for the calculation of the Jacobian matrix (Equations (1) and (2)).

Singular value decomposition (SVD) was applied for the inverse calculation of the Jacobian matrix [8]. We repeated inverse calculations 100 times until the variance of resulting Jacobian entries did not differ signficiantly ($p < 0.05$) from the previous calculation. We normalised the medians of each Jacobian entry to the square of its fluctuation which was expressed by the interquartile distance comprising 50% of all data points. As described in our previous publication, we build the differential Jacobian matrix dJ representing the $\log_2$-ratio of two different conditions a and b (Equation (5)):

$$dJ_{ij} = \log_2\left(abs\left(\frac{J_{a,ij}}{J_{b,ij}}\right)\right) \tag{5}$$

In this way we compared the ratio of Jacobian matrices derived from leaf samples of plants which were not exposed (non-acc), 2 days exposed (2d), 8 days exposed (8d), 14 days expose (14d) and 18 days exposed (18d) to 4°C and increased light intensity. We found a major impact of cold exposure on primary metabolism after 2 days and 14 days at 4°C pointing to a short-term and longterm effect of cold on plant primary metabolism [7]. The data indicated that the short-term response represents a fast protection against rapid cell damage while the longterm response is due to an increasing regulatory interaction with secondary metabolism resulting in an increase of flavonoids and other energyand oxidative stress dissipating molecules.

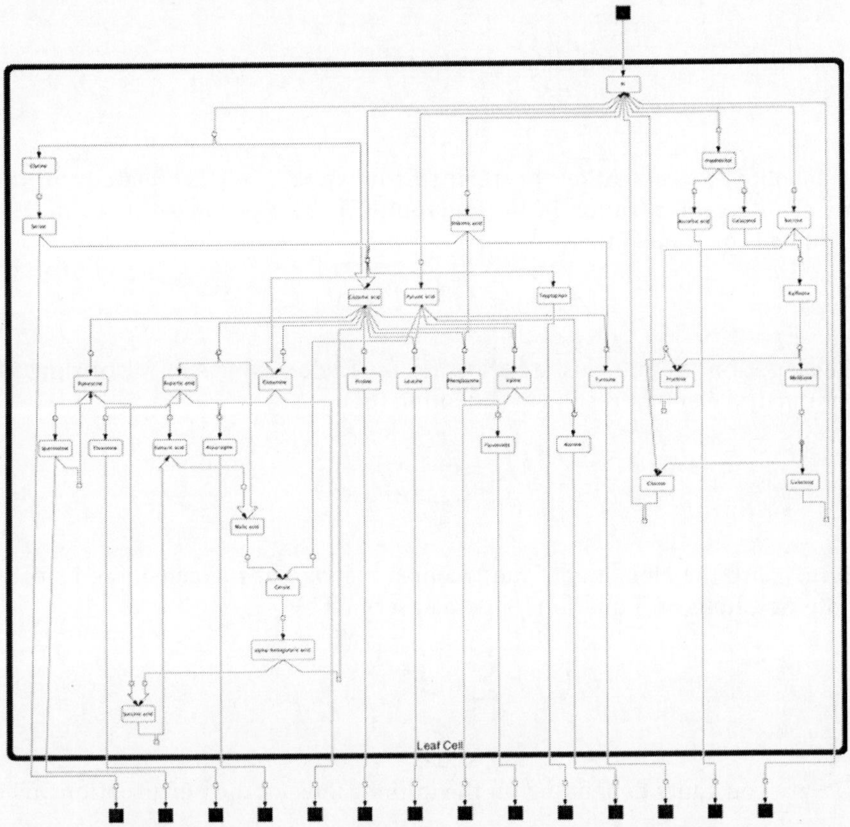

Figure 1. Model structure of primary metabolism in Arabidopsis thaliana leaf cells. Black squares represent dead ends of the model, i.e. imported and exported pools of metabolites which have not been analysed experimentally. Arrows represent steps of metabolic interconversion defining the metabolic interaction matrix, N_I, for inverse Jacobian calculation.

Deriving Eigenvalues of Jacobian Entries

Based on our results of inverse calculations, we derived the eigenvalues of the Jacobian entries as described in the following steps. All of the compared Jacobian matrices represent solutions for a certain steady state, i.e. an infintesimal time unit for which all changes in the systems equations can be assumed to equal zero (Equation (6)):

$$\frac{\mathrm{d}c_i(t)}{\mathrm{d}t} = f_i\left(M_1^*,\cdots,M_n^*\right) = 0 \tag{6}$$

Metabolite concentration at this steady state, M_i^*, is related to the metabolite concentration M_i in (Equation (2)) by a perturbation term $\theta_i(t)$ yielding (Equation (7)):

$$M_i(t) = M_i^* + \theta_i(t) \tag{7}$$

Linearization around this steady state by Taylor expansion discarding all but the first element results in (Equation (8)):

$$\frac{\mathrm{d}\theta_i(t)}{\mathrm{d}t} = \sum_{k=1}^{m} j_{ik}\theta_k(t) \tag{8}$$

Here, j_{ik} are the elements of the Jacobian matrix J at the considered steady state. Solutions of (Equation (8)) are described by

$$\theta_i(t) = \sum_{k=1}^{m} \Gamma_{ik}\mathrm{e}^{\lambda_k t} \tag{9}$$

Γ_{ik} are constants depending on the initial values of the perturbations. The constants λ_k represent the eigenvalues of the Jacobian matrix J describing the behaviour of the system. Solutions for λ_k can be derived from the characteristic equation:

$$\left(J - \lambda I\right)\theta(t) = 0 \tag{10}$$

Here, I is the m ´ m unit matrix. In general, eigenvalues are complex numbers, $\lambda = a + ib,$ composed of a real part and an imaginary part. From (Equation (9)) it becomes obvious that real parts < 0 result in an exponential decay of the perturbation matrix $\theta(t)$ while real parts > 0 induce exponential increase indicating instable system behaviour. The imaginary part of the eigenvalues determine the sinusoidal oscillation which becomes clear when rewriting λ in polar coordinates instead of Cartesian coordinates $(\lambda = \rho(\cos\varphi + i\sin\varphi))$.

To analyse whether solutions of inverse calculations may indicate a difference of system stability properties in leaf metabolism when exposed to perturbed environmental conditions, we compared the real parts of eigenvalues. All calculations and graphical representations were performed in MATLAB® (V7.12.0 R2011a). Histograms and scatterplots indicate the presence of positive real parts in all samples following the environmental perturbation (Figure 2). This points to instabilities in solution sets. In the histograms which are arranged diagonally in the scatterplots (**Figure 2**), the distribution of real parts of eigenvalues are shown for each environmental condition. Only for the samples from not cold exposed plants ("na") all real parts of eigenvalues were found to be negative. When comparing results for samples of 2 days and 8 days of cold exposure ("2d" and "8d"), the number of positive real parts of eigenvalues indicates a destabilization of the system after 2 days, followed by a restabilization after 8 days. The solution set for samples after 14 days of cold exposure ("14d") shows an increase of instabilities, i.e. positive real parts of eigenvalues. The number and extend of positive real parts declines again after 18 days indicating an increase of stable solution sets.

3. DISCUSSION

Inverse problems may often be characterized as ill-conditioned because either solutions do not exist, or solutions are not unique, or solutions are not data-dependent in a uniformly continuous way [14]. Hence, solution sets gained by an inverse approach have to be carefully evaluated and tested by further experiments. This is necessary particularly to prevent any misinterpretation of subsequent analysis which is based on the results of the inverse calculation. In the present study, we assume the realistic output of inverse calculation of the Jacobian matrix due to good agreement with a multitude of former experimental studies on plant-environment interaction [7]. It is desirable to be able to derive the Jacobian matrix directly from experimental data because of its central importance in the field of ematical modeling of complex natural systems.

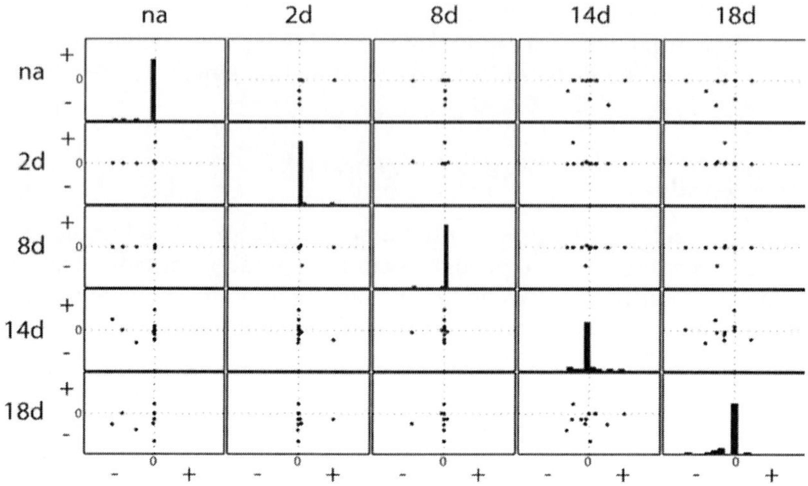

Figure 2. Histogram and scatterplot of real parts of eigenvalues. Histograms show the distribution of eigenvalues for each of the sampling conditions. In the scatterplots, eigenvalue distribution is compared between two conditions. na: non-acclimated samples (0 days at 4°C); 2d: samples after 2 days at 4°C; 8d: samples after 8 days at 4°C; 14d: samples after 14 days at 4°C; 18d: samples after 18 days at 4°C. "+" and "–" indicate the positive and negative real parts of the eigenvalues.

Even more, characterizing a biochemical system of interest by the Jacobian matrix enables the application of methods from systems theory which is inevitable for the comprehensive and systematic analysis of large metabolic networks [10]. Motivated by our previous findings based on numerical approximation of eigenvalues of Jacobian matrices in a small metabolic network [15] we now extended this approach to analyse plant-environment interactions in a larger network. The finding of changing distributions of real parts of eigenvalues allows for speculation on different metabolic constitutions. It is known from former studies that there are shortand long-term acclimatory responses in primary metabolism of Arabidopsis thaliana [16] which were also indicated by our eigenvalue calculations. The appearance of positive real parts after 2 days at changed environmental conditions may indicate the shift of the metabolic constitution towards the more stable homeostasis after 8 days. In agreement with the finding that interaction between primary and secondary metabolism is permanently induced from 14 days on [7], instable system behaviour was again observed at this time point declining until day 18 (**Figure 2**). Although the inverse calculations of the Jacobians comprise data variance which is an artefact from the Gaussian noise in the fluctuation matrix (Equations (3) and (4)), we do not assume that our findings about eigenvalues are related artefacts. First, this is due to the fact that for "na" samples only negative real parts of eigenvalues were determined. Secondly, we have normalised the medians of calculated Jacobian entries by the fluctuation of the inverse calculation. Finally, our results agree with previous findings of independent studies. However, because our experimental measurements only provide qualitative information about a metabolic homeostasis, no realistic information about the absolute values of both our Jacobian entries as well as the eigenvalues can be expected. Instead, only comparative analysis, here exemplified in scatterplots, should be applied. This problem can be circumvented by integrating metabolomics approaches which are able to measure absolute and not relative concentrations of metabolites. Like the approach of modeling photosynthesis [5,6], here we present the applicability of our approach not only to model on a molecular but also on an ecosystem level involving environmental stochastic perturbation [17]. Future studies will have to evaluate if an absolute quantification of metabolic intermediates allows for the calculation of absolute Jacobian entries and associated eigenvalues. This would be a further important step in dissecting the complexity of biological systems and their interaction with the environment by strategies of ematical modeling.

4. ACKNOWLEDGEMENTS

Thomas Nägele, Wolfram Weckwerth (GA 2010-264474) for financial support of TN.

REFERENCES

1. J. A. Morgan and D. Rhodes, "ematical Modeling of Plant Metabolic Pathways," Metabolic Engineering, Vol. 4, No. 1, 2002, pp. 80-89.
2. W. Weckwerth, "Green Systems Biology—From Single Genomes, Proteomes and Metabolomes to Ecosystems Research and Biotechnology," Journal of Proteomics, Vol. 75, No. 1, 2011, pp. 284-305.
3. G. D. Farquhar, S. V. Caemmerer and J. A. Berry, "A Biochemical Model of Photosynthetic CO_2 Assimilation in Leaves of C_3 Species," Planta, Vol. 149, No. 1, 1980, pp. 78-90.
4. G. D. Farquhar, "Models Describing the Kinetics of Ribulose Biphosphate Carboxylase-Oxygenase," Archives of Biochemistry and Biophysics, Vol. 193, No. 2, 1979, pp. 456-468.
5. C. J. Bernacchi, J. E. Bagley, S. P. Serbin, U. M. RuizVera, D. M. Rosenthal and A. Vanloocke, "Modeling C_3 Photosynthesis from the Chloroplast to the Ecosystem," Plant, Cell and Environment, 2013.
6. W. Cramer, D. W. Kicklighter, A. Bondeau, B. Moore, G. Churkina, B. Nemry, A. Ruimy, A. L. Schloss and P. P. N. M. Intercompariso, "Comparing Global Models of Terrestrial

Net Primary Productivity (NPP): Overview and Key Results," Global Change Biology, Vol. 5, 1999, pp. 1-15.

7. H. Doerfler, D. Lyon, T. Nägele, X. Sun, L. Fragner, F. Hadacek, V. Egelhofer and W. Weckwerth, "Granger Causality in Integrated GC-MS and LC-MS Metabolomics Data Reveals the Interface of Primary and Secondary Metabolism," Metabolomics, Vol. 9, No. 3, 2013, pp. 564- 574.[Citation Time(s):5]
8. X. Sun and W. Weckwerth, "COVAIN: A Toolbox for Uniand Multivariate Statistics, Time-Series and Correlation Network Analysis and Inverse Estimation of the Differential Jacobian from Metabolomics Covariance Data," Metabolomics, Vol. 8, No. 1, 2012, pp. 81-93.
9. I. Thiele and B. Ø. Palsson, "A Protocol for Generating a High-Quality Genome-Scale Metabolic Reconstruction," Nature Protocols, Vol. 5, 2010, pp. 93-121.
10. W. Weckwerth, "Unpredictability of Metabolism—The Key Role of Metabolomics Science in Combination with Next-Generation Genome Sequencing," Analytical and Bioanalytical Chemistry, Vol. 400, No. 7, 2011, pp. 1967- 1978.
11. R. Steuer, J. Kurths, O. Fiehn and W. Weckwerth, "Observing and Interpreting Correlations in Metabolomic Networks," Bioinformatics, Vol. 19, No. 8, 2003, pp. 1019- 1026.
12. N. G. van Kampen, "Stochastic Processes in Physics and Chemistry," Elsevier, Amsterdam, 1992.
13. E. Reznik and D. Segrè, "On the Stability of Metabolic Cycles," Journal of Theoretical Biology, Vol. 266, No. 4, 2010, pp. 536-549.
14. C. Eck, H. Garcke and P. Knabner, "ematische Modellierung," Springer, Berlin Heidelberg, 2008.
15. S. Henkel, T. Nägele, I. Hörmiller, T. Sauter, O. Sawodny, M. Ederer and A. G. Heyer, "A Systems Biology Approach to Analyse Leaf Carbohydrate Metabolism in Arabidopsis thaliana," EURASIP Journal on Bioinformatics & Systems Biology, 2011, p. 2.
16. T. Nägele and A. G. Heyer, "Approximating Subcellular Organisation of Carbohydrate Metabolism during Cold Acclimation in Different Natural Accessions of Arabidopsis thaliana," New Phytologist, Vol. 198, No. 3, 2013, pp. 777-787.
17. R. M. May, "Stability and Complexity in Model Ecosystems," 2nd Edition, Princeton University Press, Princeton, 1974.

CHAPTER 7

A Novel Analytical Method for Structural Characteristics of Gene Networks and Its Application

Shudong Wang, Yuanyuan Zhang, Kaikai Li, Dazhi Meng

College of Information Science and Engineering, Shandong University of Science and Technology, Qingdao, China

ABSTRACT

Analyzing gene network structure is an important way to discover and understand some unknown relevant functions and regulatory mechanisms of organism at the molecular level. In this work, mutual information networks and Boolean logic networks are constructed using the methods of reverse modeling based on gene expression profiles in lung tissues with and without cancer. The comparison of these network structures shows that average degree, the proportion of non-isolated nodes, average betweenness and average coreness can distinguish the networks corresponding to the lung tissues with and without cancer. According to the difference of degree, betweenness and coreness of each gene in these networks, nine structural key genes are obtained. Seven of them which are related to lung cancer are supported by literatures. The remaining two genes AKT1 and RBL may have important roles in the formation, development and metastasis of lung cancer. Furthermore, the contrast of these logic networks suggests that the distributions of logic types are obviously different. The structural differences can help us to understand the mechanism of formation and development of lung cancer.

Keywords: Systems Biology; Gene Network; Logic Network; Structural Parameter; Lung Cancer

1. INTRODUCTION

Lung cancer is one of the most common and lethal malignancy carcinomas, and its formation, development and metastasis is an extremely complex polygene regulatory process [1]. Studies have shown that formation of lung cancer involves activation of oncogenes, inactivation of cancer suppressor genes and mutation of apoptosis-related genes resulting in cell proliferation and abnormal death. Therefore, finding the genes critical to the formation and development of a disease from potentially disease-related genes is of significance to the diagnosis and cure of the disease and drug design. This is an important component in the research of bioinformatics [2]. Now, many researchers recognized genes closely related to lung cancer classification, early diagnosis, prognosis and treatment using microarray technology. Battacharjee et al. [3] analyzed gene expression of tissues with and without cancer using Affymetrix U195A oligonucleotide microarray, and divided pulmonary adenocarcinoma into four subtypes. Hayes et al. [4] determined three pulmonary adenocarcinoma subtypes with clinical significance using DNA microarray. Borczuk et al. [5] built class prediction models for lung cancer histology and for cancer outcome and found that gene expression profiles of biopsy specimens of lung cancers can identify unique tumoral signatures that provide information about tissue morphology and prognosis. In order to identify a gene signature predictive of recurrence in primary lung adenocarcinoma, Larsen et al. [6] analyzed gene expression profiles in a training set of 48 node-negative tumors (stage I-II) and compared tumors from cases who remained disease-free for a minimum of 36 months with those from cases whose disease recurred within 18 months of complete resection, and obtained a 54-gene signature that predicts the risk of recurrent disease independently of tumor stage. Also, Chen et al. [7] identified sixteen genes correlated with survival among patients with nonsmall-cell lung cancer (NSCLC) by analyzing microarray data and risk scores and found five-gene (DUSP6, MMD, STAT1, ERBB3, and LCK) signature is closely associated with relapse-free and overall survival among patients with NSCLC. Tsai et al. [8] studied overexpression or knockdown of gene HLJ1 in human lung adenocarcinoma CL1-5 cells and discovered that HLJ1 expression inhibits lung cancer cell proliferation, anchorage-independent growth, tumorigenesis, cell motility, and invasion, and slowed cell cycle progression.

In post-genomic era, biologists paid more attention to functions and dynamic laws of whole genome, further understood the essence of life from an integrative and systemic perspective [9]. In recent years, scientists have developed many models of gene regulatory network. For instance, Boolean network, Bayesian network, linear model and differential equation model [10-13]. Jang et al. [14] applied decision tree method to construct gene network for disease tissues. They focused mainly on the relationships between coactions of genes and pathogenesis of a complex disease and explored a new way to study genetic complexity of disease from the view of relationship and pathway. Bowers et al. [15] proposed a computational approach-logic analysis of phylogenetic profiles (LAPP) to identify detailed triplet relationships among proteins on the basis of genomic data. It can help biomedical researchers to understand potential logical relationships and discover novel biological mechanism. Now, this method has been studied in both theories and applications [16- 21].

In this work, according to the viewpoint that the structures of matters can determine their functions, we can catch disease-risk genes through analyzing the differential structure of gene network in lung tissues with and without cancer. Different from other methods [10-13], we can find the structural difference of gene networks between normal and lung cancer easily. First, mutual information networks of carcinoma-related genes are constructed using their expression profiles under the contexts of normal and two types of lung cancer (adenocarcinoma and small cell lung cancer, abbreviated as AC and SCLC respectively). From the analysis of several structural parameters for these networks, it shows that average degree, the proportion of non-isolated nodes, average betweenness and average coreness can distinguish significantly the networks corresponding to normal and diseased tissues. Computing the degree, betweenness and coreness of each gene in the three gene networks, we find nine structural key genes. Seven of them which are related to lung cancer are supported by literatures. The validity of this approach is about 78%, while the experimental methods only can determine one or two disease-risk genes. Furthermore, we predict the remaining two genes AKT1 and REL may have important roles in the formation, development and metastasis of lung cancer. Second, gene logic networks of normal and two types of lung cancer are constructed. Comparing the distributions of 2-order logic types in these networks, we find that logic types in gene logic network of normal tissues include logic types 1, 3, 5, 8, while logic types 5, 8 and logic type 8 do not appear in that of AC and SCLC respectively. Also, logic types 2, 6 and 2, 4, 6 emerge in that of AC and SCLC respecttively. The difference of the distribution of 2-order logic types in the above gene logic networks of normal and two types of lung cancer may be a reason or result of the formation and development of lung cancer. It provides some enlightenment roles to study the intrinsic mechanism of lung cancer.

2. REVERSE NETWORK MODELING STRUCTURAL PARAMETERS OF NETWORK

Now, there are two ways to build a network model for a biological system, forward and reverse network modelings. The former is based on studies of individual components of the system and, hence, is appropriate to a system for which there exist lots of experiment data in literature. However, reverse network modeling is based on the analysis of high-throughput data (from DNA chips and some other recently developed techniques) to mine regulatory mechanisms among the components of the system. Recently, reverse network modeling has been broadly applied to the studies of various biological systems [10-14,22-24]. In 2006, Werhli et al. [22] compared the accuracy of reconstructing gene regulatory networks with three different modeling and inference paradigms Relevance networks, graphical Gaussian models and Bayesian networks through constructing Raf signal transduction network, and found that Bayesian networks and graphical Gaussian models outperform Relevance networks. Using the method of reverse network modeling, Perkins et al. [23] investigated gene regulatory network of drosophila and explained the activation of genes. Wang et al. [24] constructed gene logical network for Arabidopsis subject to external stimuli and further analyzed dynamical behaviors of the logical networks.

2.1. Mutual Information

A gene expression profile is a vector whose components are its expressions in many different experiments. For convenience, we denote the gene expression profiles by their corresponding genes. We describe the relationships between genes with mutual information. For example, the mutual information of genes A and B means that of their expression profiles and is quantified as follows:

$$I(A;B) = H(A) + H(B) - H(A,B) \tag{1}$$

where

$$H(X) = -\sum_{x \in X} p(x) \log_2 p(x)$$

is the Shannon entropy of vector X, H(A,B) is the joint entropy of genes A and B. Larger values of I(A;B) imply closer interrelation between genes' expression profiles. In case of I(A;B) = 0, genes' expression profiles are independent.

2.2. LAPP Method

Bowers et al. [15-17] utilized uncertainty coefficient (abbreviated as U) to measure the logical relationships and identify logic types of 1-, 2-order among genes A, B, C. The relationships between/among two and three genes are called 1-, 2-order one, respectively. In this work, we identify logical relationships among genes by LAPP method based on 0 - 1 expression profile data of carcinoma-related genes.

LAPP method can determine logical relationships of 1-, 2-order by calculating the uncertainty coefficient of proper functions of every order among genes. The 1-order proper relationship between genes A and B is determined in the following formula:

$$U\left(B \mid f_1(A)\right) = \frac{H(B) + H\left(f_1(A)\right) - H\left(B, f_1(A)\right)}{H(B)} \tag{2}$$

where H refers to the entropy of the individual or joint distributions. $U(Y/X)$ denotes the uncertainty coefficient of the influence of X on Y. The size of U value denotes the statistical possibility of the uncertainty logic of 1-order of X to Y. f_1 is one of 1-order proper functions of A to B. The 1-order proper functions are divided into two types (**Table 1**). The first is B = A, namely, the presence of A leads to the one of B (equivalent with $\neg B = \neg A$); the second is $B = \neg A$, namely, the presence of A leads to the absence of B (equivalent with $\neg B = A$).

Similarly, the uncertainty coefficient of 2-order proper function is calculated in the following expression:

$$U\left(C \mid f_2(A,B)\right) = \left(H(C) + H\left(f_2(A,B)\right) - H\left(C, f_2(A,B)\right)\right) / H(C) \tag{3}$$

where f_2 is one of 2-order proper functions. There are ten 2-order proper functions and eight logic types (**Table 1**).

Let u_i denote thresholds of i-order uncertainty coefficients (i = 1, 2) and d' a small number and a difference threshold. When the difference of the uncertainty coefficients $U\left(B \mid f_1(A)\right)$ and $U\left(A \mid f_1(B)\right)$ is greater than a given value d', namely: only if

$$U\left(B \mid f_1(A)\right) - U\left(A \mid f_1(B)\right) > d'$$

Table 1. List of 1-, 2-order proper functions and logic types. Obviously, 1-order proper functions and logic types are the same.

order	Proper function and its serial number	Logic type and its serial number
1-order	1. $B = A$	1. $B = A$
	2. $B = \neg A$	2. $B = \neg A$
2-order	1. $C = A \wedge B$	1. $C = A \wedge B$
	2. $C = \neg(A \wedge B)$	2. $C = \neg(A \wedge B)$
	3. $C = A \vee B$	3. $C = A \vee B$
	4. $C = \neg(A \vee B)$	4. $C = \neg(A \vee B)$
	5. $C = A \wedge \neg B$	5. $C = A \wedge \neg B$ (5-1)
	6. $C = \neg A \wedge B$	$C = \neg A \wedge B$ (5-2)
	7. $C = A \vee \neg B$	6. $C = A \vee \neg B$ (6-1)
	8. $C = \neg A \vee B$	$C = \neg A \vee B$ (6-2)
	9. $C = \neg(A \leftrightarrow B)$	7. $C = \neg(A \leftrightarrow B)$
	10. $C = \neg(A \leftrightarrow B)$	8. $C = \neg(A \leftrightarrow B)$

and

$$U\left(B \mid f_1\left(A\right)\right) > u_1,$$

the relationship between genes A and B is A→B. We set threshold u_2 to judge the existence of 2-order logical relationships in the following formula:

$$\begin{cases} U\left(C \mid f_2\left(A,B\right)\right) > u_2 \\ U\left(C \mid f_1\left(A\right)\right) < u_1 \\ U\left(C \mid f_1\left(B\right)\right) < u_1 \\ u_2 > u_1 \end{cases} \qquad (4)$$

According to the above description, we can determine all the logical relationships and logic types among the investigated genes. We regard all of the logical relationships as a gene network with 1-, 2-order logics, where each node in the network represents a gene.

2.3. Structural Parameters of Network

Now, let us briefly introduce some structural parameters of network that are used to describe network structures in this work (see [25,26] for detailed explanations). Set $G=(V,E)$ be a complex network with node-set $V=\{1,2,\cdots,N\}$ and edge-set E.

1) Average degree (K)

The degree of a node is the number of nodes adjacent to it. The average degree is the average value of degrees of all nodes, denoted by K.

2) Average path length (L)

The distance between nodes i and j in V is defined as the smallest number of edges that connect nodes i to j, denoted by d_{ij}. The average path length L of the network is defined as

$$L=\sum_{i>j}d_{ij}\Big/\left(0.5N(N-1)\right)$$

3) Average clustering coefficient (C)

The clustering coefficient of node i, denoted by C_i, is equal to the proportion of the edges among its adjacent nodes in the possible edges. The average clustering coefficient is the average value of clustering coefficients of all the nodes.

4) Modularity (Q)

In a complex network, the concept of community is a good tool to describe network structures and provides better understanding of network functions. Newman [26] proposed modularity to measure the probability of a network having communities. Suppose that network G contains l communities G_1, G_2, ⋯ G_l. Define symmetric matrix $M=(m_{ij})_{l\times l}$, where h_{ij} represents the ratio of the number of edges between two communities G_i, G_j to the total number of edges of the network. The modularity is defined as:

$$Q=\sum_i Q_i=\sum_i\left(m_{ii}-a_i^2\right)$$

where a_i is the sum of all elements in the ith row of M, representing the ratio of the number of edges adjacent to community G_i to that of all edges.

5) Proportion of non-isolated nodes (R)

Proportion of non-isolated nodes, denoted by R, is the ratio of the number of the nodes with degree not 0 to that of all nodes.

6) Average betweenness (B)

$$B_n = \sum_{(i,j)} d_{inj} / d_{ij}$$

is the betweenness of the node n, where d_{ij} presents the number of the shortest paths between the node i and j, and d_{inj} presents that of paths which pass the node n. The betweenness of a node reflects its influence in the network. The average value of the betweennesses of all nodes in the network is average betweenness of the network.

7) Average coreness (H)

k-core of a graph is the subgraph after getting rid of the nodes whose degrees are smaller than or equal to k repeatedly. If a node exists in a k - core, but is removed in (k + 1)-core, the coreness of the node is k. The average value of the corenesses of all nodes is average coreness of the network.

3. MATERIAL AND NUMERICAL EXPERIMENTS

3.1. Data Sources and Processing

1) Data sources The gene expression data we work on are all from GPL96 and GPL570 in NCBI. The sample data set for normal tissue, AC and SCLC includes 73, 68 and 39 samples respectively. These data are from GSE2771, GSE12667 and GSE15240 respectively. For description convenience, we refer to the networks corresponding to these three data sets as normal network, adenocarcinoma network (abbreviated as AC network) and small cell lung cancer network (abbreviated as SCLC network). Almost all of these data sets include p-values and P-M-A (P, A and M stand for presence, absence and margin respecttively) except for some samples. We process these three data sets (corresponding to normal tissue, AC and SCLC)) by MAS 5.0 of Expression Console, and obtain values, p-values and P-M-A for each sample of each gene. Then we represent P with 1, A and M with 0 in these three processed data sets. Our work is based on these three databases (corresponding to normal, AC and SCLC group)[1].

2) Selection of carcinoma-related genes It is out of our scope to construct and analyze the networks for all genes in the databases (each of the three databases contains expressions of more than 20,000 genes). Our main focus is on carcinoma-related genes [27]. From the databases we choose 301 genes that are known to be linked with cancers. These genes correspond to 801 probes. In the cases where several probes correspond to one gene, the

highest expression value is chosen to form the gene's expression profile. We obtain 283 genes in normal group and 286 genes in AC and SCLC group. Choosing the intersection of these genes, and using Bonferroni test to compare the difference of genes' expression profiles between normal and lung cancer groups respectively, we choose the significant level 0.0001 and obtain 91 genes which both have significant difference between normal and two types of lung cancer. We focus on mutual information and logical networks of these genes in each database[2].

3) Data processing To calculate mutual information between arbitrary two genes, we discretize the p-values in each database as follows: a) Select the ranges [Min, Max] for p-values in the intervals [0,0.05] and (0.05,1] and divide them into 10 portions respectively such that each portion contains almost the same number of p-values. Order the portions in the number order and denote them by 1st, 2nd, ··· 20th interval, respectively; b) Replace the p-values in an interval by its labeling value. Obviously, the granularity of our discretization is finer than that of 0 - 1 discretization and hence our discretization loses less information contained in the 0 - 1 discretization. Therefore, it is reasonable to believe that the mutual information networks based on our finer discretization better reflect the nature of the gene regulatory system.

3.2. Establishment of Mutual Information Gene Network

Based on above discretized data, we can calculate mutual information values of any two genes by formula (1) and hence obtain a complete network of all genes in each database. Note that the ranges for mutual information values for our three databases are different. For the purpose of comparing the networks, we normalize the mutual information values for each database as

$$x' = (x - \min)/(\max - \min)$$

where x and x' represent the original and normalized mutual information values. max and min are the maximum and minimum values of the mutual information values before normalization respectively.

In order to highlight the structural characteristics of the networks so that valuable biological conclusions can be drawn, it is necessary to choose a threshold value to carry out coarse graining on normalized mutual information. We build three gene networks (with normalized mutual information values as edge weights), corresponding to the cases of normal, AC and SCLC group. For each network, we compute seven structural parameters: average degree K, average path length L, average clustering coefficient C, modularity Q, proportion of non-isolated nodes R, average betweenness B and average coreness H. These structural parameters are plotted versus threshold values T in **Figure 1**. From **Figure 1**, we can see that these structural parameters K, R, B and H can distinguish normal network from AC and SCLC networks in a broad range of the threshold variation. It shows that the difference between normal and lung cancer which do not dependent on the choosing of threshold.

We define the degree of differentiation of each structural parameter as the proportion of the distinguishing interval of threshold in [0,1], denote the degree of differentiation that the values of structural parameters of normal tissues are greater or smaller than that of lung cancer by G differentiation or S differentiation respectively. The results are show in **Table 2**. In order to show the specific of these networks, we randomize the genes' expression profiles keeping the equal number of p-values in [0, 0.05] and (0.05, 1]. We obtain the changes of those structural parameters versus thresholds in randomized networks (**Figure 2**) and compute the degrees of differentiation of those structural parameters (**Table 2**). Compared with these results in real data, these structural parameters can distinguish normal and lung cancer in their structures. According to the view that system's structure decides its function, it shows that the functions of gene networks have significant difference in normal and lung cancer.

We search genes which have largest contribution to these differences through calculating the difference of the degree, betweenness and coreness of each gene in the three gene networks. We get the top ten genes whose degrees, betweennesses or corenesses are largest (**Table 3**), which are called structural key genes. We get the intersection of these genes and find that there are seven genes contributing to the difference of structures of gene networks of AC and normal, also there are seven genes between SCLC and normal tissues. Genes FN1 and REL are special between AC and normal tissues, and genes TOP2A and RBL1 between SCLC and normal tissues.

Table 2. Comparison the degree of differentiation of real data with that of randomized data.

		K	*L*	*C*	*Q*	*R*	*B*	*H*
Real data	*G* differentiation	83%	39%	27%	30%	81%	81%	71%
	S differentiation	0	26%	5%	21%	0	0	0
Randomized data		12%	17%	13%	12%	0	14%	18%

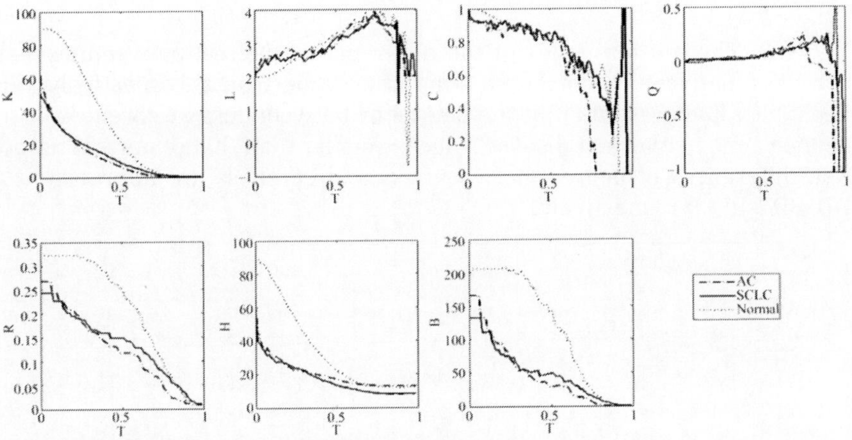

Figure 1. Plots of seven structural parameters versus threshold values T in real networks.

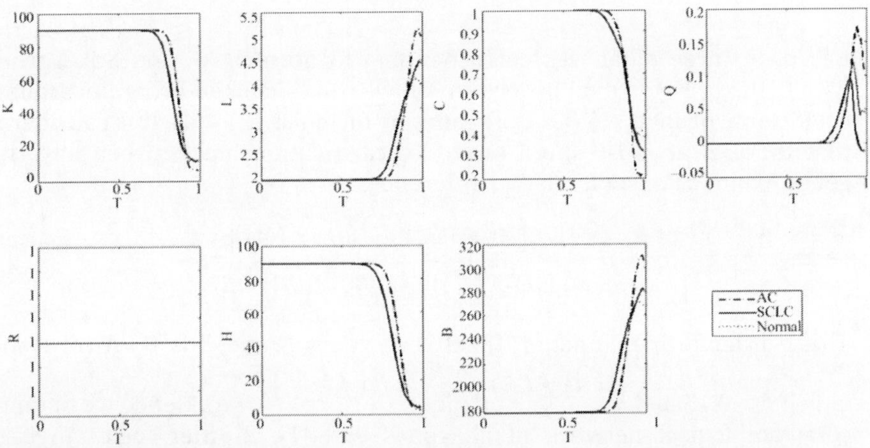

Figure 2. Plots of seven structural parameters versus threshold values T in randomized networks.

3.3. Construction of Gene Logical Networks

Looking for the regulatory relationships among genes can help us to understand the regulatory pathways related to diseases and discover the regulatory mechanisms of organism at the molecular level. Next, we will construct and analyze gene logical networks for normal, AC and SCLC respectively.

Table 3. The top ten genes of the difference of degree, betweenness and coreness between normal and lung cancer networks respectively. The negative value presents that the degree or betweenness or coreness of the gene in gene network of disease stage is smaller than that of normal tissues. The differences of degree, betweenness and coreness are abbreviated as DD, BD and CD respectively.

Genes (AC and normal)	DD	Genes (SCLC and normal)	DD	Genes (AC and normal)	BD	Genes (SCLC and normal)	BD	Genes (AC and normal)	CD	Genes (SCLC and normal)	CD
FN1	-52	PIK3CA	-54	PIK3CA	-930	PIK3CA	-955	FN1	-41	PIK3CA	-41
MSH2	-50	MSH2	-50	FN1	-399	EMP1	354	MSH2	-40	MAPK9	-41
PIK3CA	-49	NRAS	-49	MAPK9	-256	NRAS	-320	REL	-40	RBL1	-41
NRAS	-48	TOP2A	-49	FGF2	234	TOP2A	-263	NRAS	-39	TOP2A	-41
REL	-47	RBL1	-47	RBL1	-216	RBL1	-244	AKT1	-38	MSH2	-40
AKT1	-46	MAPK9	-46	MSH2	-211	MAPK9	-211	PIK3CA	-37	NRAS	-40
MAPK9	-45	AKT1	-46	AKT1	-180	AKT1	-185	PTCH1	-36	AKT1	-40
CXCL2	-43	CHEK1	-45	NRAS	-150	CHEK1	-179	EMP1	-36	MAFG	-40
ECT2	-42	SUPT4H1	-44	TOP2A	-145	SUPT4H1	-160	KDR	-35	PTCH1	-40
AXL	-42	PTCH1	-43	REL	-128	PTCH1	131	MAPK9	-35	REL	-39

Based on three databases corresponding to normal, AC and SCLC group with 0-1 expression data in Section 3.1, we can calculate i-order uncertainty coefficients of any I + 1 (i = 1, 2) genes by formulas (2) - (4). It is considered that there is at most one 1-order logical relationship between any two genes A and B. We set a relative difference:

$$d = \frac{\left| U\left(A \mid f_1(B)\right) - U\left(B \mid f_1(A)\right) \right|}{\left[\left(A \mid f_1(B)\right) + U\left(B \mid f_1(A)\right)\right]/2}$$

The 1-order proper function f_1 between genes A and B is B→A (or A→B) only if d > 0.15 and $U\left(A \mid f_1(B)\right)$ (or $U\left(B \mid f_1(A)\right)$) > u_1. Hence we obtain a complete logical network of all genes with 1-, 2-order logics in each database.

In order to eliminate the difference of 2-order logic types owing to choosing thresholds, we compare the distribution of 2-order logic types in the three networks at different threshold pairs (**Figure 3**). It shows that the distribution of 2-order logic types nearly doesn't change as the threshold pairs. In normal gene logic networks, there are logic types 1, 3, 5, 8, while in that of AC, logic types 5, 8 do not appear and logic types 2, 4, 6 emerge, also in that of SCLC, logic types 8 does not appear and logic 2, 6 emerge. The difference of 2-order logic types between normal tissues and lung cancer may be a reason or result of the formation of lung cancer.

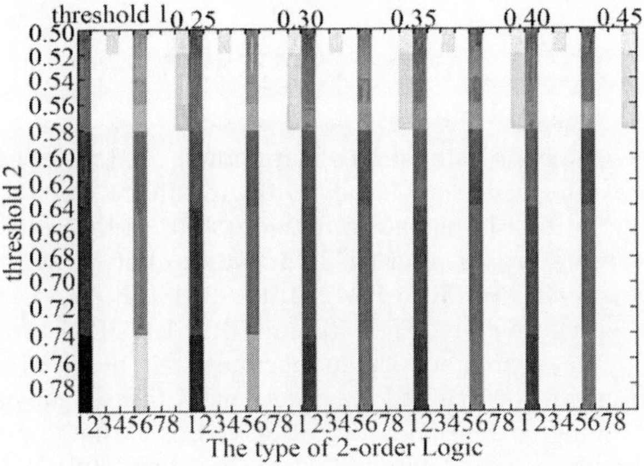

Normal

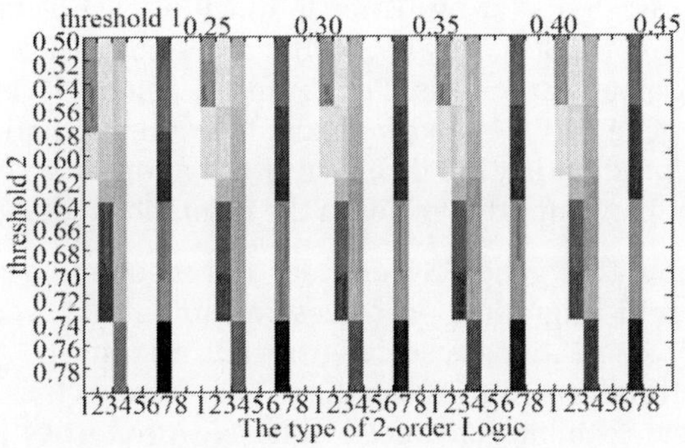

AC

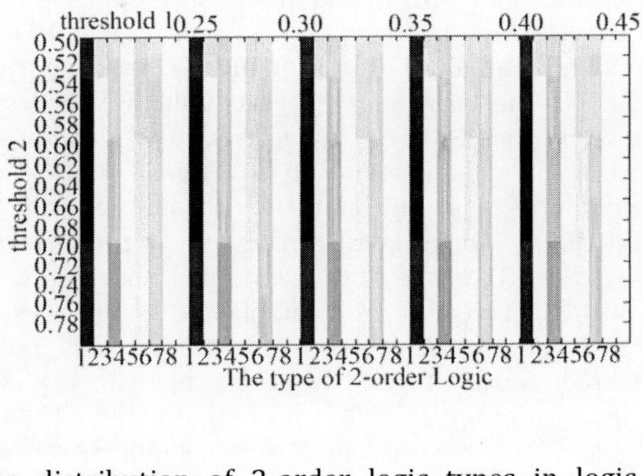

SCLC

Figure 3.The distribution of 2-order logic types in logic networks at different thresholds. The darker the color is the higher proportion of the number of a logic type has .

4. CONCLUSIONS AND DISCUSSIONS

Based on the gene expression data of carcinoma-related genes expressed in normal tissues and diseased tissues with AC and SCLC, we construct the corresponding mutual information networks. From the comparison of the structural parameters of mutual information networks, we find nine structural key genes: PIK3CA, FN1, MAPK9, MSH2, NRAS, TOP2A, RBL1, AKT1 and REL, where genes FN1, REL and RBL1, TOP2A are special genes in comparing the gene network of normal with those of AC and SCLC respectively. In the constructed gene networks, these genes are related to each other with higher mutual information values in that of normal (greater than 0.7) than in those of AC and SCLC (smaller than 0.05). We consider that these genes may play important roles in maintaining the normal functions of genome. Through analyzing Gene Ontology of each genes, we find that five genes of them PIK3CA, MAPK9, MSH2, AKT1 and TOP2A are involved in the regulation of cell apoptosis. For example, gene FN1 is related to cell adhesion and gene RBL1 is a key regulator in the regulation of cell cycle. Therefore, we predict that the disorder of cell apoptosis, cell adhesion and cell cycle might play important roles in the formation of lung cancer.

Researches show that gene FN1 encodes fibronectin which is involved in cell adhesion and migration processes including embryogenesis, wound healing, blood coagulation, host defense, and metastasis. Anastellin binds fibronectin and induces fibril formation. This fibronectin polymer, named superfibronectin, exhibits enhanced adhesive properties. Both anastellin and superfibronectin inhibit tumor growth, angiogenesis and metastasis. Anastellin activates p38 MAPK and inhibits lysophospholipid signaling. Nashino et al. [28] found that the disappearanceof FN reflects the aggressiveness of the tumor and presumably plays an important role in the prognosis of AC of the lung. So gene FN1 is an ACsuppressor gene. The protein encoded by gene RBL1 is similar in sequence and possibly function to the product of the retinoblastoma 1 (RB1) gene. RB1 gene product is a tumor suppressor protein that appears to be involved in cell cycle regulation, as it is phosphorylated in the S to M phase transition and is dephosphorylated in G1 phase of the cell cycle. Both the RB1 protein and the product of this gene can form a complex with adenovirus E1A protein and SV40 large T-antigen, with the SV40 large T-antigen binding only to the unphosphorylated form of each protein. Due to the sequence and biochemical similarities with the RB1 protein, it is thought that the protein encoded by gene RBL1 may also be a tumor suppressor. Baldi et al. [29] found an inverse correlation between the histological grading of the tumors, the development of metastasis, and the level of expression of Rb2/p130 which is structurally and functionally similar to the product of gene RBL1. So gene RBL1 may be a tumor suppressor gene of lung cancer.

Gene TOP2A encodes a DNA topoisomerase which functions as the target for several anticancer agents and a variety of mutations in this gene have been associated with the development of drug resistance. Jensen et al. [30] studied the resistant to the inhibitory action of F 11782 in SCLC patients and found that TOP2A is a target of F 11782 in vivo, and that F 11782 may act as a novel topoisomerase II poison. It shows the function of gene TOP2A in SCLC. Gene PIK 3CA encodes the p110alpha catalytic subunit, Samuels et al. [31] found that PIK3CA was frequently mutated in cancers of the colon, breast, brain and lung. The majority of mutations clustered near two positions within the PI3K helical or catalytic domains and at least one hotspot mutation appeared to increase kinase activity. The protein encoded by gene MAPK9 is a member of MAP kinase family. MAP kinases act as an integration point for multiple biochemical signals, and are involved in a wide variety of cellular processes such as proliferation, differentiation, transcription regulation and development. This gene and MAPK8 are also known as c-Jun Nterminal kinases. This kinase blocks the ubiquitination of tumor suppressor p53, and thus it increases the stability of p53 in nonstressed cells. Nitta et al. [32] investigated the role of the JNK2α (the alias of MAPK9) isoform in NSCLC formation by examining its expression in primary tumors and by modulating its expression in cultured cell lines and found Cell lines deficient in JNK2α had decreased cellular growth and anchorage-independent growth, and the tumors were four-fold smaller in mass. Gene MSH2 is a DNA mismatch repair gene, Xinarianos et al. [33] found a relationship between p53 and MSH2, there is evidence for p53 being a transcriptional activator of MSH2, and suggested a hypothesis that MSH2 acts as a DNA-damage signaller triggering p53 overexpression. Sasaki et al. [34] found NRAS mutation in 1 of 195 patients, although the rate is small, we consider that the mutation of NRAS may be a reason of the formation of lung cancer. The relationships between MSH2, NRAS and SCLC are not clear, but owing to they have important roles in structures of gene networks between normal and SCLC, we predict these two genes are related to the formation and development of SCLC. Besides, the remaining two genes AKT1 and REL are open. We predict that they may have significant roles in the formation of lung cancer based on the view that system's structure decides its function.

Through constructing gene logic networks corresponding to normal, AC and SCLC, we find the significant difference of the distribution of 2-order logic types. Relative to normal gene logic network, logic types 5, 8 do not appear in that of AC, and logic type 8 does not appear in that of SCLC. Also, logic types 2, 6 and 2, 4, 6 emerge in that of AC and SCLC respectively. The difference of the distribution of 2-order logic types in the above gene logic networks of normal and two types of lung cancer may be a reason or result of the formation of lung cancer. It provides some enlightenment roles to study the intrinsic mechanism of lung cancer.

Finally, we have applied this approach to the gene detection of other cancers, such as breast cancer and brain tumor, and Arabidopsis thaliana response to the special external stimulus. Through analyzing the results, we find most of genes can be supported by literatures. It further approves the validity of our approach.

5. ACKNOWLEDGEMENTS

Shudong Wang, Yuanyuan Zhang, Kaikai Li, Dazhi Meng (Nos: 60874036, 61170183), Scientific Research Foundation for Excellent Middle-Aged and Youth Scientists of Shandong Province of China (BS2011SW025), SDUST Research Fund of China (2010 KYJQ104) for the support to this work.

REFERENCES

1. T. Mistudomi and T. Takahashi, "Genetic Abnormalities in Lung Cancer and Their Prognostic Implication," Gan to Kagaku Ryoho, Vol. 23, 1996, pp. 990-996.
2. E. S. Lander and R. A. Weinberg, "Genomics: Journey to the Center of Biology," Science, Vol. 287, No. 5459, 2000, pp. 1777-1782.
3. A. Bhattacharjee, W. G. Richards, J. Staunton, C. Li, S. Monti, P. Vasa, C. Ladd, J. Beheshti, R. Bueno, M. Gilette, M. Loda, G. Weber, E. J. Mark, E. S. Lander, W. Wong, B. E. Johnson, T. R. Golub, D. J. Sugarbaker and M. Meyerson, "Classification of Human Lung Carcinomas by mRNA Expression Profiling Reveals Distinct Adenocarcinoma Subclasses," Proceedings of the National Academy of Sciences of the United States of America, Vol. 98, No. 24, 2001, pp. 13790-13795.
4. D. N. Hayes, S. Monti, G. Parmigiani, C. B. Gilks, K. Naoki, A. Bhattacharjee, M. A. Socinski, C. Perou and M. Meyerson, "Gene Expression Profiling Reveals Reproducible Human Lung Adenocarcinoma Subtypes in Multiple Independent Patient Cohorts," Journal of Clinical Oncology, Vol. 24, No. 31, 2006, pp. 5079-5090.
5. A. C. Borczuk, L. Shah, G. D. Pearson, K. L. Walter, L. Wang, J. H. Austin, R. A. Friedman and C. A. Powell, "Molecular Signatures in Biopsy Specimens of Lung Cancer," American Journal of Respiratory and Critical Care Medicine, Vol. 170, No. 2, 2004, pp. 167-174.
6. J. E. Larsen, S. J. Pavey, L. H. Passmore, R. V. Bowman, N. K. Hayward and K. M. Fong, "Gene Expression Signature Predicts Recurrence in Lung Adenocarcinoma," Clinical Cancer Research, Vol. 13, No. 10, 2007, pp. 2946- 2954.
7. H. Y. Chen, S. L. Yu, C. H. Chen, G. C. Chang, C. Y. Chen, A. Yuan, et al., "A Five-Gene Signature and Clinical Outcome in Non-Small-Cell Lung Cancer," The New England Journal of Medicine, Vol. 356, No. 1, 2007, pp. 11- 20.
8. M. F. Tsai, C. C. Wang, G. C. Chang, C. Y. Chen, C. L. Chen, Y. P. Yang, et al., "A New Tumor Suppressor DnaJ-Like Heat Shock Protein, HLJ1, and Survival of Patients with Non-Small-Cell Lung Carcinoma," Journal of the National Cancer Institute, Vol. 98, No. 12, 2006, pp. 825-838.
9. R. Brent, "Genomic Biology," Cell, Vol. 100, No. 1, 2000, pp. 169-183.
10. T. Akutsu, S. Miyano and S. Kuhara, "Identification of Genetic Networks from a Small Number of Gene Expression Patterns under the Boolean Network Model," Pacific Symposium on Biocomputing,1999, pp. 17-28,

11. D. Husmeier, "Reverse Engineering of Genetic Networks with Bayesian Networks," Biochemical Society Transactions, Vol. 31, 2003, pp. 1516-1518.
12. E. P. Van Someren, L. F. Wessels and M. J. Reinders, "Linear Modeling of Genetic Networks from Experimental Data," International Conference on Intelligent Systems for Molecular Biology, Vol. 8, 2000, pp. 355-366.
13. T. Chen, H. Y. He and G. M. Church, "Modeling Gene Expression with Differential Equations," Pacific Symposium on Biocomputing, 1999, pp. 29-40.
14. W. Jiang, C. Wu, J. K. Xu, Y. Y. Yang and X. Li, "Using Decision Forest to Construct Gene Networks for Complex Disease," Chinese Journal of Biomedical Engineering, Vol. 28, 2009, pp. 263-269.
15. P. M. Bowers, S. J. Cokus, D. Eisenberg and T. O. Yeates, "Use of Logic Relationships to Decipher Protein Network Organization," Science, Vol. 306, No. 5705, 2004, pp. 2246-2249.
16. X. Zhang, S. C. Kim, T. Wang and B. Chitta, "Joint Learning of Logic Relationships for Studying Protein Function Using Phylogenetic Profiles and the Rosetta Stone Method," IEEE Transactions on Signal Processing, Vol. 54, No. 6, 2006, pp. 2427-2435.
17. J. X. Zhang, X. Ji and L. Zhang, "Extracting Three-Way Gene Interactions from Microarray Data," Bioinformatics, Vol. 23, No. 21, 2007, pp. 2903-2909.
18. M. W. Anderson, "Functional Genomics Annotation: It's Logical! Symbolic Logic Helps Unravel Complex Protein Networks," Scientist, Vol. 19, No. 5, 2005, p. 33.
19. V. Spirin, M. S. Gelfand, A. A. Mironov and L. A. Mirny, "A Metabolic Network in the Evolutionary Context: Multiscale Structure and Modularity," Proceedings of the National Academy of Sciences of the United States of America, Vol. 103, No. 23, 2006, pp. 8774-8779.
20. G. Glazko, M. Coleman and A. Mushegian, "Similarity Searches in Genome-Wide Numerical Data Sets," Biology Direct, Vol. 1, No. 1, 2006, pp. 1-14.
21. P. M. Bowers, B. D. O'Connor, S. J. Cokus, E. Sprinzak, T. O. Yeates and D. Eisenberg, "Utilizing Logical Relationships in Genomic Data to Decipher Celluar Processes," FEBS Journal, Vol. 272, No. 20, 2005, pp. 5110-5118.
22. A. V. Werhli, M. Grzegorezyk and D. Husmeier, "Comparative Evaluation of Reverse Engineering Gene Regulatory Networks with Relevance Networks Graphical Gaussian Models and Bayesian Networks," Bioinformatics, Vol. 22, No. 20, 2006, pp. 2523-2531.
23. T. J. Perkins, J. Jaeger, J. Reinitz and L. Glass, "Reverse Engineering the Gap Gene Network of Drosophila melanogaster," PLOS Computational Biology, Vol. 2, No. 5, 2006, pp. 417-428.
24. S. D. Wang, Y. Chen, Q. Y. Wang, E. Y. Li, Y. S. Su and D. Z. Meng, "Analysis for Gene Networks Based on Logical Relationships," Journal of Systems Science and Complexity, Vol. 23, No. 5, 2010, pp. 999-1011,
25. A. Réka and A. L. László, "Statistical Mechanics of Complex Networks," Reviews of Modern Physics, Vol. 74, 2002, pp. 47-97.
26. M. E. J. Newman and M. Girvan, "Finding and Evaluating Community Structure in Networks," Physical Review. E, Statistical, Nonlinear, and Soft Matter Physics, Vol. 69, No. 2, 2004, pp. 6113-6127.
27. L. S. Si and X. Li. "Oncogenes, Cancer Suppressor Genes, Cancer-Related Genes," Shanxi Technology and Science Press, Xi'an, 2002.
28. T. Nishina, T. Ishida, K. Yasumoto and K. Sugimachi, "Distribution of Fibronection in Adenocarcinoma of the Lung: Classification and Prognostic Significance," Journal of Surgical Oncology, Vol. 43, No. 2, 1990, pp. 94- 100.
29. A. Baldi, V. Esposito, A. De Luca, FuY, I. Meoli, G. G. Giordano, M. Caputi, F. Baldi and A. Giordano, "Differential Expression of Rb2/p130 and p107 in Normal Human Tissues and in Primary Lung Cancer," Clinical Cancer Research, Vol. 3, No. 10, 1997, pp. 1691-1697.
30. L. H. Jensen, A. Renodon-Corniere, K. C. Nitiss, B. T. Hill, J. L. Nitiss, P. B. Jensen and M. Sehested, "A Dual Mechanism of Action of Anticancer Agent F 11782 on Human

Topoisomerase II Alpha," Biochemical Pharmacology, Vol. 66, No. 4, 2003, pp. 623-631.
31. Y. Samuels and V. E. Velculescu, "Oncogenic Mutations of PIK3CA in Human Cnacers," Cell Cycle, Vol. 3, No. 10, 2004, pp. 1221-1224.
32. R. T. Nitta, C. A. Del Vecchio, A. H. Chu, S. S. Mitra, A. K. Godwin and A. J. Wong, "The Role of the c-Jun NTerminal Kinase 2-α-Isoform in Non-Small Cell Lung Carcinoma Tumorigenesis," Oncogene, Vol. 30, No. 2, 2011, pp. 234-244.
33. G. Xinarianos, T. Liloglou, W. Prime, G. Sourvinos, A. Karachristos, J. R. Gosney, D. A. Spandidos and J. K. Field, "P53 Status Correlates with the Differential Expression of the DNA Mismatch Repair Protein MSH2 in Non-Small Cell Lung Carcinoma," International Journal of Cancer, Vol. 101 No. 3, 2002, pp. 248-252.
34. H. Sasaki, K. Okuda, O. Kawano, K. Endo, H. Yukiue, T. Yokoyama, M. Yano and Y. Fujii, "Nras and Kras Mutation in Japanese Lung Cancer Patients: Genotyping Analysis Using LightCycler," Oncology Reports, Vol. 18, No. 3, 2007, pp. 623-628.

NOTES

[1]http://cise.sdkd.net.cn/institute/isbbc/data/AGNLC/raw_database.rar

[2]http://cise.sdkd.net.cn/institute/isbbc/data/AGNLC/working_database.ra
r

CHAPTER 8

Use of FFT in Protein Sequence Comparison under Their Binary Representations

Jayanta Pal[1], Soumen Ghosh[2], Bansibadan Maji[3], Dilip Kumar Bhattacharya[4]

[1]*Department of Computer Science & Engineering, Narula Institute of Technology, Kolkata, India*

[2]*Department of Information Technology, Narula Institute of Technology, Kolkata, India*

[3]*Department of Electronics & Communication Engineering, National Institute of Technology, Durgapur, India*

[4]*Department of Pure ematics, Calcutta University, Kolkata, India*

ABSTRACT

The paper considers Voss type representation of amino acids and uses FFT on the represented binary sequences to get the spectrum in the frequency domain. Based on the analysis of this spectrum by using the method of inter coefficient difference (ICD), it compares protein sequences of ND5 and ND6 category. Results obtained agree with the standard ones. The purpose of the paper is to extend the ICD method of comparison of DNA sequences to comparison of protein sequences. The topic of discussion is to develop a novel method of comparing protein sequences. The main achievements of the work are that the method applied is completely new of its kind, so far as protein sequence comparison is concerned and moreover the results of comparison agree with the previous results obtained by other methods for the same category of protein sequences.

Keywords: Voss Type Representation, Inter-Coefficient Difference (ICD) Method, Distance Matrix, Phylogenetic Tree, Fast Fourier Transform (FFT), ND5 and ND6 Category of Protein

1. INTRODUCTION

Among the numerous available amino acids only 20 are generally found in living beings and every protein sequence is expressed by these 20 amino acids. The representation of protein in terms of its amino acids is called its primary sequence. Based on this primary sequence representation, protein sequence comparison involves basically two types of methods: 1) Alignment Based Method and 2) Alignment Free Method. Protein sequence comparison was primarily done by different alignment based methods [1] - [3] . But especially due to execution time and comparatively difficult procedure, alignment free methods were preferred subsequently. So far as alignment free methods are concerned, a good literature up to 2003 is available in [4] . So we start with highlighting some of the most important contribution in protein sequence comparisons by alignment free methods from 2003 onwards [5] - [25] . Obviously in most cases, protein sequence comparison also follows similar approach as is considered in genome sequence analysis, because the role of four nucleotides is the same as the role of 20 amino acids in a protein sequence. In details, first of all, numerical representations of the protein sequences are obtained from the numerical values given to the individual amino acids, then graphical representation of the protein sequences is obtained; from these graphs descriptors are derived. These are finally used in comparing protein sequences. All the papers from [7] - [24] involve graphical representations. But another completely different approach is also followed in protein sequence comparison. These are based on classification of amino acids in different groups with different cardinality [8] [26] [27] . Again application of Discrete Fourier Transform in Bioinformatics is also well known. Discrete Fourier Transform (DFT) is nicely used in signal and image processing [28] - [34] . The main areas of its application in DNA research are found in gene prediction, hierarchical analysis and such others [35] - [40] . It is effectively used in identification of protein coding regions, because a DFT spectrum of a DNA sequence reflects the distribution and periodic pattern of the sequence [41] . Use of DFT on binary sequence is found in [42] , where the binary sequence is generated from genome sequences by Voss type of representation. Naturally to find similar use of DFT in protein sequence analysis, corresponding Voss type representation of amino acids is to be known priori. Fortunately Voss representation of DNA sequences involving

4 nucleotides has already been generalized to Voss type representation of 20 amino acids in protein sequences [43] . Such representation of amino acids has already been used in obtaining fuzzy representation of amino acids [43] . These are found to be effective in classification of amino acids in 6 different groups. Finally protein sequence classification has been obtained based on such classified groups of amino acids [8] [25] - [27] [44] . Thus Voss type representation of amino acids is an important contribution in protein sequence analysis. But use of FFT on the binary representations of protein sequences generated by such Voss type representation of amino acids has not yet been attempted in protein sequence comparison. This is the motivation of the paper to consider such binary sequences in comparing protein sequences.

2. METHODOLOGY

2.1. Voss Type Binary Representation of Amino Acid

20 amino acids are taken in the following order:

Alanine (A), Cysteine (C), Aspartic acid (D), Glutamic acid (E), Phenylalanine (F), Glycine (G), Histidine (H), Isoleucine (I), Lysine (K), Leucine (L), Methionine (M), Asparagine (N), Proline (P), Glutamine (Q), Arginine (R), Serine (S), Tyrosine (T), Valine (V), Tryptophan (W) and Threonine (Y).

Each amino acid is represented by a 20 component vector of which one bit is 1 and others are 0. But the representation follows the order of amino acid taken. For example amino acid Alanine(a) is represented by 10000000000000000000. The same rule is applied for other amino acids also, so that the last amino acid Threonine (Y) is represented by 00000000000000000001.

From each protein sequence S we get 20 different representations corresponding to 20 different amino acids by putting in the protein sequence 1 for the particular amino acid considered and the rest all 0 for the remaining amino acids. Thus 20 different binary representations viz., U_A, U_C, U_D, U_E, U_F, U_G, U_H, U_I, U_K, U_L, U_M, U_N, U_P, U_Q, U_R, U_S, U_T, U_V, U_W and U_Y are obtained.

2.2. ICD Method for Protein Sequence Analysis

The ICD method of DNA sequence and Protein sequence analysis basically remains the same as both deals with binary sequence only. So we describe ICD method as described in [43] . First of all FFT is applied on the binary represented protein sequences of length N say. In the Fourier spectrum the amplitudes are taken, which are N/2 distinct numbers. We normalize these N/2 components by their lengths. On these N/2 normalized components, we take absolute value of the inter coefficient difference (ICD) by calculating the differences of the succeeding terms from the preceding ones. Thus we get (N/2 – 1) distinct elements corresponding to each amino acid. Now 20 such (N/2) – 1 distinct components are concatenated to give a descriptor of length 20*((N/2) – 1)). From such descriptors distance matrix is formed by considering Euclidian Distance measures as follows.

If $x=(x_1,x_2,\cdots,x_n)$ and $y=(y_1,y_2,\cdots,y_n)$ are two sequences for two proteins X and Y, then the distance between X and Y is given by

$$d(x,y)=\left(\sum_{i=1}^{n}(x_i-y_i)^2\right)^{1/2}$$

This is the Euclidean distance between X and Y. The smaller is the distance; more similar are the protein sequences. On the basis of this formula the distances between pair of proteins are calculated and they are used to form the diagonal matrix. Due to similarity, only the lower half of the matrix is taken. Now using the UPGMA software on this matrix the Phylogenetic Tree for all the species is obtained. For comparison of protein sequences of different lengths the question of making all the lengths same does not arise normally in FFT. But if necessary, the length may be manually adjusted by putting additional zeros. For example, suppose two protein sequences are of lengths M and N. Then the descriptors for the first and second sequences are of lengths 20*((N/2) – 1) and 20*((M/2) – 1) respectively. As the descriptors are of unequal lengths, so comparison becomes infeasible. Hence if M = N – 2, say, then we first make the lengths of both the sequences equal to N, by putting two additional zeros to the second sequence. But there is no problem in doing so, as the Fourier transform of zeros gives zero spectrum.

3. SEQUENCES FOR COMPARISON

We have used the NADH dehydrogenase subunit 5 (ND5) and subunit 6 (ND6) protein sequences of nine species for comparison as shown in Table 1.

Table 1. List of nine species with their versions and lengths.

Sl. No.	Species	ND5		ND6	
		NCBI Reference	Length	NCBI Reference	Length
1	HUMAN	AP-000649.1	603	AP-000650.1	174
2	GORILLA	NP-008222.1	603	NP-008223.1	174
3	COMMON CHIMPANZEE	NP-008196.1	603	NP-008197.1	174
4	PYGMY CHIMPANZEE	NP-008209.1	603	NP-008210.1	174
5	FIN WHALE	NP-006899.1	606	NP-006900.1	175
6	BLUE WHALE	NP-007066.1	606	NP-007067.1	175
7	RAT	AP-004902.1	610	AP-004903.1	172
8	MOUSE	NP-904338.1	607	NP-904339.1	172
9	OPOSSUM	NP-007105.1	602	NP-007106.1	168

4. RESULTS AND DISCUSSIONS

4.1. Results

Distance matrix obtained by applying our method for 9 protein sequences of ND5 and ND6 category have been presented in Table 2 and Table 3 respectively. Phylogenetic tree obtained from these data have been presented in Figure 1 and Figure 2 for ND5 and ND6 category respectively.

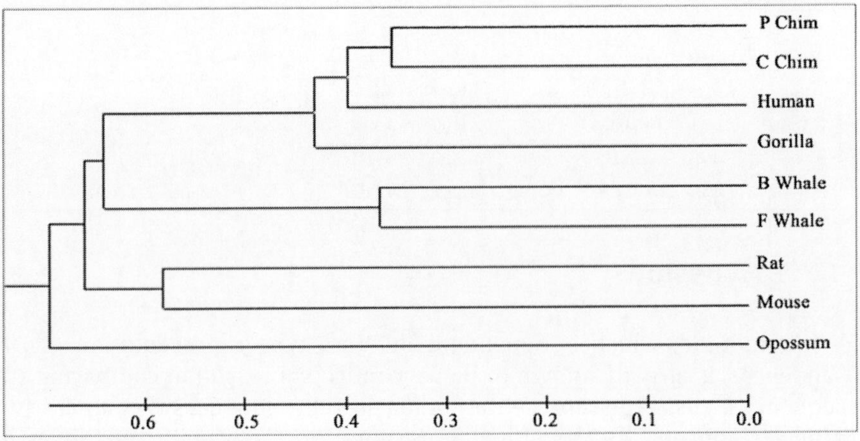

Figure 1. Phylogenetic tree obtained for 9 protein sequences of ND5 category.

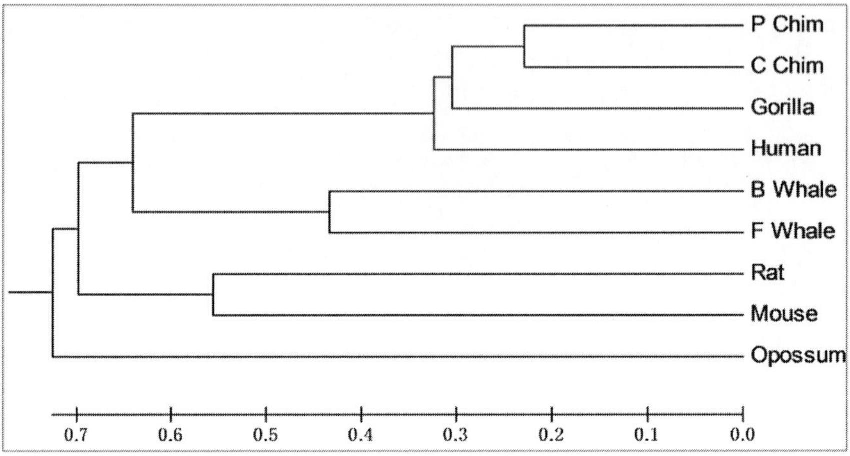

Figure 2. Phylogenetic tree obtained for 9 protein sequences of ND6 category.

Table 2. Distance matrix (lower triangular) for 9 protein sequences of ND5 category.

	Human	Gorilla	P. Chim	C. Chim	Rat	Mouse	B_Whale	F_Whale	Opossum
Human	0								
Gorilla	0.88011	0							
P. Chim	0.78392	0.83493	0						
C. Chim	0.80377	0.86392	0.70748	0					
Rat	1.33343	1.32515	1.32102	1.33353	0				
Mouse	1.31723	1.3021	1.29413	1.31185	1.16305	0			
B_Whale	1.28719	1.28518	1.27026	1.29161	1.32947	1.3423	0		
F_Whale	1.27388	1.2825	1.26707	1.29284	1.31818	1.32474	0.73081	0	
Opossum	1.39053	1.38974	1.37859	1.3812	1.39501	1.3896	1.4018	1.40041	0

4.2. Discussion

Ÿ ICD method, which is dependent on Voss type representation of DNA sequences, is already known to be very much successful in comparing DNA sequences. Voss type representation for protein sequences is comparatively a newer concept. As Voss type representation for protein sequences has been applied recently in different areas and found to be very much successful there, so it is expected that this type of representation might be useful in protein sequence comparison also. This is why; in our paper ICD method based on Voss type representation for protein sequences has been developed and used for protein sequence comparison. No doubt that the present method is a new contribution to the literature of protein sequence comparison.

Ÿ Obviously ICD method, may be for DNA sequence comparison or Protein sequence comparison, is comparatively easier and straight forward to apply.

Ÿ To compare our results with those obtained earlier by other methods on the same species, we first mention them as far as possible. The phylogenetic tree obtained in [25] for 9 species of ND5 category is given in Figure 3. Similarly the phylogenetic trees obtained in [26] for 9 species of ND5 category and ND6 category are given in Figure 4 and Figure 5 respectively and the phylogenetic trees obtained in [44] for 9 species of ND5 category and ND6 category are given in Figure 6 and Figure 7 respectively.

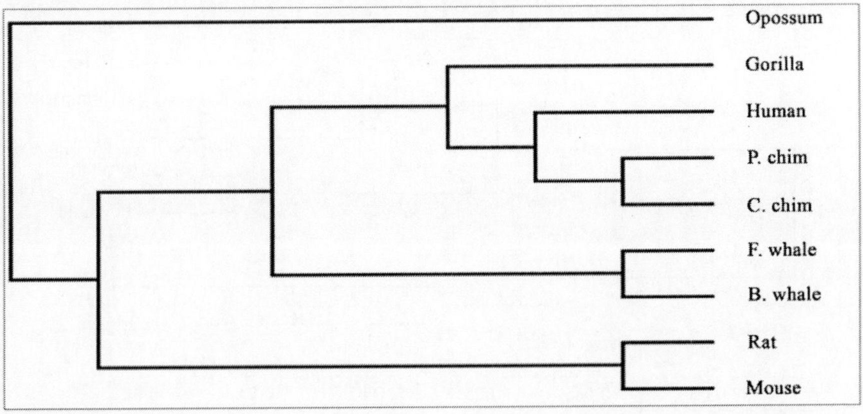

Figure 3. Phylogenetic tree obtained in [25] for 9 species of ND5 category.

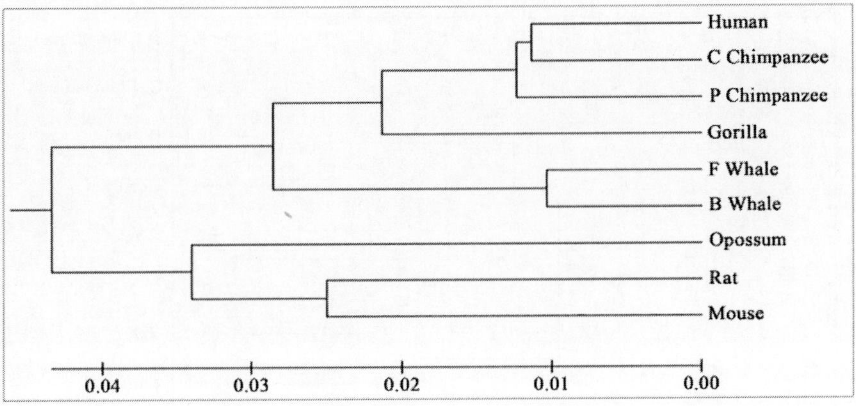

Figure 4. Phylogenetic tree obtained in [26] for 9 species of ND5 category.

Table 3. Distance matrix (lower triangular) for 9 protein sequences of ND6 category.

	Human	Gorilla	P. Chim	C. Chim	Rat	Mouse	B_Whale	F_Whale	Opossum
Human	0								
Gorilla	0.61527	0							
P. Chim	0.66602	0.60809	0						
C. Chim	0.65876	0.60947	0.45713	0					
Rat	1.40056	1.3828	1.37694	1.38812	0				
Mouse	1.44023	1.43475	1.43451	1.44952	1.10948	0			
B_Whale	1.29625	1.28337	1.26906	1.27875	1.35032	1.37165	0		
F_Whale	1.28281	1.28369	1.265	1.28498	1.35476	1.36975	0.86639	0	
Opossum	1.48984	1.46302	1.4753	1.47082	1.37682	1.43112	1.46181	1.44397	0

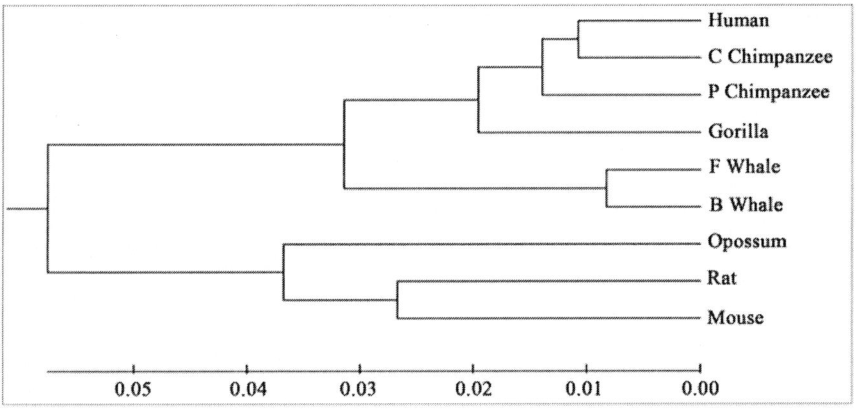

Figure 5. Phylogenetic tree obtained in [26] for 9 species of ND6 category.

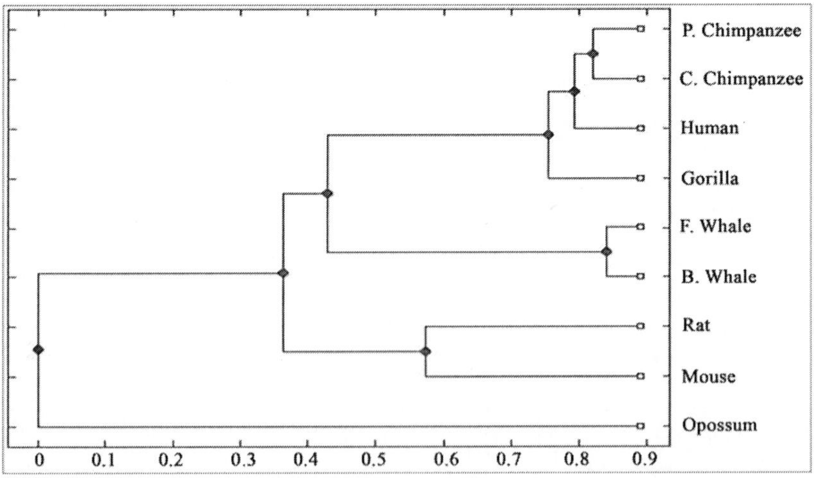

Figure 6. Phylogenetic tree obtained in [44] for 9 species of ND5 category.

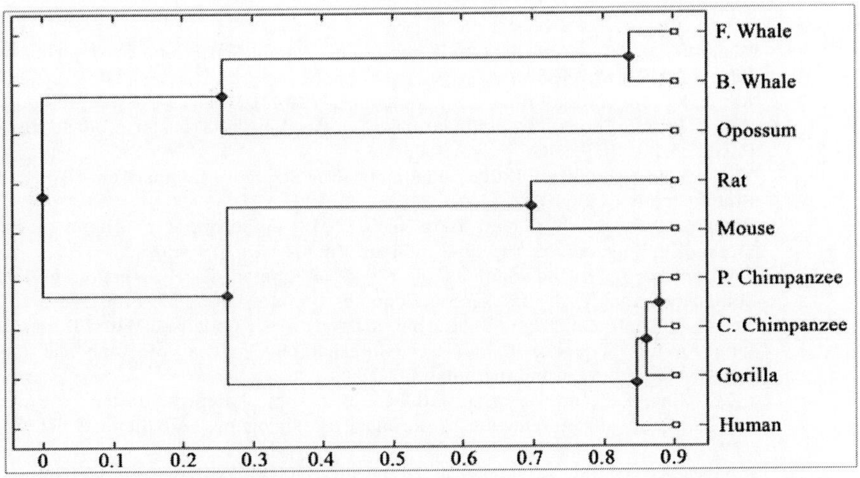

Figure 7. Phylogenetic tree obtained in [44] for 9 species of ND6 category.

From the above phylogenetic trees obtained for ND5 and ND6 categories of protein, it is revealed that in both the cases the phylogenetic trees obtained by our method almost agree with the earlier phylogenetic trees obtained by other methods.

5. CONCLUSION

Our method is effective and easier to apply in protein sequence comparison.

CITE THIS PAPER

Jayanta Pal, Soumen Ghosh, Bansibadan Maji, Dilip Kumar Bhattacharya (2016) Use of FFT in Protein Sequence Comparison under Their Binary Representations. *Computational Molecular Bioscience*,**06**,33-40.

REFERENCES

1. Phillips, A., Janies, D. and Wheeler, W. (2000) Multiple Sequence Alignment in Phylogenetic Analysis. Molecular Phylogenetics and Evolution, 16, 317-330.
2. Thompson, J.D., Higgins, D.G. and Gibson, T.J. (1994) CLUSTAL W: Improving the Sensitivity of Progressive Multiple Sequence Alignment through Sequence

Weighting, Position-Specific Gap Penalties and Weight Matrix Choice. Nucleic Acids Research, 22, 4673-4680.
3. Katoh, K., Misawa, K., Kuma, K. and Miyata, T. (2002) MAFFT: A Novel Method for Rapid Multiple Sequence Alignment Based on Fast Fourier Transform. Nucleic Acids Research, 30, 3059-3066.
4. Vinga, S. and Almeida, J. (2003) Alignment-Free Sequence Comparison—A Review. Bioinformatics, 19, 513-523.
5. Pinello, L., Lo Bosco, G. and Yuan, G.-C. (2013) Applications of Alignment-Free Methods in Epigenomics. Briefings in Bioinformatics, 15, 419-430.
6. Domazet-Loso, M. and Haubold, B. (2011) Alignment-Free Detection of Local Similarity among Viral and Bacterial Genomes. Bioinformatics, 27, 1466-1472.
7. Ghosh, S., Pal, J., Maji, B. and Bhattacharya, D.K. (2016) Condensed Matrix Descriptor for Proteinb Sequence Comparison. International Journal of Analytical Mass Spectrometry and Chromatography, 4, 1-13.
8. Li, C., Xing, L.L. and Wang, X. (2008) 2-D Graphical Representation of Protein Sequences and Its Application to Coronavirus Phylogeny. BMB Reports, 41, 217-222.
9. Randic, M., Mehulic, K., Vukicevic, D., Pisanski, T., Vikic-Topic, D. and Plavsic, D. (2009) Graphical Representation of Proteins as Four-Color Maps and Their Numerical Characterization. Journal of Molecular Graphics and Modelling, 27, 637-641.
10. Bai, F. and Wang, T. (2006) On Graphical and Numerical Representation of Protein Sequences. Journal of Biomolecular Structure and Dynamics, 23, 537-545.
11. Randic, M. (2007) 2-D Graphical Representation of Proteins Based on Physico-Chemical Properties of Amino Acids. Chemical Physics Letters, 440, 291-295.
12. Ghosh, A. and Nandy, A. (2011) Graphical Representation and ematical Characterization of Protein Sequences and Applications to Viral Proteins. Advances in Protein Chemistry and Structural Biology, 83, 1-42.
13. Randic, M., Zupan, J. and Vikic-Topic, D. (2007) On Representation of Proteins by Star-Like Graphs. Journal of Molecular Graphics and Modelling, 26, 290-305.
14. Wen, J. and Zhang, Y. (2009) A 2D Graphical Representation of Protein Sequence and Its Numerical Characterization. Chemical Physics Letters, 476, 281-286.
15. Liao, B., Sun, X. and Zeng, Q. (2010) A Novel Method for Similarity Analysis and Protein Sub-Cellular Localization Prediction. Bio-Informatics, 26, 2678-2683.
16. Novic, M. and Randic, M. (2008) Representation of Proteins as Walks in 20-D Space. SAR and QSAR in Environmental Research, 19, 317-337.
17. Yu, H.-J. and Huang, D.-S. (2012) Novel 20-D Descriptors of Protein Sequences and Its Applications in Similarity Analysis. Chemical Physics Letters, 531, 261-266.
18. He, P.-A., Wei, J., Yao, Y. and Tie, Z. (2012) A Novel Graphical Representation of Proteins and Its Application. Physica A: Statistical Mechanics and Its Applications, 391, 93-99.
19. Randic, M., Novic, M. and Vracko, M. (2008) On Novel Representation of Proteins Based on Amino Acid Adjacency Matrix. SAR and QSAR in Environmental Research, 19, 339-349.
20. Abo-Elkhier, M.M. (2012) Similarity/Dissimilarity Analysis of Protein Sequences Using the Spatial Median as a Descriptor. Journal of Biophysical Chemistry, 3, 142-148.
21. Randic, M., Zupan, J. and Balaban, A.T. (2004) Unique Graphical Representation of Protein Sequences Based on Nucleotide Triplet Codons. Chemical Physics Letters, 397, 247-252.
22. El-Lakkani, A. and El-Sherif, S. (2013) Similarity Analysis of Protein Sequences Based on 2D and 3D Amino Acid Adjacency Matrices. Chemical Physics Letters, 590, 192-195.
23. Feng, Z.-P. and Zhang, C.-T. (2002) A Graphic Representation of Protein Sequence and Predicting the Sub-Cellular Locations of Prokaryotic Proteins. International Journal of Biochemistry and Cell Biology, 34, 298-307.

24. Yao, Y.H., Kong, F., Dai, Q. and He, P.-A. (2013) A Sequence-Segmented Method Applied to the Similarity Analysis of Long Protein Sequence. MATCH: Communications in ematical and in Computer Chemistry, 70, 431-450.
25. He, P.-A., Li, X.-F., Yang, J.-L. and Wang, J. (2011) A Novel Descriptor for Protein Similarity Analysis. MATCH: Communications in ematical and in Computer Chemistry, 65, 445-458.
26. Ghosh, S., Pal, J., Das, S. and Bhattacharya, D.K. (2015) Differentiation of Protein Sequence Comparison Based on Biological and Theoretical Classifications of Amino Acids in Six Groups. International Journal of Computer Science and Software Engineering, 5, 695-698.
27. Zhang, Y.S. and Yu, X.T. (2010) Analysis of Protein Sequence Similarity. IEEE, 1255-1258
28. Wu, Y.-L., Agrawal, D. and El Abbadi, A. (2000) A Comparison of DFT and DWT Based Similarity Search in Time-Series Databases. Proceedings of the 9th International Conference on Information and Knowledge Management, McLean, 6-11 November 2000, 488-495.
29. Anastassiou, D. (2000) Frequency-Domain Analysis of Bimolecular Sequences. Bioinformatics, 16, 1073-1081.
30. Vaidyanathan, P. and Yoon, B.-J. (2004) The Role of Signal Processing Concepts in Genomics and Proteomics. Journal of the Franklin Institute, 341, 111-135.
31. Brigham, E.O. and Morrow, R.E. (1967) The Fast Fourier Transform. IEEE Spectrum, 4, 63-70.
32. Lyons, R.G. (2004) Understanding Digital Signal Processing. Pearson Education, Upper Saddle River.
33. Oppenheim, A.V. and Schafer, R.W. (2010) Discrete-Time Signal Processing. 3rd Edition, Prentice Hall, Upper Saddle River.
34. Akhtar, M., Epps, J. and Ambikairajah, E. (2008) Signal Processing in Sequence Analysis: Advances in Eukaryotic Gene Prediction. IEEE Journal of Selected Topics in Signal Processing, 2, 310-321.
35. Yin, C.C. and Yau, S.S.-T. (2007) Prediction of Protein Coding Regions by the 3-Base Periodicity Analysis of a DNA Sequence. Journal of Theoretical Biology, 247, 687-894.
36. Tiwari, S., Ramchandran, S., Bhattacharya, A., Bhattacharya, S. and Ramaswami, R. (1997) Prediction of Probable Genes by Fourier Analysis of Genome Sequences. Computer Applications in the Biosciences, 13, 263-270.
37. Afreixo, V., Bastos, C.A., Garcia, S.P. and Ferrieira, P.J. (2009) Genome Analysis with Inter-Nucleotide Distances. Bioinformatics, 25, 3064-3070.
38. Abu-Zahhad, M., Ahmed, S.M. and Abd-Elrahman, S.A. (2012) Genomic Analysis and Classification of Exon and Intron Sequences Using DNA Nu-merical Mapping Techniques. International Journal of Information Technology and Computer Science (IJITCS), 8, 22-36.
39. Sitansu, S.S. and Panda, G. (2010) A DSP Approach for Protein Coding Region Identification in DNA Sequence. International Journal of Signal and Image Processing, 1, 75-79.
40. Saberkari, H., Shamsi, M., Sedaaghi, M. and Golabi, F. (2012) Prediction of Protein Coding Regions in DNA Sequences Using Signal Processing Methods. IEEE Symposium on Industrial Electronics and Applications (ISIEA), Bandung, 23-26 September 2012, 355-360.
41. Hoang, T., Yin, C.C., Zheng, H., Yu, C.L. and He, R.L. (2015) A New Method to Cluster DNA Sequences Using Fourier Power Spectrum. Journal of Theoretical Biology, 372, 135-145.
42. King, B.R., Aburdene, M., Thompson, A. and Warres, Z. (2014) Application of Discrete Fourier Inter-Coefficient Difference for Assessing Genetic Sequence Similarity. EURASIP Journal on Bioinformatics and Systems Biology, 2014, 8.

43. Ghosh, S., Pal, J. and Bhattacharya, D.K. (2014) Classi-fication of Amino Acids of a Protein on the Basis of Fuzzy Set Theory. International Journal of Modern Sciences and Engineering Technology, 1, 30-35.
44. Jafarzadeh, N. and Iranmanesh, A. (2015) A New Measure for Pairwise Comparison of Protein Sequences. MATCH: Communications in ematical and in Computer Chemistry, 74, 563-574.

CHAPTER 9

Factors That Affect the Computational Prediction of Hot Spots in Protein-Protein Complexes

Jianping Lin[1*], Pi Liu[2], Hua-Zheng Yang[2], Nagarajan Vaidehi[3*]

[1]*College of Pharmacy, Nankai University, Tianjin, China*

[2]*State Key Laboratory of Elemento-Organic Chemistry, Elemento-Organic Chemistry Institute, Nankai University, Tianjin, China*

[3]*Department of Immunology, Beckman Research Institute of the City of Hope, Duarte, USA*

ABSTRACT

Protein-protein complexes play an important role in the physiology and the pathology of cellular functions, and therefore are attractive therapeutic targets. A small subset of residues known as "hot spots", accounts for most of the protein-protein binding free energy. Computational methods play a critical role in identifying the hotspots on the proteinprotein interface. In this paper, we use a computational alanine scanning method with all-atom force fields for predicting hotspots for 313 mutations in 16 protein complexes of known structures. We studied the effect of force fields, solvation models, and conformational sampling on the hotspot predictions. We compared the calculated change in the protein-protein interaction energies upon mutation of the residues in and near the protein-protein interface, to the experimental change in free energies. The AMBER force field (FF) predicted 86% of the hotspots among the three commonly used

FF for proteins, namely, AMBER FF, Charmm27 FF, and OPLS-2005 FF. However, AMBER FF also showed a high rate of false positives, while the Charmm27 FF yielded 74% correct predictions of the hotspot residues with low false positives. Van der Waals and hydrogen bonding energy show the largest energy contribution with a high rate of prediction accuracy, while the desolvation energy was found to contribute little to improve the hot spot prediction. Using a conformational ensemble including limited backbone movement instead of one static structure leads to better predicttion of hotpsots.

Keywords: Hotspot Prediction; Computational Mutagenesis; Concoord; Ensemble; Protein-Protein Complexes

1. INTRODUCTION

Protein-protein complexes are involved in various physiological processes and are also challenging targets for various pathological conditions. Protein-protein interaction surfaces are large and hence designing small molecule inhibitors that would disrupt the complex formation is a major challenge. It is known that a small subset of residues known as "hot spots" accounts for most of the protein-protein binding free energy [1]. Hot spots are shown to be surrounded by energetically less important residues forming a "hydrophobic O-ring" responsible for bulk solvent exclusion [2]. Identification of hotspots in protein-protein complexes is still challenging both experimentally and computationally, yet important to understand protein-protein binding and complex stability. It has been found that hot spots cannot be unequivocally defined by any single attribute such as location of the residue in the protein complex, charge, protein shape, or hydrophobicity [3-6]. Experimentally, alanine scanning mutagenesis has been used successfully on few proteinprotein complexes to probe the effect of these mutations on the stability of the protein-protein complexes to identify hot spots [7-11]. But alanine scanning mutagenesis is both laborious and expensive because hundreds of mutants need to be analyzed by biophysical methods [12]. Moreover, there are hotspots that are not in the proteinprotein interface. Such "allosteric hotspots" are more tedious to locate using experimental techniques. Therefore computational alanine scanning methods that can reliably predict hotspots and quantify the binding free energy is desirable [13-16].

In the past decade, a number of computational alanine scanning methods have been developed to identify hot spots. Simple empirical energy-based computational methods [17,18], linear interaction energy methods [19], Monte Carlo methods [20], and FOLDEF whose energy terms are derived for predicting protein stability [21] are computationally feasible, while still providing good prediction for more than 70% of the mutants. All these methods use a single protein-protein complex conformation from the crystal structure and empirical energy terms to calculate the change in protein-protein binding energy upon mutation. Protein-protein complexes are particularly dynamic and it is important to account for this flexibility while calculating the possible hotspots [22]. The protein-protein binding free-energy calculation using an all-atom forcefield (FF) combined with molecular dynamics simulations, has been used to improve the prediction of hot spots [16,23-26]. These methods provide good qualitative prediction of results, but they are still computationally intensive for many large protein-protein complexes [15]. There is change in binding free energy measured for several single point mutants in sixteen protein-protein complexes with known crystal structures of the complexes [17]. This serves as a gold standard for validating any computational method to predict hotspots.

In this paper, we have used alanine scanning method with all-atom forcefields (FF) and conformational sampling, for identifying hotspots in 16 different protein-protein complexes. We have compared the performance of various commonly used FF such as Charmm27 [27], AMBER9 (ff03) [28], and OPLS (2005) [29], towards prediction of hotspots in 16 protein-protein complexes with known crystal structures. We have also studied the role of implicit solvation methods such as Delphi [30], and Adaptive Poisson-Boltzman solvation APBS [31], in the hotspot predictions. Protein-Protein complex interfaces are dynamic and hence conformational sampling could play an important role in identifying hotspots. To keep the conformational sampling procedure computationally viable, we used a fast conformational sampling method based on geometric constraints Concoord [32], and studied the effect of sampling on the calculated protein-protein binding energies. Our results show that the Charmm27 FF predicts 74% of correct positives with a low percentage of false positives, while the AMBER force field (FF) performed better for the hot spot prediction (86% correct out of a total of 115 mutations), while showing a high rate of false positives. Desolvation energy was found to contribute little to improve the hot spot prediction efficiency. This is understandable given that most of the hotspots are dominated by small and hydrophobic residues [33]. Conformational sampling with Concoord improves the prediction efficiency of hot spots from 85% to 90%.

2. COMPUTATIONAL METHODS

2.1. Datasets

Datasets of single mutations and mutational data for 16 chosen protein complexes were taken from reference [17] and references therein, www.rtc.riken.go.jp/jouhou/pro therm/protherm.html, and http://mullinslab.ussf.edu/kurt /hotspot/index.php. The 16 protein complexes are shown in Table S1 of the supporting information.

2.2. Preparation of the Protein-Protein Complexes and Mutations

The coordinates of the 16 protein-protein complexes (shown in Table S1) were downloaded from the PDB databank. The program VMD [34] was used to add hydrogens. The potential energy of each protein-protein complex was minimized for 500 steps using conjugate gradient method in NAMD [35] (with Charmm force field [27]) and then annealed to 300K temperature. The annealed structures were minimized for 200 steps of conjugate gradient method in AMBER [28], Maestro 7.5 [36] upon which the minimization converged.

2.3. Methods for Calculating Protein-Protein Binding Free Energy and Alanine Scanning Mutations

We have calculated the protein-protein binding free energy using [24]

$$\Delta G_{\text{binding}} = \Delta G_{\text{complex}} - \left(\Delta G_{\text{partner A}} + \Delta G_{\text{partner B}}\right) \tag{1}$$

where

$\Delta G = E_{gas} + \Delta G_{solvation}$

$E_{gas} = E_{internal} + E_{electrostaic} + E_{vdw}$

$E_{internal} = E_{bond} + E_{angle} + E_{torsion}$

The E_{bond}, E_{angle}, $E_{torsion}$ are contributions to the internal energy from the bonds, angles, torsion angles. The $E_{electrostatic}$, E_{vdw} are the electrostatic and van der Waals interaction energies, respectively. E_{gas} was calculated using NAMD energy plugin in VMD [34] (for the Charmm FF [27]), MM_PBSA for the AMBER 9 FF (ff03 force field) [28], and Maestro 7.5 [36] for the OPLS2005 FF [29]). $\Delta G_{solvation}$ is the polar contribution to the solvation energy and was calculated using two methods, Delphi [30] (with 0.5 Å spacing grid, 1000 iterations, a dielectric constant of 2 for solute, 80 for exterior medium), and Adaptive Poisson-Boltzman Solvation APBS [31] (with 0.5 Å spacing grid, 1000 iterations, a dielectric constant of 2 for solute, 80 for exterior medium). Mutations were modeled using a mutation plugin in VMD [34]. And no more further optimization was performed [26].

The change in binding energy of the two monomers of a protein-protein complex, due to mutation to alanine is given by

$$\Delta\Delta G_{binding} = \Delta G^{mut}(binding) - \Delta G^{wt}(binding) \quad (2)$$

where "mut" represents mutants, WT represents wild type structures. $DDG_{binding}$ is the change in binding energy of the two monomers upon mutation of a single residue in the complex to alanine.

2.4. Hydrogen Bonding Energy

Since none of the FF tested here have an explicit hydrogen bond term, we have examined the effect of hydrogen bonding potential on the protein-protein binding energy, using a stand alone hydrogen bond potential [37]. The hydrogen bond potential we have used is given by:

$$E_{hbond} = \varepsilon\left[(\sigma/R_{D\text{-}A})^6 - (\sigma/R_{D\text{-}A})^4\right]\cos^4(\theta-\theta_0)SW(R_{D-A}) \quad (3)$$

where ε is a weight of the hydrogen bond interaction; σ is given by s = $1.73R_0$, R_0 is the optimal distance between hydrogen donor D and acceptor A at minimal potential, E_{min}; θ and θ_0 are the D-A-AA and target D-A-AA angles, respectively (D indicates donor; A , acceptor; AA, acceptor antecendent). SW is a switching function [38] to smoothly decrease E_{hond} to zero beyond a cut-off distance.

2.5. Concoord Conformational Sampling

Conformational sampling was performed using Concoord [32], a method that starting from a crystal structure, generates protein conformations that fulfill a set of upper and lower interatomic distance limits which depends on distances measured in experimental structures and the strength of the interatomic interactions. For each wild type protein complex structure, we generated 60 conformers [from reference 39, 60 is a reasonable number of conformers for convergence] and then performed alanine scanning mutation for each of the 60 conformers. For each conformer, mutations were modeled using a mutation plugin in VMD [34]. And no more further optimization was performed [26]. The binding energy was calculated as the average of all the binding energy for each conformation in the ensemble.

3. RESULTS AND DISCUSSION

The change in binding free energy (interaction energy between partners of protein complex + desolvation energy) as defined in the methods section was calculated for a total of 313 mutations in 16 protein-protein complexes. This test set was chosen due to availability of the experimentally measured change in binding free energies upon several alanine mutations. Additionally, these 16 protein complexes have known crystal structures. In our calculations, hot spots are defined as residues that show a change in calculated or observed binding affinities of more than 1 kcal/mol upon alanine replacement, and "neutral residues" are those positions which upon mutation to alanine show less than 1 kcal/mol difference in predicted or observed protein-protein binding affinities [17]. The number of total hotspots is a combination of those hotspots present in the protein-protein interaction interface, and the hotspots that have allosteric effect on protein-protein interactions. Neutral residues do not affect the protein-protein interaction upon mutation.

3.1. Comparison between Different Force Fields

Table 1 shows the % correct predictions for various FF and also the various energy components in each of these FF. The first three rows of **Table 1** shows the percentage of hotspots and neutral residues predicted correctly with various all-atom FF. The AMBER FF performed better among the three, with 86% correct prediction of total hotspots and 93% for the hotspots in the interface, while showing a high percentage of false positives or low percentage of correct prediction of neutral residues as shown in **Figure 1**. The Charmm27 FF on the other hand shows good prediction for both hotspots (74%) and for neutral residues (61%) and the OPLS force field predicted the least number of hot spots among the three FF. All of the three force fields predicted more than 60% of the hot spots. Interfacial hot spots have higher percentage of correct prediction than the allosteric ones.

3.2. Effect of Solvation on Prediction of Hotspots

Two continuum solvent methods solving the PoissonBoltzman equations, namely the Adaptive Poisson-Boltzman solvation (APBS) [31] and Delphi

[30] were tested in the prediction of hot spots. From **Table 1**, it is seen that inclusion of solvation effects using Delphi improved

Table 1. Comparison of the performance of different all-atom force fields and solvation models for hot spot prediction.

Force Field + desolvation	Neutral residues (%)	Total hot spots (%)	Hot spots in interface (%)
Charmm	61	74	80
OPLS	60	61	63
AMBER	39	86	93
Charmm + delphi	64	79	82
OPLS + delphi	55	55	59
AMBER + delphi	42	83	85
Charmm + APBS	61	48	52
OPLS + APBS	64	37	40
AMBER + APBS	51	73	81
Charmm-vdw	68	76	82
Charmm-diel-4.0	66	80	90
Charmm-vdw + Hbond	61	81	88
Charmm-concoord-4.0	50	85	95

Charmm-vdw represents calculated binding affinities using van der Waals interaction energy only from Charmm force field; Charmm-diel-4.0 represents prediction from Charmm force field with a dielectric constant of 4.0; Charmm-vdw + Hbond represents prediction from Charmm van der Waals interaction plus hydrogen bonding interaction; Charmm-concoord-4.0 represents prediction from Charmm force field using concoord conformational sampling with a dielectric constant of 4.0.

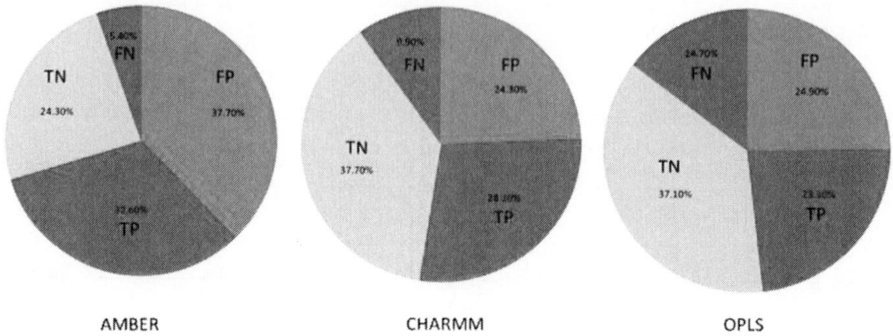

Figure 1. Pie chart of the prediction of hot spots and neutral residues for different force fields. TP is the true positive, neutral residues are true negative (TN). FP represents false positive, and FN represents false negative.

both the hotspot (from 74% to 79%) and neutral residue (from 61% to 64%) predictions for the Charmm force field. For other forcefields, inclusion of the solvation energies, decreased the percentage of hotspot residues predicted. Overall the effect of solvation is minor on hotspot predictions since most of the residues in the interface are hydrophobic and hence burial desolvation penalty is low. This is consistent with the analysis of the nature of residues in the interface of protein-protein complexes [33] and the O-ring hypothesis [2] that most of the hot spot residues are small and hydrophobic or that the hot spot residues are surrounded by hydrophobic residues (form an O-ring) to block the interactions between hot spot and solvent.

We have further analyzed the effect of solvation on prediction of mutation of polar versus non-polar residue hotspots. Tables S2-S4 of the supplementary material shows this data. It is seen that Amber FF performed the best with 85% of polar mutants and 88% of non-polar mutants for hotspots predicted correctly. This percentage is not affected substantially by inclusion of solvation. This is because the polar residues still stay partially solvated even upon formation of protein-protein interface. The protein-protein interface is dynamic and has a substantial solvent exposed surface area. Charmm FF with Delphi solvation predicted 68% of true positives with 73% of polar hot spots, and 84% of non-polar hot spots. The prediction of hotspots worsened by including solvation with OPLS force field. From this section onwards we have pursued further analysis with Charmm FF only, since Charmm FF with solvation showed the best prediction of hotspots.

3.3. Effect of Electrostatics and Van Der Waals Interactions on Hot Spot Predictions

To understand the role of various energy components such as electrostatics and van der Waals interactions in the protein-protein interface, we separated the van der Waals interaction energies and calculated the percentage of mutation residues predicted using just the van der Waals and electrostatic interactions. As seen in **Table 1**, using van der Waals interactions alone we find that the Charmm FF predicts about 2% more hotspots than using the complete energy function in Charmm. This indicates that the van der Waals interaction is critical to the protein-protein interface interactions. The study of proteinprotein interactions by other groups have shown that van der Waals interactions are important [39,40] in hot spot predictions. From Table S5, including Delphi solvation energy with Charmm-VDW reduced the prediction of hotspots dramatically. This indicated that solvation energy compliments the electrostatic interaction, and should not be included if the electrostatics are omitted [26,39].

Baker and coworkers showed that [17], hydrogen bonding term provides a better description of polar interactions in protein interfaces than electrostatics with a prediction of 76% over 56% [17]. Accurate calculation of electrostatics is notoriously difficult and the protein internal dielectric constant has been shown to be important in binding energy prediction [26,39]. This is a tricky parameter when it comes to calculating protein-protein interaction energies since these surfaces could be solvent exposed. Therefore we studied the effect of a range of internal dielectric constant varying from 1 to 80 for the calculation of electrostatic energies. We use the mutations available for the protein-protein complex (1JCK) as an example to optimize the internal dielectric constant to be used for the hot spot predictions. We chose this system since there are only nine mutants including 8 hot spots with polar, non-polar, charged amino acids. **Figure 2** shows the variation of the calculated change in binding energy as a function of the dielectric constant and it shows that the electrostatic energy difference and the binding energy difference become stable when the dielectric constant is greater than 4. We examined the effect of electrostatic interactions on hotspot predictions with dielectric 4.0 in Charmm FF. From **Table 1**, we can see that the prediction of hot spots increased by 4%, and the prediction of interfacial hot spots increased by 10%.

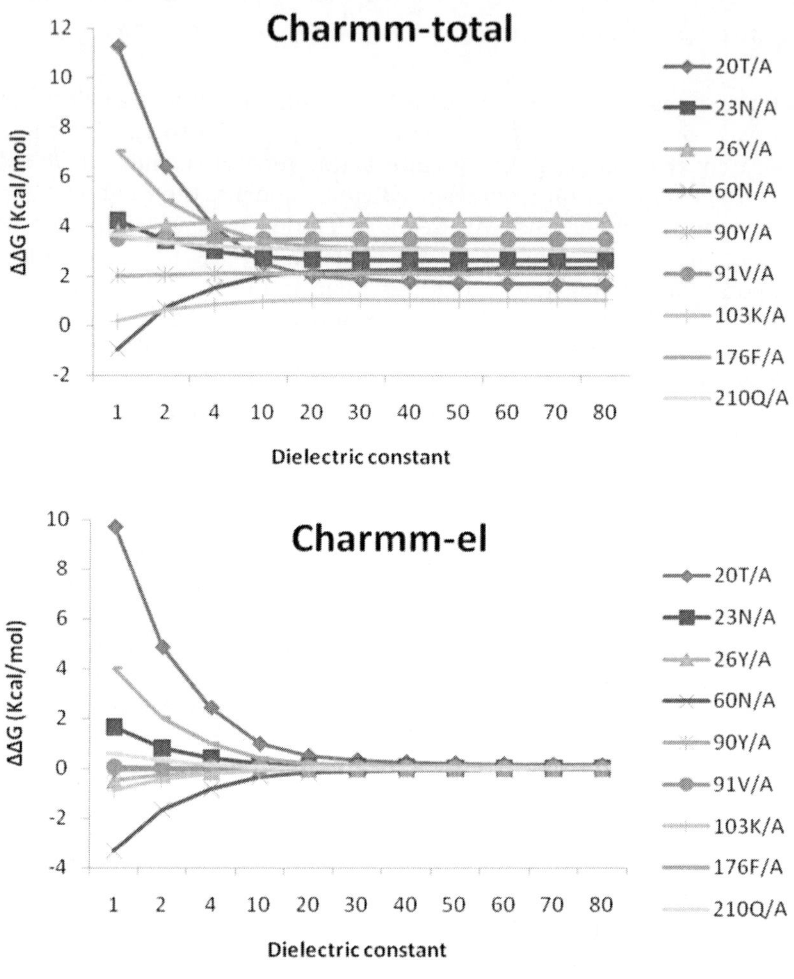

Figure 2. Effect of the dielectric constant on the difference in the total binding energy and electrostatic energy.

3.4. Effect of Hydrogen Bond Energy on Hot Spot Predictions

Since the electrostatic energies are notoriously difficult to calculate accurately especially at the protein-protein interface, we added an explicit hydrogen bond function to examine if we could eliminate the electrostatic component of the energy. Moreover, the all-atom forcefields tested in this study do not have an explicit hydrogen bond term in the potential energy function. To be able to test the influence of the hydrogen bonds on protein-protein interactions we used an explicit hydrogen bond function described in detail in the Methods section (Section 2.4). We calculated the hydrogen bond energy using this function and added the van der Waals energy from Charmm FF. It is seen from **Table 1**, that the hot spot prediction improves by 5% upon inclusion of hydrogen bond energies, compared to using only van del Waals energies. This correlates with the fact that residues that most frequently form a hot spot are tryptophan (21%), arginine (13%), and tyrosine (12%) [15], those show both hydrogen bonding as well hydrophobic interaction with other residues. If hydrogen bonding energy alone was used to predict the hot spot, only 9.8% of the hot spots (data not shown in the table) were predicted correctly. This indicates that hydrogen bonding energy is important in predicting hotspots, but is not as important as van der Waals energy. Since the intermolecular electrostatic and polar solvation free energies cancel each other upon proteinprotein binding [26,41], one needs only hydrogen bonding and van der Waals energies to predict the hot spots.

3.5. Effect of Conformational Sampling on Hotspot Predictions

Because protein-protein interface is dynamic, making mutations in the interface could lead to conformational changes in the neighborhood of the mutations or in some cases even show allosteric conformational changes. Here we have examined the effect of including conformational sampling (both backbone and side chains) in the prediction of hotspots. For faster computations, we have used a distance constraints based method, known as Concoord [32]. It has been shown that the sampling done by Concoord method is equivalent to MD simulations [32]. We have used Concoord method to generate a conformational ensemble for each of the 15 protein-protein complexes. One of the 16 protein-protein complexes was omitted because it is a multimer and not feasible with Concoord. The RMSD in coordinates of the C_α atoms in the ensemble of conformations generated by Concoord for the various protein complexes, varies from 1.8 Å to 6.0 Å. The variation in conformations in the ensemble for each protein comes mostly from the loop regions. The RMSD spread in the protein-protein interface is between 0.6Å and 2.6 Å. Compared to the entire proteins, the protein-protein interface is more rigid. **Figure 3** (a) shows the fluctuation of the interfacial residues for 60 conformers of protein 1JCK and **Figure 3**(b) shows the RMSD by residues (RMSF) of the protein. **Figure 3** shows that the conformations generated by Concoord [32] show variability in the interface. The RMSF figure shows that the interfacial residues are more rigid than the rest residues of the proteins (see Figure S2 for RMSF of other proteins). From **Table 1** it is clear that including conformational sampling for the protein-protein complexes leads to 85% correct prediction of the hotspots and 95% of interfacial hot spots predicted which is a substantial increase compared to using single static conformation. 3.6 Comparison of calculated change in binding energy to the experimental values.

Although Charmm FF function with dielectric constant 4.0 predicted 90% of the interfacial hotspots correctly, the accuracy of the predictions for different protein interfaces range from 50% to 100% (see Table S1 of supporting material). The change in binding energy calculated using Charmm FF function with dielectric constant 4.0 for the proteins 1VFB, 1AHW are shown in **Figure 4**. We have chosen these proteins since they have a wide range of mutations showing both neutral residues and hotspots. The correlation coefficient between the experimental free energies and calculated change in binding energy ranges from 0.7 to 0.8. For 1VFB, Figure S1 of supporting material shows the correlation for the neutral and hotspot residues separately. The correlation coefficient for the hotspots is 0.7 while for the neutral residues, the correlation is poor with a coefficient of 0.3.

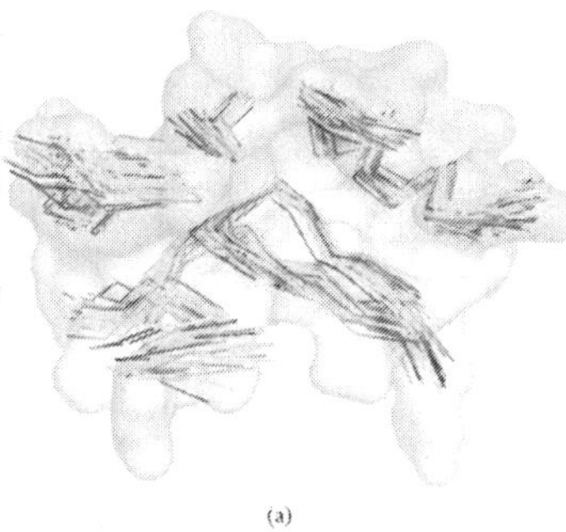

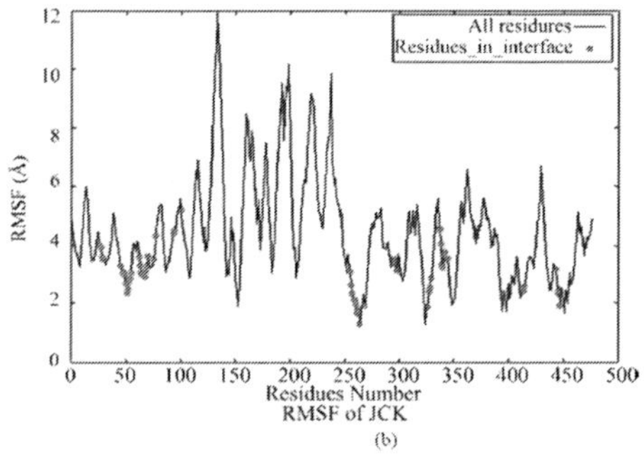

Figure 3. (a) Overlay of the donor-acceptor interfaces of 60 conformers of protein 1JCK. Pink represents the surface of the interface of one monomer, and pale cyan represents the surface of the other monomer. The interfacial residues are represented in trace representation and each color represents one conformer; (b) RMSF of Cα atoms for each residue averaged 60 conformers for protein 1JCK. Black solid line represents all residues, and red dots are residues in the interface.

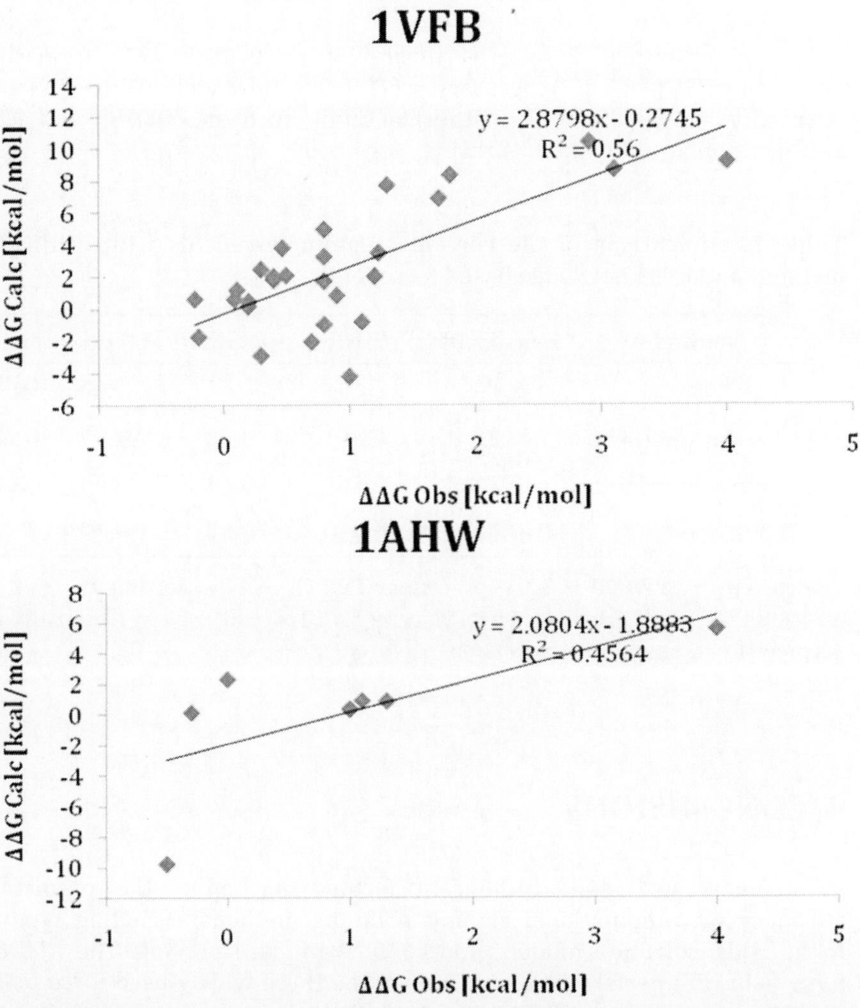

Figure 4. Correlation between the experimentally measured free energies of binding of the monomers in protein-protein complex 1AHW, and the calculated free energies of binding.

3.6. Comparison of Performance of All-Atom Forcefield with Other Methods

Table 2 shows comparison of our results with Robetta that uses a knowledge based forcefield for alanine scanning [17]. The performance of Charmm force field with dielectric constant 4.0 and inclusion of hydrogen bonds is similar to that of Robetta method. We have calculated the sensitivity and accuracy of the prediction from the number of true positives (TP), true negatives (TN), false positives (FP) and false negatives (FN). The following terms are defined as sensitivity = TP/(TP + FN); precision precision = TP/(TP + FP); accuracy = (TP + TN)/(TP + FP + TN + FN) and Specificity = TN/(TN + FP). Other methods such as FOLDEF and KFC perform similar to Robetta [42,43].

Table 2. Comparison of the current all-atom forcefields with ROBETTA method that uses knowledge based forcefields.

Method	Sensitivity	Precision	Specificity	Accuracy	F1
Charmm	0.74	0.54	0.61	0.66	0.62
Charmm-diel-4.0	0.80	0.59	0.66	0.71	0.68
Charmm-vdw + Hbond	0.81	0.56	0.61	0.68	0.66
Robetta	0.70	0.73	0.84	0.79	0.71

Sensitivity = TP/(TP + FN); precision=TP/(TP + FP); accuracy = (TP + TN)/(TP+ FP + TN + FN); Specificity = TN/(TN + FP); F1 = 2Xsensitivity Xprecision/(sensitivity + precision)

4. CONCLUSIONS

In summary, we studied different factors that affect the prediction efficiency of computational alanine scanning methods, including type of force fields, solvation models, and conformational sampling. The AMBER force field (FF) performed better among the three widely used force fields for hot spot predictions (86% correct prediction), while showing a high rate of false positives since the percentage of neutral residues correctly predicted was low with AMBER FF. The Charmm FF however showed an evenly balanced prediction of both hot spots and neutral residues. Dielectric constant of 4.0 for electrostatic interaction calculation improves the hot spot prediction efficiency dramatically. Among all the energy terms, van der Waals energy seems to play a critical role followed by the hydrogen bond energy. Conformational sampling with Concoord also predicts the hot spots better than using a single conformation.

Using these techniques we identified an allosteric inhibitor for the formation of the protein-protein complexSTAT3 dimer that has been published elsewhere [44].

5. ACKNOWLEDGEMENTS

Jianping Lin, Pi Liu, Hua-Zheng Yang, Nagarajan Vaidehi* JL thanks Dr. Gouthaman Balaraman for sharing the script to calculate the hydrogen bond function. We thank the STAT3 drug design project team at Beckman Research Institute of the City of Hope for instigating this project with the goal of designing small molecules for disrupting the STAT3 dimer.

REFERENCES

1. T. Clackson and J. A. Wells, "A Hotspot of Binding Energy in a Hormone Receptor Interface," Science, Vol. 267, 1995, pp. 383-386.
2. A. A. Bogan and K. S. Thorn, "Anatomy of Hot Spots in Protein Interfaces, Review," Journal of Molecular Biology, Vol. 280, 1998, pp. 1-9.
3. O. Keskin, B. Ma, et al., "Hot Regions in Protein-Protein Interactions: The Organization and Contribution of Structurally Conserved Hot Spot Residues," Journal of Molecular Biology, Vol. 345, No. 5, 2005, pp. 1281-1294.
4. W. L. DeLano, "Unraveling Hot Spots in Binding Interfaces: Progress and Challenges," Current Opinion in Structural Biology, Vol. 12, No. 1, 2002, pp. 14-20.
5. L. Lo Conte, C. Chothia, et al., "The Atomic Structure of Protein-Protein Recognition Sites," Journal of Molecular Biology, Vol. 285, No. 5, 1999, pp. 2177-2198.
6. S. N. Jones, A. T. Sands, et al., "The Tumorigenic Potential and Cell Growth Characteristics of p53-Deficient Cells Are Equivalent in the Presence or Absence of Mdm2," Proceedings of the National Academy of Sciences USA, Vol. 3, No. 24, 1996, pp. 14106-14111.
7. O. Y. Tikhomirov and J. W. Thomas, "Alanine Scanning Mutants of Rat Proinsulin I Show Functional Diversity of Anti-Insulin Monoclonal Antibodies," Journal of Immunology, Vol. 165, No. 7, 2000, pp. 3876-3882.
8. B. C. Cunningham and J. A. Wells, "Rational Design of Receptor-Specific Variants of Human Growth Hormone," Proceedings of the National Academy of Sciences USA, Vol. 88, No. 8, 1991, pp. 3407-3411.
9. Z. Hu, B. Ma, et al., "Conservation of Polar Residues as Hot Spots at Protein Interfaces," Proteins, Vol. 39, No. 4, 2000, pp. 331-342.
10. B. Ma, T. Elkayam, et al., "Protein-Protein Interactions: Structurally Conserved Residues Distinguish between Binding Sites and Exposed Protein Surfaces," Proceedings of the National Academy of Sciences USA, Vol. 100, No. 10, 2003, pp. 5772-5777.
11. K. S. Thorn and A. A. Bogan, "ASEdb: A Database of Alanine Mutations and Their Effects on the Free Energy of Binding in Protein Interactions," Bioinformatics, Vol. 17, No. 3, 2001, pp. 284-285.
12. B. Jayaram, K. McConnell, et al., "Free-Energy Component Analysis of 40 Protein-DNA Complexes: A Consensus View on the Thermodynamics of Binding at the Molecular Level," Journal of Computational Chemistry, Vol. 23, No. 1, 2002, pp. 1-14.
13. Y. Gao, R. Wang, et al., "Structure-Based Method for Analyzing Protein-Protein Interfaces," Journal of Molecular Modeling, Vol. 10, No. 1, 2004, pp. 44-54.
14. G. M. Verkhivker, D. Bouzida, et al., "Computational Detection of the Binding-Site Hot Spot at the Remodeled Human Growth Hormone-Receptor Interface," Proteins-Structure Function and Genetics, Vol. 53, No. 2, 2003, pp. 201-219.

15. I. S. Moreira, P. A. Fernandes, et al., "Hot Spots—A Review of the Protein-Protein Interface Determinant Amino-Acid Residues," Proteins, Vol. 68, No. 4, 2007, pp. 803-812.
16. I. S. Moreira, P. A. Fernandes, et al., "Computational Alanine Scanning Mutagenesis—An Improved Methodological Approach," Journal of Computational Chemistry, Vol. 28, No. 3, 2007, pp. 644-654.
17. T. Kortemme and D. Baker, "A Simple Physical Model for Binding Energy Hot Spots in Protein-Protein Complexes," Proceedings of the National Academy of Sciences USA, Vol. 99, No. 22, 2002, pp. 14116-14121.
18. T. Kortemme, D. E. Kim, et al., "Computational Alanine Scanning of Protein-Protein Interfaces," Science's STKE, Vol. 2004, No. 219, 2004, p. l2.
19. J. Aqvist, C. Medina, et al., "A New Method for Predicting Binding Affinity in Computer-Aided Drug Design," Protein Engineering, Vol. 7, No. 3, 1994, pp. 385-391.
20. G. M. Verkhivker, D. Bouzida, et al., "Monte Carlo Simulations of the Peptide Recognition at the Consensus Binding Site of the Constant Fragment of Human Immunoglobulin G: The Energy Landscape Analysis of a Hot Spot at the Intermolecular Interface," Proteins, Vol. 48, No. 3, 2002, pp. 539-557.
21. R. Guerois, J. E. Nielsen, et al., "Predicting Changes in the Stability of Proteins and Protein Complexes: A Study of More than 1000 Mutations," Journal of Molecular Biology, Vol. 320, No. 2, 2002, pp. 369-387.
22. E. L. Humphris and T. Kortemme, "Prediction of Protein-Protein Interface Sequence Diversity Using Flexible Backbone Computational Protein Design," Structure, Vol. 16, No. 12, 2008, pp. 1777-1788.
23. P. A. Kollman, et al., "Calculating Structures and Free Energies of Complex Molecules: Combining Molecular Mechanics and Continuum Models," Accounts of Chemical Research, Vol. 2000, No. 33, 2000, p. 889.
24. I. Massova and P. A. Kollman, "Computational Alanine Scanning to Probe Protein-Protein Interactions: A Novel Approach to Evaluate Binding Free Energies," Journal of the American Chemical Society, Vol. 121, No. 36, 1999, pp. 8133-8143.
25. S. Huo, I. Massova, et al., "Computational Alanine Scanning of the 1:1 Human Growth Hormone-Receptor Complex," Journal of Computational Chemistry, Vol. 23, No. 1, 2002, pp. 15-27.
26. I. S. Moreira, P. A. Fernandes, et al., "Unraveling the Importance of Protein-Protein Interaction: Application of a Computational Alanine-Scanning Mutagenesis to the Study of the IgG1 Streptococcal Protein G (C2 Fragment) Complex," Journal of Physical Chemistry B, Vol. 110, No. 22, 2006, pp. 10962-10969.
27. A. D. Mackerell Jr., et al., "All-Artom Empirical Potential for Molecular Modeling and Dynamics Studies of Proteins," Journal of Physical Chemistry B, Vol. 102, 1998, pp. 3586-3616.
28. D. A. Case, T. A. Darden, T. E. Cheatham III, C. L. Simmerling, J. Wang, R. E. Duke, R. Luo, K. M. Merz, B. Wang, D. A. Pearlman, M. Crowley, S. Brozell, V. Tsui, H. Gohlke, J. Mongan, V. Hrnak, G. Cui, P. Beroza, C. Schafmeister, J. W. Caldwell, W. S. Ross and P. A. Kollman, "AMBER 9," University of California, San Francisco, 2004.
29. W. L. Jorgensen, D. S. Maxwell, et al., "Development and Testing of the OPLS All-Atom Force Field on Conformational Energetics and Properties of Organic Liquids," Journal of the American Chemical Society, Vol. 118, No. 45, 1996, pp. 11225-11236.
30. B. Honig and A. Nicholls, "Classical Electrostatics in Biology and Chemistry," Science, Vol. 268, No. 5214, 1995, pp. 1144-1149.
31. N. A. Baker, D. Sept, et al., "Electrostatics of Nanosystems: Application to Microtubules and the Ribosome," Proceedings of the National Academy of Sciences USA, Vol. 98, No. 18, 2001, pp. 10037-10041.
32. B. L. de Groot, D. M. van Aalten, et al., "Prediction of Protein Conformational Freedom from Distance Constraints," Proteins, Vol. 29, No. 2, 1997, pp. 240-251.

33. O. Keskin, B. Y. Ma and R. Nussinov, "Hot Regions in Protein—Protein Interactions: The Organization and Contribution of Structurally Conserved Hot Spot Residues," Journal of Molecular Biology, Vol. 345, 2005, pp. 1281-1294.
34. W. Humphrey, A. Dalke, et al., "VMD: Visual Molecular Dynamics," Journal of Molecular Graphics, Vol. 14, No. 1, 1996, pp. 27-38.
35. J. C. Phillips, R. Braun, et al., "Scalable Molecular Dynamics with NAMD," Journal of Computational Chemistry, Vol. 26, No. 16, 2005, pp. 1781-1802.
36. Maestro, "Schrödinger," Version 7.5, LLC, New York, 2005.
37. F. Fabiola, R. Bertram, et al., "An Improved Hydrogen Bond Potential: Impact on Medium Resolution Protein Structures," Protein Science, Vol. 11, No. 6, 2002, pp. 1415-1423.
38. A. T. Brunger, "Free R Value: A Novel Statistical Quantity for Assessing the Accuracy of Crystal Structures," Nature, Vol. 355, No. 6359, 1992, pp. 472-475.
39. I. S. Moreira, P. A. Fernandes, et al., "Hot Spot Occlusion from Bulk Water: A Comprehensive Study of the Complex between the Lysozyme HEL and the Antibody FVD1.3," Journal of Physical Chemistry B, Vol. 111, No. 10, 2007, pp. 2697-2706.
40. I. D. Kuntz, K. Chen, et al., "The Maximal Affinity of Ligands," Proceedings of the National Academy of Sciences USA, Vol. 96, No. 18, 1999, pp. 9997-10002.
41. A. Benedix, C. M. Becker, et al., "Predicting Free Energy Changes Using Structural Ensembles," Nature Methods, Vol. 6, No. 1, 2009, pp. 3-4.
42. N. Tuncbag, A. Gursoy, et al., "Identification of Computational Hot Spots in Protein Interfaces: Combining Solvent Accessibility and Inter-Residue Potentials Improves the Accuracy," Bioinformatics, Vol. 25, No. 12, 2009, pp. 1513-1520.
43. K. I. Cho, D. Kim, et al., "A Feature-Based Approach to Modeling Proteinprotein Interaction Hot Spots," Nucleic Acids Research, Vol. 37, No. 8, 2009, pp. 2672-2687.
44. R. Buettner, R. Corzano, et al., "Alkylation of Cysteine 468 in Stat3 Defines a Novel Site for Therapeutic Development," ACS Chemical Biology, Vol. 6, No. 5, 2011, pp. 432-443.

CHAPTER 10

Antibody-Like Phosphorylation Sites in Focus of Statistically Based Bilingual Approach

Jaroslav Kubrycht[1*], Karel Sigler[2], Pavel Souček[3], Jiří Hudeček[4]

[1]*Department of Physiology, Second Faculty of Medicine, Charles University, Prague, Czech Republic*

[2]*Laboratory of Cellular Biology, Institute of Microbiology, Academy of Sciences of the Czech Republic, Prague, Czech Republic*

[3]*Toxicogenomics Unit, National Institute of Public Health, Prague, Czech Republic*

[4]*Department of Biochemistry, Charles University, Prague, Czech Republic*

ABSTRACT

In accordance with previous reports, the sequences related to phosporylated protein segments occur in conserved variable domains of immunoglobulins including first of all certain N-terminally located segments. Consequently, we look here for the sequences 1) composing human and mouse proteins different from antigen receptors, 2) identical with or highly similar to nucleotide sequence representatives of conserved variable immunoglobulin segments and 3) identical with or closely related to phosphorylation sites. More precisely, we searched for the corresponding actual pairs of DNA and protein sequence segments using five-step bilingual approach employing among others a) different types of BLAST searches, b) two in-principle-different machine-learning methods predicting phosphorylated sites and c) two large databases recording existing phosphorylation sites. The approach identified seven existing phosphorylation sites and thirty-seven related human and mouse segments achieving limits for several predictions or phylogenic parameters. Mostly serines phosporylated with ataxia-telangiectasia-related kinase (involved in regulation of DNA-double-strand-break repair) were indicated or predicted in this study. Hypermutation motifs, located in effective positions of the selected sequence segments, occurred significantly less frequently in transcribed than non-transcribed DNA strands suggesting thus the incidence of mutation events. In addition, marked differences between the numbers and proportions of human and mouse cancer-related sequence items were found in different steps of selection process. The possible role of hypermutation changes within the selected segments and the observed structural relationships are discussed here with respect to DNA damage, carcinogenesis, cancer vaccination, ageing and evolution. Taken together, our data represent additional and sometimes perhaps complementary information to the existing databases of empirically proven phosphorylation sites or pathogenically important spots.

Keywords: Ataxia Telangiectasia-Mutated-Protein (i.e. Kinase ATM, Whose Pathogenic Mutation Is Responsible for Early Death of People), Complementarity Determining Region 1 (of Immunoglobulins, i.e. CDR1 or Hypervariable Region 1), Database (of Functional Structures), Hypermutation (i.e. Mutation of DNA Sequences Mediated by Enzymes), Immunoglobulin (i.e. Ig or Antibody), Phosphorylation (Enzyme Mediated Modification Concerns Here Mostly Protein Sequences)

1. INTRODUCTION

Protein kinase (PK) mediated phosporylation of proteins occurs specifically on phosporylated protein segments (PPS), whereas down-regulation of phosphorylated sites is specifically mediated by phosphatases. PPS can markedly influence the reactivity or interactivity of distinctly located functional sites of the same molecule or molecular complex via allosteric effects [1] - [4] . In addition, PPS represent functionally important structural elements of regulatory cascades and complex interconnected networks due to their presence in PK and phosphatases or molecules involved in triggering or modulation of the same enzymes [5] . Though the lengths of PPS mostly do not exceed ten amino acid residues (aa), these segments are specifically recognized by different PK even in cases of small sequence differences between PPS (cf. [6] [7]).

Comparable diversified specificity of short chains is also necessary for interactions of complementary determining (hypervariable) regions (CDR) of variable immunoglobulin domains (IgV) of immunoglobulins. In accordance with this fact, it is interesting that N-terminal parts of conserved IgV (called here PPSIg) containing hypervariable region CDR1 (and also related parts of certain conserved constant immunoglobulin domains) exhibited sequence relationship to commercially accessible peptidic substrates or inhibitors of protein kinases (cf. PKSI in previous papers [8] - [10]). The described sequence relationships were further supported by the result of successful prediction of PPS within various BLAST-accessible conserved IgV of antigenic receptors (AR). The superior prediction scores were mostly observed in two sites of these IgV. One site was located in PPSIg (the corresponding paper [11] contains also short recent summary of the corresponding investigation).

Due to their phylogeny, IgV domains of AR are adapted to effects of somatic sequence changes including hypermutation [12] [13] . In addition, these domains contain at least several types of structures or candidate structures supporting these somatic changes (cf. following paragraph). This raises the question whether at least some of the two strands of translated NS (in fact NS denotes here nucleotide sequences of the corresponding cDNA) closely similar to NS-representatives of (conserved) PPSIg but encoding PPS composing molecules distinct from AR can perform IgV-related somatic changes with functional consequences. To gradually answer to this question, newly focused bilingual approach searching for PPSIg-related NS encoding peptidic PPS was proposed. More precisely, we look here for NS 1) encoding PPS in proteins distinct from AR, 2) forming sufficiently high and dense similarities when compared with representative mRNA segments highly similar to PPSIg- related segments and 3) containing sites necessary for functionally important hypermutation.

Hypermutation mentioned above represents in fact an enzyme-mediated processes causing alteration of DNA sequences. First of all, APOBEC family member AID (activation-induced cytidine deaminase) has been extensively investigated in studies of Ig hypermutation [14] . AID exhibits a tumorigenic effect, when it is transfected and constitutively expressed and it can be found in current somatic cells different from lymphocytes in response to certain activating signals [15] - [18] . In accordance with its tumorigenic effect, AID hypermutates also genes encoding proteins different from AR [19] [20] . AID-mediated hypermutation occurs selectively at sites of certain primary and secondary DNA structures [21] [22] . First of all, this concerns well-known hypermutation motifs (HM) of sequence WRCH [21] [23] and specifically distant WRCH-pairs (W-pairs) which were frequently found in NS encoding immunoglobulins [24] [25] . In addition to this restriction of substrate specificity, initiating selective alignment between AID and DNA is also necessary for AID activity. Hence, this event does not occur at sites with sole HM, but only in G-loops [20] , and perhaps also in the other secondary DNA structures [22] or W-pairs proposed to trigger bi-bi-random mechanism of deamination [25] . Besides AID effects, mutations mediated by other APOBEC family members were observed in various non-lymphoid tissues currently not expressing AID. The research determined HM with the TCW sequence as an important target of APOBEC1, APOBEC3A, APOBEC3B, APOBEC3F and APOBEC3H reactions accompanying carcinogenesis [26] - [30] .

In summary, the goals of this paper consisted in 1) description of structurally and functionally interesting PPSIg-related PPS or PPSIg-related a posteriori reselected sequences predicted as PPS, 2) record of substantial common and extreme features of these molecules and 3) statistical evaluation of their group-related interrelationships. The corresponding search for topical PPSIg-related items occurred in five main steps (Figure 1).

About eighty percent of the predicted or existing PPSIg-related PPS found here comprised phosphorylation sites with serine processed by Ataxia telangiectasia mutated molecule (ATM), a key regulatory kinase of the DNA double-strand-break response [31] - [33] . All displayed PPSIg-related oligonucleotides contained HM WRCH and TCW at positions critical with respect to aa alterations. These critically located HM mostly occurred in non- transcribed DNA strands. The list of displayed segments contained molecules involved in carcinogenesis, cell division or specific regulatory functions.

2. METHODS

2.1 Bioinformatic Tools

Programs of BLAST family [34] including conserved domain searches [35] were mostly accessed through publicly available NIH gate. Multiple-sequence alignment (MSA) was performed with Clustal W 2.1 present in the server of European Bioinformatic Institute [36] (for older version of the MSA record displayed here, see paper [11]). The server with FFAS03 program [37] [38] was used for scanning based on the shortest possible profile- profile alignments enabling identification of conserved fold regions (FFAS-scan; cf. WP5.1). Hypermutation motifs were searched in the preformed multi-segment constructs using EMBOSS program Fuzznuc present on the web page of Pasteur Institute [25] [39] . Only PPS candidates simultaneously achieving the score 0.800 in two different machine learning

Figure 1. Short methodological description of the employed heuristic bilingual approach. Five main procedural steps including both serial and parallel selection procedures determined here human, mouse and FFAS-scan-derived or NITR-related MPL. *, ** in the same box-two immediately subsequent steps (scheme simplification); repeating * in the same box-parallel procedures or common comments to these procedures; combined presence # and ∩-simultaneous occurrence of the selected items in records primarily obtained with the adjacent: 1) query sequences and 2) adjustments of BLAST searches; AR-antigen receptor(s); HM*-hypermutation motif(s) located at NS position critical with respect to possible aa alteration; NS-nucleotide sequence(s) of the corresponding both strands of cDNA. For CSB1, CSB2, Erα, MNSQ, MPL, MRNS, MSA, MSEP and SEM see inner part of this table. For additional details or abbreviations see supplementary files SF1 (cf. WF- and WP-related references), SF2 (dictionary of abbreviations) and SF3 (supplements to **Figure 1**). All three supplementary files are accessible on the web page address www.papersatellitesjk.com or via email jkub@post.cz.

programs (neural-network-based NetPhos2.0 and support-vector-machine-related KinasePhos2.0 [6] [7]) were selected before the final reselection (cf. Figure 1 and sections WP2.4.1-2). On the other hand, alternative usage of databases Phospho.ELM a Phosida enabled us to find experimentally confirmed PPS [40] - [42] .

2.1.1. Typical Sequences

1) Conserved domain sequences (CDS; cf. [35]), 2) phylogenically interesting actual sequence exhibiting superior similarities with CDS (IgW of clone AAB03680; igw; Figure 2; [11] [13]) or 3) actual sequence achieving repeating superior similarities in initial searches for MNSQ units (consensus-like sequence, i.e. cls, of heavy chain with clonal names AF273898.1 and AAK20241.1; Figure 2) were all denoted here as typical sequences. Besides cls, derived in addition, the typical sequences assembled initial MSA record including two important conserved sequence blocks (CSB1 and CSB2; see Figure 2, sections 2.3 and WP2.2.2).

2.1.2. Statistical Enumerations

These enumerations were performed in accordance with the textbook of Zvárová [43] . In accordance with basic Bayesian approximation [44] , all groups of values with zero observations presented in 2 × 2 tables were increased by one and then the modified odds ratio value was enumerated as $OR^*[0]$. Fisher exact probability of 2 × 2 tables was enumerated using Active web page [45] . This web page was used to statistically evaluate the validity of 1) currently enumerated odds ratios (OR) and 2) original 2 × 2 tables determining $OR^*[0]$. Current search limits for long sequence similarities, i.e. 40 bits (a bit score limit for middle range BLAST similarities present in all headings of BLAST records) and $p < 0.005$ (indeed 5×10^{-3}; cf. [34]) restricted the validity of a) conserved segments of initial MSA record (Figure 2; cf. $BS^{\#}$ and $E^{\#}$ in Section 2.2.3) and b) searches for NS representatives of AR (Figure 1; sections 2.3 and WP2.4.2-4). We have to note that it holds if $p < 0.05$, then $p < E$ and $p \approx E$ including $E = E^{\#}$ (cf. a formula of Dembo [46]). Specific statistically derived bit score restrictions of searches for short conserve-domain-related and MNSQ-derived similarities (MNSQ denotes multi-nucleotide sequence queries; see below) of lengths comparable with subsequently searched PPS are described in Figure 1 and section WP2.3.2.

2.1.3. Overall Scheme of Employed Procedures

For the scheme including all main procedures described in this chapter see Figure 1 and SF3.

```
CLUSTAL 2.1 multiple sequence alignment[a]
Positions of SBC        123456789012345678901234567890123456789012345678901234567890 60
cd04983 (TCR_ALL)       ---TQSPQSLSVQEGENVTLNCNYSTS--TFYYLFWYRQYPGQGPQFLIYIS--SN-GEE 53
cd07706 (TCR_D)         --VTQAQPDVSVQVGEEVTLNCRYETS-WTNYYIFWYKQLPSGEMTFLIRQK--SI-YGN 55
cd05899 (TCR_B)         --VTQSPRYLIKGRGQSVTLRCSQTSG---HDNMYWYRQDPGKGLQLLFYSNGGSL-NEE 54
cd04984 (L_K IgV)       ------------SPGETVTITCTGSSGNISGNYVNWYQQKPGSAPRYLIYEDSD----RP 47
cd04980 (L_L IgV)       IVLTQSPATLSVSPGESATISCKASQS-VSSNYLAWYQQKPGQAPKLLIYGAST----LA 55
cd00099 (IgV)           ---------LSVSEGESVTLSCTYSGS-FSSYYIFWYRQKPGKGPELLIYISSNGS-QYA 49
cd04981 (H_IgV)         --LVESGPGLVKP-GQSLKLSCKASGFTFTSYGVNWVRQAPGKGLEWIGFINPGGGETYY 57
smart00406 (IgV)        ----------------SVTLSCKFSGSTFSSYYVSWVRQPPGKGLEWLGYIGSNGS-SYY 43
AAB03680_igv (IgV)      IVLTQPESVVKKP-GETVRLSCGVTGFDIDTHYITWVKQVPGKGLEWLLYHDSRPQ--EF 77
cd04982 (TCR_G)          --LEQPQLSITREESKSVTISCKVSGIDFSTTYIHWYRQKPGQALERLLYVSSTSTQRKL 58
                                             : *          : * :* *.        :

similarities and consensi[b,c,d]
consensus               IVLTQSPPSLSVαEGESVTLSCKYSGSNFSSYYIYWYRQKPGKGLEWLIYINSβSNENγδ
fip                     102381nnn4nnnn989999991n911n141398n9999n9937140929n15n0nnnnn
fin                     ppp-pp359p3765pppppppppp6ppp3ppp+pp5pppp6pp-pppppp3pp4p81788
F ≥ 6                   ----Q---------GESVTLSC--S-------YI-WYRQ-PG-G---L-Y---------
p < 0.005               ----Q---------GESVTLSC--S-------Y--WYRQ-PG-----L-Y---------
p < 10^-6               --------------------C-------------W--Q-P------L-Y---------
common aa               --------------------C-------------W--Q-P------------------
pm3 consensus           --------------GESVTLSCKYSTS-FSSYYIFWYRQKPG-----------------
experimentally supplemented sequence queries (i.e. hybrid queries and cls)[e]
cdxal1                                GESVTLSCTYSGSTFSSYYIFWYRQKPGKGPELLIYISS
igw1                                  GESVTLSCGVSGFDIDTHYITWYRQVPGKGLEWLLYHDS
smx1                                    SVTLSCKFSGSTFSSYYISWYRQPPGKGLEWLGYIGS
cls (AF273898.1)                      GESVTLSCTVSGFSMSSYYMHWIRQKPGKGLEWIGYIDT
-----------------------------------------------------------------------------------
Positions of SBC        123456789012345678901234567890123456789012345678901234567890 120
cd04983 (TCR_ALL)       K--EKGRFSATLDKSRKSSSLHISAAQLSDSAVYFCA----------------------- 88
cd07706 (TCR_D)         A--TKGRYSVNFQKAQKSISLTISALQLEDSAKYFCA----------------------- 90
cd05899 (TCR_B)         EGDPKDRFSASRPSLTRS-SLTIKSAEPEDSAVYLCASSLGGGADEAY--FGPGT----- 106
cd04984 (L_K IgV)       SG-IPDRFSGSK--SGNTASLTISGAQTEDEADYYCQV---WDSNSYV--FGGGTKLTV- 98
cd04980 (L_L IgV)       SG-TPSRFSGSG--SGTDFTLTISRVEPEDAAVYYCQQ---YGTFPYT--FGGGTKLEI- 106
cd00099 (IgV)           GG-VKGRFSGTRDSSKSSFTLTISSLQPEDSAVYYCAVSLSGGTYKLY--FGQGTRLTV- 105
cd04981 (H_IgV)         ADSVKGRFTITRDTSKSTVYLQLNSLTPEDTAVYYCARGLGGYGYGYFDYWGQGTLVTVS 117
smart00406 (IgV)        QESYKGRFTISKDTSKNDVSLTISNLRVEDTGTYYCA----------------------- 80
AAB03680_igv (IgV)      APGIEGRFTPS--VVSNTAYLEITSLSVTDTAIYYCARIYTGALAWVFDYWGNGTFVEVT 135
cd04982 (TCR_G)         SGGTKNKFEARKDVGKSTSTLTIQNLEKEDSATYYCAYWESGS-SYYIKVFGSGT----- 112
                              .::                  * :          *  .  * *

similarities and consensi[b,c,d]
consensus               AGGεKGRFSASRDTSKNζηSLTISSLQPEDSAVYYCARθLSGGTYWYYDYFGQGTλLTVT
fip                     n0nn87997n7n3n5n24n49897554n997929999nnnnnnnnn0nn059n99n0n4n
fin                     1p43ppppp2p5+4p1pp8pppppppp2ppppppppp874357559p31ppp7pp7p2p2
F ≥ 6                   ----KGRFS-S---------LTIS----EDSA-YYCA--------------G-GT-----
p < 0.005               ----K-RFS-----------LTI-----ED-A-YYCA--------------G-GT-----
p < 10^-6               ------RF------------L-I-----D---YYC-------------------------
common aa               --------------------L-------D---Y-C-------------------------
```

Figure 2. Initial multiple sequence alignment, its analysis and sequence derivatives. [a]The set of sequences present in this table comprises variable domain of shark IgW sequence AAB03680 (pre-selected as representative of Ig-ancestor- related sequences of Elasmobranchii origin [11] [13]). The included conserved Ig domains:_AL,_B,_G,_D-TCR related domains, i.e. alpha-like, beta, gamma and delta, respectively; K_L, L_L, H-kappa or lambda light chain and heavy chain domains, respectively. For further comments see sections 3.1 and 4.1. [b]Two main conserved segments of sequence blocks (CSB1 and CSB2) are restricted here using bold characters of the corresponding consensus segments. The positions of the MSEP block segment (related to MNSQ1, MNSQ2) are indicated by gray background. Greek alphabets in the consensus denote aa differences in LE-/σ- related consensi determined here: α-P/S; β-N/S; γ-Y/E; δ-Y/A; ε-I/T, ζ-S/T; η-I/A; θ-W/S, λ-R/K. As follows, CSB1 related part of MSA-record determines unique consensus (consensus 1). [c]fin, fip-fuzzy related intervals in the ranges of negative or positive values, respectively. Numbers 1 - 9 in fip-/fin-related rows-values denoting hierarchy of fuzzy-related intervals, which represent the degrees of SBC similarities/dissimilarities (cf. Figure 3 and Section 2.2.3); n, p-negative and positive F-values described in Figure 3, respectively; plus and minus present in the fin row under fip value F = 3-presence or absence of the LE-value at least 1.5 necessary for candidate CBS edge, respectively (cf. Sections 2.2.4). For significance levels of SBC see sections WP4.1. [d]The consensus pm3 is described in Sections 3.1 and WP5.1. Gray, dark gray proofs of pm3 segments localize positions corresponding specifically to CDR1 of light chains (CDR1light) or generally to all CDR1 (CDR1all), respectively. [e]For details concerning the displayed structures see Sections 2.3, 3.1 and WP2.2.2.

2.2. Two Types of Consensus Sequences

We distinguish here two types of consensi (see also section WP2.2.1): 1) statistically important σ-consensus containing aa achieving the highest mean column score and 2) LE-consensus composed of aa determining length equivalents (LE) related to individual sequence block column (SBC). Both these types of consensi were differently used in enumerations described below and performed with minicomputer Casio Algebra 2 PLUS (cf. section WP2.1.4).

2.2.1. BLAST Derived Enumeration of LE

Each aa species present in evaluated SBC determined a non-invasive LE_i^* value, i.e. i-th aa-related length equivalent candidate (cf. sections WP2.1.1 and WP5.2). LE_i^* was defined as integer or non-integer height of artificial SBC composed only of i-th aa, provided that the probabilities of this artificial SBC and actual SBC (evaluated with respect to the selected i-th aa) were the same [9] . In score-related BLAST-derived evaluation of LE, we used conditioned Expect values (to include compared blank values) instead of probabilities (for details see section WP2.1.1). This evaluation finally determined the consistent formula:

$$LE = \mathrm{Max}_{i=1\ \mathrm{to}\ h} LE_i^* = \mathrm{Max}_{i=1\ \mathrm{to}\ h}\left\{S_i / Abs\left(s\left[x_i, x_i\right]\right)\right\}$$
$$= \mathrm{Max}_{i=1\ \mathrm{to}\ h}\left\{\left(C_i - D - \gamma\right) / Abs\left(s\left[x_i, x_i\right]\right)\right\}, \quad (1)$$

where Abs() enumerates absolute value; D is coefficient of column diversity; h is SBC height equal to the number of chains in MSA record; S_i represents i-th aa related score in evaluated SBC; $s[x_i, x_i]$ is score of i-th aa identity; x_i is the number of pointing column and row positions of scores in the employed substitution matrix (PAM30 in our case, cf. limiting SBC heights of ten in Figure 2 and section WP2.1.1); γ denotes column related gap penalty. The following formulas further characterize the C_i, D and γ:

$$C_i = \sum_{j=1 \text{ to } h} s\left[x_i, x_j\right], \tag{2}$$

$$D = (1/\lambda) \times \ln\left\{\left[\operatorname{sgn}(g) \times h + (h-g) \times d\right]/h\right\}, \tag{3}$$

$$\gamma = 10 + \text{Min}(\text{DEL}, \text{INS}) = 10 + \text{Min}(g, h-g) = 10 + h/2 - Abs(h/2 - g), \tag{4}$$

where j denotes positions of the compared aa in SBC; λ = 0.294 is BLAST constant; g is the number of gaps; d is the number of different aa in the same column; DEL and INS are the numbers of deletions and insertions deduced from the number of gaps, respectively; Min selects the minimum enumerated value in agreement with parsimony-related attempts of our approach. For explanation of $s\left[x_i, x_j\right]$ see $s\left[x_i, x_i\right]$ and j-values.

2.2.2. Enumeration of Numbers Determining the Degree of Fuzzy-Related Intervals (F or F-Values)

Each SBC-related F value was determined based on the fuzzy-related system (shown in Figure 3, see also WP5.3), implicating a two-step enumeration of F value based on υ value:

$$\upsilon = \operatorname{sgn}(LE) \times \text{int}\left\{\left[Abs(LE) + 0.25\right]/0.5\right\}, \tag{5}$$

if $\{Abs(\upsilon) \leq 9\}$, then $\{F = \upsilon\}$, else $\{F = \operatorname{sgn}(\upsilon) \times 9\}$. (6)

where int denotes an integer value of the enumerated number.

2.2.3. Restriction and Double-Sequence-Related Evaluation of the Conserved Segments of Sequence Blocks (CSB) Proposed for IgV-Related Sequences

The presence of a sequence pattern (briefly pattern determined by SBC containing unique aa species) was required in each proposed CSB including otherwise 16 - 50 neighbor SBC and achieving LE ≥1.5 in both its edge SBC. In the proposed approximating fuzzy-related approach, six rules concerned F values: 1)-3) at least 50% and 20% or at most 30% of SBC kept F ≥ 4, F ≥ 6, and F < 0, respectively. Among the state-of-the-art conditions, we assume that 4)-5) each list including F-values related to ten or five neighbor SBC

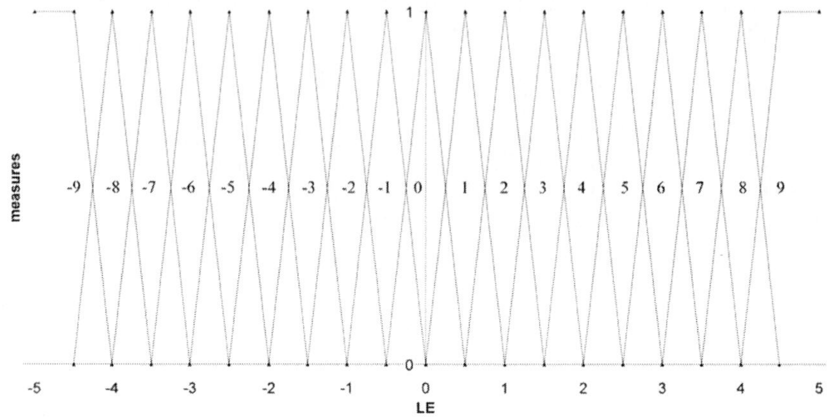

Figure 3. Fuzzy-related system based on length equivalents (LE). Nineteen integer F-values from −9 to 9 are displayed here in the middle part of the graph. The following fuzzy-related description holds for the displayed positive F-values: 0—absence of similarity; 1—rampart of randomness; 2—area of "promising" noise; 3—quasi-similar (minimum but weak column similarities); 4—similar (close to "deterministic" double sequence aa identity); 5—quasi-cohesive; 6—cohesive (corresponds to aa identity in three compared chains); 7—quasi-rigid; 8—rigid (aa identity in four compared chains); 9—improved-rigid, i.e. maximum F-value in our single character system. Negative F-values approximate here more than random extent of diversity/variability (cf. sections the first section of Results and WP5.3). For MSA-record-compatible two-row entry of LE-derived fuzzy-related intervals indicating extents of SBC similarities see Figure 2.

had to contain at least three $F \geq 3$ or at least two $F \geq 1$, respectively, and 6) at least one of the three following alternative rules holds for the set of SBC composing evaluated CSB: mean(υ_i) ≥ 3.5, mean(F_i) ≥ 3.5 or mean(LE_i) ≥ 1.75 (cf. Figure 3 or formulas 5 and 6). The selected candidates of CSB were then proved using current BLAST statistics [46] [47] . Hence, the values of double sequence similarity equivalents of bit score ($BS^{\#}$) and Expect value ($E^{\#}$) were enumerated, when assuming a simplified equal validity of all model chains in the evaluated reference MSA record (more general CSB-chain-related linear combination can be also used, when employing "empirical", i.e. Monte Carlo-related and database comparison-derived coefficients). The topical corresponding formulas follow:

$$BS^{\#} = \{\lambda \times \rho_{CSB}(S) - \ln K\}/\ln 2 = BS_m - \{\lambda/(\ln 2 \times h)\} \times (\Sigma_{n=1 \text{ to } N} D_n), \quad (7)$$

$$E^{\#} = K \times (L \times h) \times N \times \exp\{-\lambda x \rho_{CSB}(S)\}, \tag{8}$$

$$\rho_{CSB}(S) = 1/h \times \{(\sum_{n=1 \text{ to } N} C_n^* - D_n) - \Theta(X^*[j,k])\}, \tag{9}$$

where $\rho_{CSB}(S)$ is score density per single chain; K, λ are BLAST constants (K = 0.11 and λ = 0.294 in given case); BS_m is equal to the mean bit score between consensus and chains forming CSB (in contrast to $BS^{\#}$, mentioned above, BS_m disregards the negative effects of column diversity, i.e. $BS_m \geq BS^{\#}$; for comparison of BS_m and $BS^{\#}$ see section 3.1); D_n denotes the coefficient of n-th column diversity; L ´ h is the length of chains present in MSA record including gaps; N is the number of SBC determining the proposed CSB; C_n^*-is σ-consensus- related sum of scores concerning n-th SBC; $\Theta(X^*[j,k])$ denoted minimized gap penalty with respect to the three considered models (cf. WP2.1.2). In case of our enumeration, we used the simplest alternative formula:

$$\Theta(X^*[j,k]) = \text{Min}_{t=1 \text{ to} 3} \Theta(X_t[j,k]) = r \times 10 + G, \tag{10}$$

where r is the number of chains with gaps and G denotes the number of all gaps in evaluated CSB-related segment of MSA record. The evaluation of gaps is the unique procedure in which differ evaluations of block-consensus and block-chain comparisons based on this section. For limits restricting valid $BS^{\#}$ and $E^{\#}$ see Section 2.1.2.

2.3. Generation of Two Initial Sequence Supersets Leading Separately to the Construction of Two Different Multi-Sequence Queries (MNSQ1 or MNSQ2) Representing NS of AR

In principle, two types of non-redundant different sequence sets (initial supersets of NS items) were generated, when using two types of TBLASTN records obtained with two different sets of query sequences (QS), each associated with different taxonomical restriction of searches. The first initial superset contained superior similarities with all "double conserved" sequences of CSB1 and consensus 1 (cf. Figure 1(d) in SF3), whereas the second superset comprised only the similarities found with consensus 1, three CSB1 sequences successfully examined in the previous paper [13] (i.e. cd00099, smart00406 and AAB03680_igv), their consensus-1-related derivatives and cls sequence found in addition during the formation of the second superset (see Sections 2.1.1, WP2.2.2 and Figure 2). The former and latter supersets contained sequences of Elasmobranchii or vertebrate origin, respectively. An iterative process of species- and query-related randomization consisted in 1) maximum number of five distinct species representatives as well as maximum overall numbers of selected items including non-redundant sequence samples in each set separately derived with individual QS and 2) existence of unlimited SIMULT set and hierarchy of QS-records both important for set-related rearrangement of promiscuously similar sequences (for details see WP2.2.3 or SF3). Initial TBLASTN searches employed BLOSUM62 matrix, whereas the following cumulative revision steps used two substitution matrices (BLOSUM62 and PAM30). For restriction of Expect and bit score values see Section 2.1.2.

2.4. Memory Problems in Some Searches with MNSQ1 and MNSQ2

Working memory problems were observed in differently advanced customer computers, when completing too extensive BLAST records obtained with initial steps of TCA searches and negatively Entrez-restricted two-step BLASTN searches limited by 35 bits (BNsup35; cf. Figure 1 and WP2.3.1). Consequently, special compromising processing of the corresponding records was necessary when downloading only well accessible but less extended records of items without reports of selected sequence alignments. These procedures comprised 1) a simplified total-bit-score-limited approach approximately substituting inaccessible frequencies of MUSAS (MNSQ- unit derived similarities with almost the same subject sequence positions; Figure 1) in the case of ternary combined approach (TCA) and 2) partial elimination of item names, immediately recognized as those determining AR, during the first step of BN35sup. For additional information see Altschul [34] and sections WP2.3.1-4, WP5.6 and WP5.7.

2.5. FINAL SELECTION AND A POSTERIORI RESELECTION OF THE DISPLAYED ITEMS

Both final selection and reselection described in Figure 1 restricted non-redundant terminal list of items displayed in Figure 5 (for details see Sections 2.1, WP2.4.1, WP2.4.2, WP3.2 and WP3.3). Due to an unknown extent of losses following from inefficient assessing of dynamic (allosterically or proteolytically mediated) accessibility of MPL-related peptides, final selection did not include prediction of accessibility.

3. RESULTS

3.1. Analysis of the Initiating MSA Record and Its Segments

Occurrence of the two conserved block segments (CSB1 and CSB2) was confirmed based on Section 2.2.3, when analyzing MSA record assembled by Clustal W 2.1 (Figure 2). CSB1 determined the same σ- and LE- consensi (consensus 1) containing pattern CX(10,13)WXXQXP. CSB1 achieved 1) $F \geq 6$ and $F < 0$ in 51.3% and 12.8% SBC, respectively, 2) double sequence similarity related bit score $BS^{\#} = 48.2$ bits and consensus 1 related mean score $BS_m = 54.9$ bits and 3) significant Expect value $E^{\#} = 1.441 \times 10^{-10}$ (for the almost equal p value see Sections 2.1.2 or WP4.1). CSB2 was characterized by 54.5% and 18.2% of columns with $F \geq 6$ and $F < 0$, respectively, and contained the pattern LX(8)DX(3)YXC. CSB2 achieved also valid values of $BS^{\#} = 44.6$ bits, $E^{\#} = 1.461 \times 10^{-9}$ and $BS_m = 50.0$ bits. 16 of the displayed 120 SBC (Figure 2) were classified as "at most anti- cohesive" (AAC; $F \leq -6$; cf. Figure 3). CDR1-, CDR2- and CDR3-related block segments contained one, four and four of sixteen AAC SBC in Figure 2, respectively. This means that CDR1 substantially differed from CDR2 and CDR3, but not from the surrounding conservative framework regions (CDR1: positions 23 - 35, RCD = 0.89; CDR2: positions 51 - 62, RCD = 3.86, CDR3: positions 97 - 110, RCD = 3.31, where RCD denotes ratio between column densities of AAC occurrence determined in a) individual hypervariable and b) all accessible segments of framework regions).

Paralogues of AR were frequently similar to N-terminal parts of CSB1-related segments of cd00099, cd07706 and cd04980, when using FFAS-scan (Sections 2.1.1 and WP5.1). The similarities between these three selected domain segments then determined the consensus pm3, which was also N-terminally located in CSB1 (Figure 2). Since pm3-derived QS only rarely achieved the required bit score limit of 40 bits (cf. Section 2.2.3) in TBLASTN searches, we do not use pm3 for MNSQ generation. Nevertheless, a pm3-related segment of MSA record was suitable as an important fold- and sequence-related conserved block segment restricting the extension of PPSIg to C-terminus (cf. section WP5.5 and PPSIg enveloping segments of MSA record, i.e. MSEP described in Figure 2).

3.2. Selection of MNSQ1 and MNSQ2

Initial selection of sequence items restricted 121 and 169 items forming starting supersets of MNSQ1 (sequences of Elasmobranchii origin) and MNSQ2 (various vertebrate sequences), respectively. Subsequently, 41 and 108 sequence items were passed through reselection with MPQ and anti-redundant procedures assembling finally MNSQ1 and MNSQ2, respectively. Although the Ig items had not any preference in our searches, sequences of T-cell receptors (TCR) composed rarely MNSQ1 (two TCR sequences) and were not present in MNSQ2.

3.3. Paths Selecting MPL

The main paths of MPL selection are described in Table 1. OR-mediated analysis of these paths revealed interesting linkages within BLASTN-derived sets (Figure 4). Each employed OR represented ratios between cancer- related-set-derived R ratio (ratio between the numbers of human and mouse items present in the corresponding elements of Table 1) and the similarly derived Q-ratios by two species-related pairs of reference sets (BN35sup and KPO; Figure 4). Eleven and one OR values indicated strong ($OR \geq 2.0$) and weak ($1.4 \leq OR < 2.0$) associations of cancer-related NS items with their human origin, respectively (Figure 4). These OR values appeared to form significantly increased set of values ($p < 0.05$) even when skeptically considering equal distribution of OR values (as discrete values) above and under the value $OR = 2.0$ as random and employing model constant distribution (cf. section WP3.2) or Dirichlet statistics. In accordance with Figure 4, some of the described individual

Table 1. Most effective paths of MPL selection.

Paths[d,e]	BLAST-derived NS[a]			Translated NS[a,b]			Selected MPL[a,b,c]		
	hu	mu	both	hu	mu	both	hu	mu	both
KPO1	17	28	45	10	13	23	3	3	6
KPO2	50	75	125	29	30	59	5	10	15
C1	12	3	15	10	2	12	2	0	2
C2	20	15	35	12	8	20	3	1	4
BN1sup35	12	16	28	9	9	18	3	3	6
BN2sup35	33	51	84	15	26	41	3	5	8
TCA1TX	5	2	7	3	1	4	3	0	3
TCA2TX	16	7	23	12	6	18	7	2	9

[a]Starting procedures included BLASTN searches or combined BLAST searches (cf. footnote d). NS—nucleotide sequences of the corresponding both strands of cDNA; hu, mu, both—sequences of human, mouse and both species origin, respectively. [b]About half of MNSQ-related mRNA segments encoded peptide segments, whereas about one seventh of these mRNA segments encoded existing or predicted PPS. [c]All enumerated NS segments with predicted or database-confirmed PPS relationship (denoted here as **MPL**) fulfilled two conditions: 1) existence of MPL-derived similarities with chains of initial MSA record overlapping CSB1, 2) presence of at least single HM* in MPL. For details see Figure 1. [d]Independent paths of MPL selection differed in their initial BLAST-derived procedures. Numbers 1 and 2 in strategy names-MNSQ1 and MNSQ2 were used as query sequences, respectively; gray and white background in left column of table elements-paths of prevailing selection of mouse or human sequences, respectively. Selections using sole BLASTN: BN1sup35, BN2sup35-molecules different from antigen receptors achieving top similarities (limit 35 bits) in set without positive Entrez restriction; C1, C2—cancer related Entrez restriction (limit 30 bits); KPO1, KPO2-Entrez restriction concerning proteins involved in phosphorylation (limit 30 bits). Combined selections: TCA1TX, TCA2TX-MNSQ1- and MNSQ2-query-derived TBLASTX variants of ternary combined approach (**TCA**), respectively. Each of these TCA variants used a) four searches using two different Entrez restrictions in two "subpaths" composed of two cumulative TBLASTX searches differently adjusted with respect to the matrices BLOSUM62 or PAM30 and word sizes three or two, respectively, b) limit for score maxima 25 bits, c) score limit for subject-sequence-related co-localized similarities (denoted as **MUSAS**) 22 bits when simultaneously requiring occurrence of at least five MUSAS in addition to segment with maximum score (*i.e.* limit six for all MUSAS). For additional information see Figure 1, sections 2.4, WP2.3.2-4 and WP5.6-7. [e]Due to independent selection paths (forming non-redundant sets), some items sometimes repeat in different sets (cf. also strategy records in Figure 5, sections WP2.3.3 and WP4.2.3).

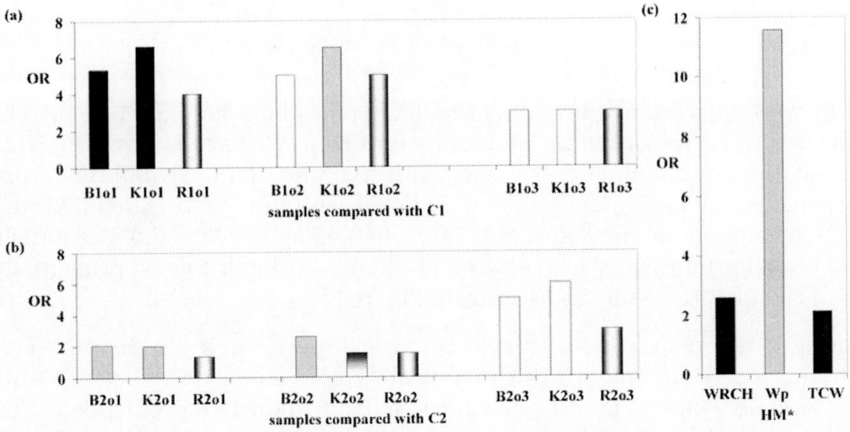

Figure 4. Odds ratio—(OR-) mediated linkage analyses. Achievement of the three limits (i.e. sufficient sample size of at least 40 NS items, significant difference $p < 0.05$ and strong association indicated by $OR \geq 2$) is recorded here by different column fillings: black, gray—three or two limits were attained, respectively; gradient of gray—sole sufficient sample size (important in cases of low or no associations); empty columns-other cases of OR; gray in flanks—reference R ratios compared with OR values (see above). (a) (b) Selected results of OR-mediated analysis concern Table 1 (cf. also the corresponding abbreviations) and are commented in section 3.3. Numbers 1 and 2 in abbreviation-NS items found with MNSQ1 and MNSQ2, respectively; suffixes o1, o2 and o3—selection steps 1, 2 and 3 mentioned in Table 1, respectively; B, K-BNsup35 and KPO sets, respectively; C1, C2—sets of cancer-related BLASTN-records containing sequences similar to MNSQ1 (part (a)) or MNSQ2 (part (b)), respectively; R-ratios between numbers of human and mouse NS items present in C1 and C2 sets. (c) Overall (robust) evaluation of strand-related HM*occurrences in MPL displayed in Figure 5. Strong and significant associations of HM* with non-transcribed DNA strands were found. Wp—W-pairs. For details see Abbreviations and Section 3.4.

strong associations were also significant ($p < 0.05$) and/or robust (sample size $s \geq 40$ NS). Higher or more valid R and OR values mostly formed upper MNSQ1-related graph indicating certain phylogenic context of this evaluation due to exclusive Elasmobranchii origin of MNSQ1 units (cf. section b in Figure 4). For additional comments see Section 4.3.

Predicted phosphorylated serines (Ser) achieved scores higher than 0.990, when using both NetPhos 2.0 and KinasePhos2.0. Similarly, NetPhos-related prediction yielded score maxima for threonine (Thr) and tyrosine (Tyr) higher than 0.980, but scores lower than 0.950 were determined during the corresponding KinasePhos- mediated predictions. In accordance with this difference in score maxima, only predicted phosphorylated Ser but not Thr and Tyr were present in set of double-predicted PPS encoded by MPL. This result led us to complete MPL set with four new MPL 1) different from originally found MPL, 2) encoding predicted Thr- and Tyr-re- lated PPS, and 3) achieving minimum score of 0.950, when using NetPhos 2.0. For details concerning the results of the parallel database search for experimentally confirmed MPL see the first section of Figure 5.

Another type of statistical reevaluation was performed on the subset of MPL-encoded peptides (MEP) immediately containing phosphorylated aa (i.e. uniquely limited by the same score of 0.800 in the two predictions). The constructed histogram and histogram-related distributions based on the product of the resulting two types of scores (s1 × s2) in fact approximated each result of performed double selection as unique value (Figure 6).

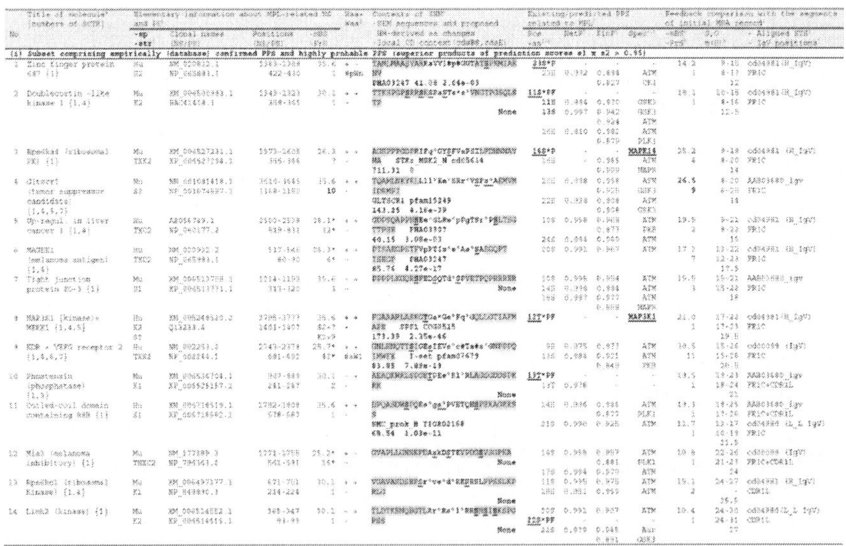

No	Title of molecule[a] (numbers of ACTR)	-sp -str	Clonal names (NM/NP)	Positions (NM/NP)	-mHS -FrS	Has- Nas	Contexts of SHM -SHM sequences and proposed HM-derived as changes -local CD context (nasHL.cdsH)	Pos +aa	NetP	KinP	Spec	-mHS -FrS	S,O m(S)	- Aligned STH - IgV positions
15	Translation factor BP (EIF4EBP1) {1,4}	Hu S1	AK312011.1 BAG34949.1	33-56 1-8 N-term	35.6 1	+ ^ #aHN #uHn	MSGg#o#abc#SQTPSRAIPAT eIF_4EBP pfam05456 139.40 3.42e-43	68 SE*P	0.444 -	0.987 -	ATM ATM	6.5 cdr1	24-33 23-34 17.5	smart00406 (IgV) CDR1a
(ii) Other items with product of prediction scores higher than empirically derived limit 0.88 (in fact 0.95 ≥ s1 x s2 > 0.88)														
16	Kalirin, RhoGEF kinase {1}	Hu TXK2	XM_006713814.1 XP_006713877.1	6838-6909 2224-2234	36.3* 9*	+ - -	KKRSTAGKKSQpaRLPQaSPRPYSSVRAGKE None	138	0.886	0.961	ATM	21.2 2	7-15 6-17 11	cd04981 (H_IgV) FR1C
17	Poly (ADP-ribose) polymerase (Parp4) {1}	Hu S2	NM_001145978.2 NP_001139450.2	5534-5559 1807-1816	35.6 1	- + -	SIKTSSQKECAPCA*a*d*QRSPVPRASLPA LQ None	138	0.827	0.974 0.839	ATM MAPK	21.0 2	8-16 7-17 12	AAB03680_igv FR1C
18	Tripartite motif -containing protein NHA-like {1}	Hu S2	XM_006508379.1 XP_006508442.2	965-988 217-230	37.4 3	- + ^	ELRKLKEEKESVVKRLGe*g*ENRLVQQRQW VR SMC_prok_B TIGR02168 43.12 7.58e-05	135	0.870	0.946 0.888	ATM CK2	24.0 2	6-17 9-18 13	AAB03680_igv FR1C
19	Rab11-BP 1B (membrane recycling) {1,4}	Hu S2	AY280969.1 AAQ18787.1	717-742 76-84	35.6 1	- + -	GAVSPVGELAAGGA*r*d*LKRq*ASSLVES KAKD None	235	0.876	0.969 0.943	ATM Aur	22.5 1	10-17 9-18 13.5	AAB03680_igv FR1C
20	A-kinase anchor protein 3 {1}	Hu K2	NM_006422.2 EDK98845.1	547-562 61-66	35.1 1	+ - -	LKKSTAGFQDSRFRpGKSSPVEKVAVPVDQ AKAP_110 smart00807 1461.79 0	198	0.834	0.888	ATM	15.5 1	12-16 11-17 14	AAB03680_igv FR1C
21	MAGEA1 (melanoma antigen) {1}	Hu C1	AK314978.1 NP_004979.3	416-338 58-65	33.7 2	+ + -	PTAGSIDIPQSPq*GaSAFPTTINFTRQRQP MAGE_N pfam12440 92.25 1.98e-23	128	0.988	0.988	ATM	14.6 3	12-18 11-19 15	cd04981 (H_IgV) FR1C
22	Potassium voltage-gated channel, Kcnh7 {1}	Hu S2	NM_133207.2 NP_573470.2	3258-3272 1014-1021	37.4 1	- + +	DEKCSDSGLQRAaWGVs*e*TESDLTYGEVE G None	179	0.966	0.939 0.834	ATM CK2	10.1 1	12-19 15-16 15.5	cd04981 (H_IgV) FR1C
23	MAGEB4 (melanoma antigen) {1,4}	Hu TXKC2	NM_002387.2 NP_002358.1	413-438 83-91	27.3* 3e*	+ + -	PQREPPTTSAAAs*Ms*cTGSDKSDESQKKE E MAGE_N pfam12440 101.50 1.19e-26	192	0.989	0.982	ATM	15.9 1 13.4 2	21-24 18-25 31-37 30-37 22.5	cd04984 (LK_IgV) FR1C+CDR1L smart00406 (IgV) CDR1a+FR2N
24	Flt4 (receptor tyrosine kinase) {1}	Hu K2	NM_008029.3 NP_032055.1	1411-1438 440-449	33.7 2	+ - -	SSRSIYSRRSRQTLTrTAYGVRQRLSVQMR None (near smart00410)	108	0.989	0.917	ATM	18.1 3	19-26 17-26 22.5	cd00099 (IgV) FR1C+CDR1L
25	Titin {1,4,6}	Hu S1	NM_133432.3 NP_597676.3	67246-314 22330-341	35.2 1	+ + ^	DKVSITTTKDAtTLTvKDgMr*GDSRVFLT LE Ig_Titin_like cd05748 88.71 2.50e-18	196	0.968	0.814 0.812	ATM PKC	19.7 6	18-27 17-27 22.5	AAB03680_igv FR1C+CDR1L
26	Zinc finger protein 619 {4}	Hu S2	NM_001004138.2 NP_001004139.2	607-678 161-171	35.6 1	+ + spWI SpWP	SSEVTDTEENPs*e*cKPGgTt#1#SSPNYPP PCS None	125	0.828	0.810	ATM	17.2 6	21-29 20-32 26	cd04981 (H_IgV) FR1C+CDR1L
27	Tyrosine kinase Ptk7 {1,6}	Hu K1	XM_006524524.1 XP_006524587.1	501-516 165-170	35.3 1	+ - -	THTVSSEEKKLTLSFaSPEDQGLYECVARN Ig2_PTK7 cd05762 167.78 1.61e-48	175	0.998	0.900 0.858	ATM PLK1	9.3 cdr1	24-27 23-27 25.5	cd04980 (IL_L_IgV) CDR1L
28	Inositol hexaphosphate kinase 3 {1}	Hu K2	BC141202.1 AAI41264.1	386-388 67-97	35.1 1	+ - -	ERGHLSLVARPLKENLEpPQVSPESRAVALM None	128	0.982	0.788	ATM	25.2 6	22-30 21-30 26	cd04981 (H_IgV) FR1C+CDR1L
29	HIPK2 (kinase) {1}	Hu K2	XM_006511686.1 XP_006555664.1	2926-2938 856-864	31.3 4	+ - -	QTIVIPDTPSPTVSVITIsSDTDEEEEQKHA None	208	0.961	0.982 0.841	ATM CK2	11.1 1	24-30 23-31 27	cd04981 (H_IgV) CDR1L*
30	HIPK2 (kinase) {1}	Hu K22	XM_006718936.1 XP_006715389.1	2837-2862 856-864	31.9 K1x2 K2+1	+ - -	QTIVIPDTPSPTVSVITIsSDTDEEEEQKHA None	209	0.961	0.982 0.841	ATM CK2	13.0 1	24-31 23-31 27.5	cd04981 (H_IgV) CDR1a*

(b)

No	Title of molecule[a] (numbers of ACTR)	-sp -str	Clonal names (NM/NP)	Positions (NM/NP)	-mHS -FrS	Has- Nas	Contexts of SHM -SHM sequences and proposed HM-derived as changes -local CD context (cdaHS.cdsH)	Pos +aa	NetP	KinP	Spec	-mHS -FrS	S,O m(S)	- Aligned STH - IgV positions
31	vav 1 oncogene {1}	Hu K2	NM_001163816.1 NP_001157288.1	1279-1310 387-394	31.9 1	+ + -	DMSTLRQIINFQiSIe*NLDQSLARYGRPKI PH_Vav cd01223 199.61 1.07e-59	168	0.995	0.918	ATM	- cdr1	25-33 25-32 28	AAB03680_igv CDR1a
32	Slowmo homolog 2 {1}	Hu S1	XM_006500000.1 XP_006500061.1	458-478 118-123	39.2 2	+ - #aKW #aHn	ETVLTQKALITVKGVELs#s#YLEGLMASTI SSNAS PRELI pfam04707 203.71 6.55e-67	168 158	0.907 0.961	0.937 0.945 0.877	ATM ATM Aur	13.8 3	35-35 29-36 36	consensus CDR1a
33	FBO1A (rhabdomyosarcoma) {1}	Hu C1	BT007456.1 Q12778.3	1083-1085 355-361	30.1 1	- + spWI	VHSMVYIFSAARMa*sRT1#PELSSISNKSDM SN None	182	0.827	0.987 0.906	ATM Aur	16.4 1	36-39 35-41 37.5	AAB03680_igv FR2N
34	Atm1 (melanoma) {1}	Hu C1	XM_006512499.1 XP_006512562.1	3895-3925 1237-1243	36.1 1	- + -	SVDADFLGLFKASSYd*PGNSFSKMSLDSMT None	135	0.985	0.939	ATM	21.2 2	36-40 34-40 38	cd04981 (H_IgV) FR2N
35	Tyrosine phosphatase PTPRZ1 {1,4}	Hu K2	XM_006716076.1 XP_006716139.1	2140-2164 582-590	30.1 1	+ + -	KSLLTSPKLDTGs*e*d*s*e*gPSPATSAIP FLSR None	168 175 298	0.812 0.886 0.987	0.887 0.884 0.888 0.978 0.803	ATM ATM PLK1 ATM Aur	28.3 2	36-41 34-43 38.5	cd04981 (H_IgV) FR2N
36	Phosphatase Phlpp1 {1}	Hu K2	NM_133821.3 NP_598582.3	4388-4412 1418-1426	31.9 1	+ - -	VSRVIKDSDSDGLAVpSASSKMASEISSEL None	118 205	0.994 0.828	0.931 0.989 0.938 0.889	ATM PKB ATM Aur	23.9 2	36-41 33-42 38.5	smart00406 (IgV) FR2N
37	TRAF2 and NCK interacting kinase {1}	Hu TXK12	NM_001161566.1 XP_006531881.1	354-376 10-17 (N-ter)	25.4* 18*	^ + -	MASDSPARSLDe*Id*1*SALRDGAGIFELVE L STKc_MAP4K4_6_N cd06636 601.23 0	88	0.993	0.932	ATM	18.6 3	36-42 36-42 39	AAB03680_igv FR2N
(iii) Items passed through terminal reselection via other requirements than the limit of score products (cf. Figure 1 and Figure 1B in SF3)														
38	Pnliprp1 (lipase related) {3,4}	Hu S2	NM_018874.2 NP_061362.1	1802-1828 449-458	35.6 1	+ + -	VQNGKENTRHNFcSe*a*IVr*EDILKELLNC K PLAT_PL cd01759 188.73 5.74e-58	177	0.986	-	-	17.3 1	8-16 9-16 12	AAB03680_igv FR1C
39	BOC (cell adhesion) {2,4,8,7}	Hu TXK12 DXK2	NM_006247893.1 XP_006247948.1	408-464 40-59	34.1* 34*	+ - -	ADLNRVDQVTVQpASTVQKpGgTVILgCVVEP PRMNVTRR IgSF cl11960 48.93 4.37e-07	156	0.808	0.889	ATM	54.3 9	5-22 4-24 13.8	AAB03680_igv FR1C
40	ret proto-oncogene {2,5}	Hu C12 K12 TKC12	X12949.1 NP_066124.1	1489-1497 517-522	33.7 K1+C1 - 8	+ - -	NSSYVASIAGCPLSCa*VSKHRLHCHHQQGL None	185	0.875	0.856	ATM	12.1 1	21-25 19-25 22	cd04983 (TCR_G) FR1C+CDR1L
41	CS0DD006YL02, cDNA (neuroblastoma) {6}	Hu TXC2	BX161420T1 CAD61894.1	1498-1528 267-279	28.8* 28*	+ + ^	PSAKEQLHLKESATIICLVTgPs*EADVFVQW MQ IgC_CH4 cd05768 194.51 3.18e-61	128 238	0.896 0.900	0.909 0.896	ATM ATM	19.8 8	16-30 16-32 23	cd00099 (IgV) FR1C+CDR1L
42	Alk (anaplastic lymphoma kinase) {4}	Hu TXC2	NM_007439.2 NP_031465.2	1498-1443 282-321	27.6* 9*	+ + spWI	PDPPCELEYSGPLh#NHGNq*sWSNRKVYSEE ASEHRLLIKIP MAM cd06263 79.73 1.28e-16	288	0.937	0.893 0.812	ATM GSK3	20.3 6	20-32 20-35 26	cd04981 (H_IgV) FR1C+CDR1a
43	Adenylate cyclase 1 (Adcy1) {3}	Hu S2	NM_009622.1 NP_033752.1	2793-2763 874-884	35.6 2	+ + -	SQVRVMFASIPNFSd*FYIe*LDSNNRGVECL K Guanylate_cyc pfam00211 284.86 3.93e-78	179	0.955	-	-	36.4 1	28-36 29-37 32	smart00406 (IgV) CDR1a+FR2N
44	Copb1 (coatomer complex) {3,4,7}	Hu S2	NM_033370.3 NP_203534.1	2323-2293 728-738	38.2 1	+ + -	ASKLNRVTQLTgPSDPVYa*e*AYSRVMQYDI V Coatomer_beta_C pfam07718 243.66 4.27e-75	187	0.878	-	-	28.3 8	33-41 33-40 37	cd04982 (TCR_G) CDR1a+FR2N
(iv) compared NITR molecules														
45	NITR12 (related to AR) {4,6,7}	Hu TX1	KJ879999.1 A1A56875.1	4-45 2-15 (N-term)	25.3* 13*	+ + -	YKpId*NFTI*tCPVs*GQMNGLPYWVE IG_like smart00410 66.37 1.36e-14	148	0.911	0.897 0.833	ATM CK2	29.9 8	12-26 12-25 19	AAB03680_igv FR1C+CDR1L

(c)

Figure 5. Existing and predicted antibody-like phosphorylation sites. [a]The MPL subsets (subdivided according to types of reselection) were arranged in the separate table segments according to the mean positions of their feedback similarities (m(S)) between MPL and initial MSA record displayed in Figure 2 (cf. footnote i and Figure 1). BOC—brother of CDON, cell adhesion associated molecule, oncogene regulated; BP-binding protein; Copb1-coatomer protein complex, subunit beta 1; EIF4EBP1—Eukaryotic translation initiation factor 4E binding protein 1; Pnliprp1-pancreatic lipase related protein 1; regul.-regulated; Rps6ka4—ribosomal protein S6 kinase, polypeptide 4; Rps6kc1—ribosomal protein S6 kinase, polypeptide 1; SCTR—successful criteria of terminal reselection (for encoding numbers see fifth part of Figure 1(b) in SF3); VEGF— vascular endothelial growth factor. [b]FrS—frequencies of limited co-localizing similarities (i.e. MUSAS mentioned in Sections 2.4, WP2.3.3-4 and WP5.7); mBS—maximum bit score of the selected BLAST similarities; NS-nucleotide sequence; PS-protein sequence; *-minimum of two alternative score maxima or MUSAS numbers was recorded in cases of two co-evaluated searches. [c]sp-species origin (current items: Hu—human, Mu—Mus musculus; reference NITR-related item: Mm-Miichthys miiuy); str-strategies yielding topical MRNS (sections WP5.6 and WP5.7). Capitals and numbers in abbreviations of Entrez restriction (section WP2.3.1): C, K-cancer- and phosphorylation-related special BLASTN searches; DX—double combined approaches comprising pairs of differently adjusted TBLASTX searches, only if they yield co-localizing similarities limited by 30 or 35 bits; S—global BLASTN searches limited by 35 bits; T—ternary combined approach (TCA, limit 25 bits); X, NX, N after T-TCA accompanying pairs of BLAST searches (selecting repeatedly co-localizing similarities; cf. WP2.3.3), i.e.: 1) two differently adjusted TBLASTX; 2) BLASTN and TBLASTX and 3) two BLASTN differently employing two MNSQ, respectively; 1, 2, 12—items selected by MNSQ1 or MNSQ2 and both MNSQ, respectively. [d]Haa + Waa-occurrence of effectively located hypermutation motifs (HM*) including W-pairs (for details see sections 3.4 and 4.4). Upper rows: left/right evaluation using plus or minus-WRCH/TCW are present or are not present at the critical positions, respectively. Bottom rows: #aWi, #pWi-W-pairs located at DNA strands anti-parallel or parallel with respect to the direction of transcription, respectively (i denotes the number of unclassified nucleotides inserted between the observed WRCH; if n substitutes the number i, then WRCWRCH is indicated); #aRW, #pRW-palindromic W-pairs RGYWRCY includes WRCY unit critical with respect to aa alteration in antiparallel or parallel (transcribed) DNA strands, respectively. [e]SEM-sequences enveloping (and including) selected MPL encoded peptides; gray background covering edges of peptide segments-chains extending central MRNS-encoded peptide sequences; underlined-predicted or empirically confirmed phosphorylated aa; CD-conserved domains; cdsBS, cdsE-conserved domain similarity-related bit scores and Expects, respectively. Lower cases (LC) in alphabets denoting peptide sequences indicate aa which could be altered during hypermutation changes via cytidines of HM (HM*-cytidines): LC and LC with—sole HM*—cytidine form WRCH and TCW, respectively; LC with *—two HM*-cytidines are present in two neighbor but antiparallelly located positions of palindromic WRCH; LC followed by dominant #—WRCH containing HM*-cytidine composes W-pair (cf. Sections 3.4 and 4.4). [f]pos + aa—phosphorylated aa is denoted by a single character, which is accompanied by the number of given aa position in the displayed peptide; NetP, KinP-scores obtained by NetPhos2.0 and KinasePhos2.0, respectively (maximum is equal to unity); spec—specificity of kinase—mediated phosphorylation (identified by KinasePhos2.0). [g]Seven experimentally proved PPS exhibited overlaps with peptides encoded by MPL (MEP), whereas phosphorylation sites of four these PPS immediately formed MEP. In addition, two proved PPS with phosphorylation sites located at position immediately neighboring to MEP were encoded by sequence containing HM* occurring within MPL. The residual (seventh) confirmed PPS (with the site outside MEP) then overlapped MEP encoded by MPL containing double-active W-pair (i.e. two HM* forming W-pair). F, P-database record of existing PPS was found using Phosida or phospho.ELM, respectively; *—indicator of empirical confirmation recorded in given databases. [h]Only nine groups of protein kinases (PK) were predicted or empirically proved to phosphorylate the displayed MEP: Aur—Aurora-related kinase; ATM—Ataxia-telangiectasia mutated (kinase); CK1, CK2-casein kinases 1 and 2, respectively; GSK3-Glycogen synthetase kinase-3; MAP3K-MAPK kinase kinase (=MEKK1); MAPK-mitogen activated PK; PKB-PK B; PKC-PK C; PLK1—polo-like kinase 1, absence of abbreviation in a database-confirmed case—only phosphopeptide was observed without knowledge about PK specificity. [i]Align versions of BLAST were used in feedback comparison (see WP2.4.1). O, S, m(S)—MSA-record-related positions of deduced sequence overlaps, feedback similarities, and mean aa position of these similarities (cf. footnote a), respectively. [j]The segments of MSA record present in Figure 2 formed feedback (BLASTX, TBLASTN or TBLASTX; cf. section WP2.4.1) similarities at defined IgV-related positions of this "initial" MSA- record (PIM): FR1C—framework region 1 including CSB1-related overlaps (PIM from 5 to 15 - 22); CDR1L—at least 50% overlap of CDR1light (PIM: 23 - 30); CDR1a—C-terminal part of CDR1 in light chains and co-localizing CDR1 of heavy chains (PIM: 31 - 35); CDR1s-prevailing overlaps of CDR1light but short of CDR1all (PIM: 23 - 35); FR2N—N-terminus of framework region 2 (PIM: 36 - 42; the most conserved part of this framework region in Figure 2). [*]—different classification follows from NS differences. For additional comments see Figure 2, Section WP4.2.2 and SF2.

Unexpectedly, the obtained graphs indicated bimodal-like distribution with a dominant peak located at range of superior score products. This distribution contrasted with the expected random exponential decrease assumed at least in upper half of score product values. Even the comparison with the model constant distribution (stricter evaluation than the assumed random exponential decrease) indicated strong and significant ($p < 0.05$) association of score products with the restricted region of dominant peak. Based on these facts and considerations concerning Figure 6, we explained the existence of the peak as the consequence of functionally conditioned structural maintenance. Based on this statistically supported explanation, we derived the limit for score products 0.88, subsequently used in the terminal reselection. For important details see sections WP2.4.1, WP3.2, WP3.3 and file SF3.

3.4. Final Bioinformatic Analysis of the Selected Pairs of MEP and MPL

About eighty percent of regularly restricted human and mouse MEP achieved superior score related to ATM. Similarly to PPSIg, part of ATM-related sites, predicted here, achieved also co-dominant prediction of PPS phosphorylated with Aurora and protein kinase B. Interestingly, MPL were not always translated in the same reading frame or orientation as the

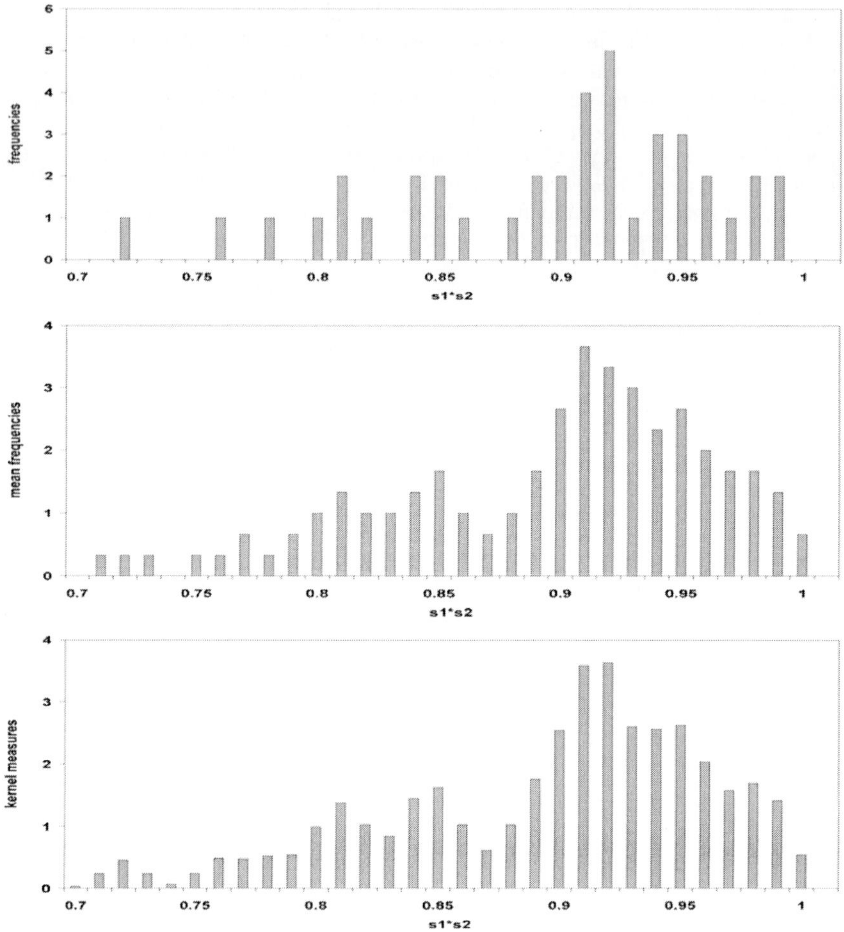

Figure 6. Distribution of score products evaluating predicted phoshorylation sites encoded by MPL. Model MPL subset of predicted sequence items was selected before overlap-, Thr-, Tyr- and database-related completions of data (cf. Figure 1 and WP2.4.1). This subset included thirty-two MPL-encoded peptide sequences containing forty predicted phosphorylation sites, which were represented here by the products of their maximum predicted scores (s1*s2). The score maxima were individually obtained with NetPhos 2.0 (s1) and KinPhos 2.0 (s2) in the searches equally restricted by the limiting minimum value 0.800 (cf. Figure 5). Axes x of the displayed graphs comprise the specific score product intervals of 0.01 extent. Upper part includes a current histogram of the score-product-related frequencies. Frequencies of the evaluated score products look like the main peak in the range of upper values. The corresponding significant prevalence to high score products ($p < 0.05$)

perhaps consisted in the existence of phylogenic pressures keeping PPS function and enabled us to empirically assess reselection limit. In addition, we simultaneously diminished effects of possible overestimations (i.e. false negativities) frequently accompanying pre-calculated restriction of combined selection. For other comments see Sections 2.1.2, 3.3 and WP3.2. Middle part contains widely used gliding mean values of frequencies enumerated here based on three neighboring (interval-related) frequencies present in the upper histogram. Each gliding mean value is displayed specifically in the positions of central (second counted) interval yielding thus the simplest form of smoothened histogram. Bottom part contains more adequate/specific form of histogram smoothening based on linear combinations formed by central and neighbor interval related frequencies and Gaussian kernel functions which enumeration takes into consideration empirical statistics of the evaluated differences between the corresponding score pairs s1 and s2 (for details see section WP3.3).

compared IgV-related segments of initial MSA record. Five MEP formed chains of valid conserved domain similarity with Ig domains. Four of them achieved feedback similarity with the segment CDR1light in initial MSA record. However, feedback similarities overlapping common CDR1all segments were found only in five cases of Ig-unrelated MEP/MPL (cf. Figure 5 and section WP4.2.2; for explanation of CDR1light and CDR1all see Figure 2). Only pairs of human and mouse orthologues of HIPK2 contained equally co-localized MPL in Figure 5.

The DNA-strand-related occurrence of well-known hypermutation motifs (HM; [21] [23] [24] [30]) was observed, when selecting HM subset (HM^*) located at MPL sequence positions critical with respect to aa alteration. Interestingly, the number of HM^* was significantly higher ($p < 0.05$) in the non-transcribed DNA strands of MPL than in transcribed ones when comparing the strand-related occurrences of HM^* with those of HM incapable to alter aa (non-HM^*). The corresponding significant increase concerned also certain WRCH-related pairs (W-pairs; $OR^*[0] = 11.6$ but an extremely low set of 22 items) determined by the patterns: RGY-WRCY, WRCWRCH, WRCHNWRCH and WRCHN(9)WRCH (see section 4.4 and previous papers [24] [25]). For the graphs of HM^*-related OR see Figure 4. The occurrences of HM^*-related W-pairs in framework-associated and CDR1-associated regions of MPL (determined by feedback comparison) also differed, determining 3.85 times higher density of such W-pairs in CDR1-associated regions. In addition, we found three types of MPL relationships interesting with respect to HM^* occurrence. This concerned: 1) knocking-out alteration of predicted phosphorylated serine in cases of twelve human and mouse MPL and one reference MPL of bony fish origin, 2) distinct relationship of the two HM^* (WRCH and TCW) to transcribed DNA strand in site encoding phosphorylated serine following from genetic code and 3) maximum number of HM* per single MPL (two different MPL triads distinctly contained four times such WRCH or such TCW). For additional details and comments see Figure 5, Sections 4.4, WP2.4.1 and WP4.2.1-3.

4. DISCUSSION

4.1. Structures Found in Initial MSA-Record

Both sequences of patterns displayed in Figure 2 exhibit certain relationship to the published data. The first two aa of CSB1-related pattern CX(10,13)WXXQXP include well known positional markers restricting CDR1. Cysteine is the current C-terminal aa of FR1 composing light chains of IgV, whereas tryptophan mostly occurs in N-terminus of FR2 in both light and heavy chains of IgV [48] [49] . Trans-cis isomerization of C-terminal aa of the same pattern, i.e. proline, is probably important for local flexibility of IgV [8] [50] . Another CSB2-related pattern identified here, LX(8)DX(3)YXC, contains the sequence DX(3)YXC identical with the pattern found in non-vertebrate molecules closely related to AR [13] .

The marked differences between the occurrence of at least anti-cohesive columns in CDR1 and other two hypervariable regions (cf. Section 3.1) indicate higher conservativeness of usually effectively hypermutating CDR1. If we also consider location of CDR1 in CSB1 and PKSIg [8] [9] [11] (cf. also Figure 2), the indicated conservativeness supports as an additional fact the model importance of N-terminally located PPSIg region for studies of antibody-like PPS.

4.2. Methodological Aspects

The bilingual approach employed here was in fact based on simultaneous evaluation of 1) occurrence of HM motifs present in MPL composing IgV-related mRNA or cDNAs segments (the first formal language; Figure 1), 2) existing or predicted phosphorylation of MPL encoded peptides extended by their immediately neighboring chains (the second formal language) and 3) possible alteration of aa composing the selected PPS via HM (site specific translation). Due to various actual reading frames of MPL, the bilingual approach determined a larger set of MPL than would simply follow from the searches for protein sequence similarities.

To unify the selection of MRNS (cf. Figure 1), we supplemented here a simplified BLAST derived fuzzy- system associated restriction of CSB (Section 2.2.3) consistent with ELEMS principles [9] [25] . The unified usage of BLAST-related evaluation in the three starting main steps (cf. Figure 1) does not mean refusing the possible prediction by means of machine-learning methods such as SVM or neural networks [51] - [55] . More likely, we assume that some results obtained in our searches represent important starting information for future machine learning. This concerns among others 1) feedback similarities and 2) reevaluation similarity extents and cooperative effects related to different elementary MNSQ units participating in or associated with MPL selection process.

4.3. Statistical and Phylogenic Contexts of Employed Search Procedures

In fact, we did not observe here NS mutations in cell clones, but only potential NS mutation. Consequently, we considered here only inherited changes of DNA, when looking for an explanation of unexpected prevailing occurrence of cancer-related NS items in the human set described in section 3.3. This consideration focused our attention first of all on 1) several events occurring in germ-line cells (GLC) or 2) parameters of these events, i.e.: a) prevention of DNA attacks by oxygen radicals in GLC, b) extents or specificity of meiotic hypermutation (comprising also the described changes in Ig-related HM [56]) and its possible effects on genome plasticity, c) percentage of mutated descendants bred by old individuals d) immune surveillance on GLC. Better understanding of the observed and potential relationships, including phylogenically interesting (Elasmobranchii-related) MNSQ1 linkage (cf. Sections 2.3, 3.3 and Figure 4), needs also more detailed analysis on a larger set of PPSIg- related molecules and MPL in future.

In accordance with our working hypothesis [13] , at least some MPL displayed here (encoding most probably ATM-phosphorylated MEP; cf. section 3.4) and PPSIg segments originate in common ancestral oligonucleotide segments of possible increased mutability. This hypothesis appears to be in agreement with the displayed strong and significant association of score products with superior score values (Figure 6; Sections 3.3 and WP3.2) and existence of four molecules containing MPL/MPL-like segments most frequently (i.e. four times) selected in terminal reselection (see the first column of Figure 5 and section WP4.3). The latter possibility concerns: 1) NITR2 of close phylogenic relationship to AR [57] otherwise only near the limits of PPS prediction (cf. WP4.3), 2) two molecules containing MPL with Ig domain context molecules (KDR of superior PPS-related scores, oncogene regulated adhesion molecule BOC); 3) MPL of glioma tumor suppressor candidate region gene 1 (Gltscr1). In contrast to other molecules, Gltscr1 did not achieve any conserved Ig domain similarity, the corresponding CDART-indicated relationship to Ig superfamily (IgSF) representatives or local PPSIg-related fold similarity in records obtained with FFAS03 program (cf. section 3.1). To explain this anomalous sequence behavior of Gltscr1, we assumed participation of well-known mechanisms such as a) transposition of short repeats, b) gene conversion and/or c) perhaps also functionally conditioned convergence of DNA segments (cf. [5]).

4.4. MPL in the Light of Hypermutation Events

Collisions of transcription apparatus (synthesizing mRNA copy of transcribed strand of DNA) with APOBEC enzymes interacting with HM present in transcribed DNA strand (CTAE) are able to enlarge hypermutation effects of these deaminases. This occurs due to 1) extended low-fidelity repair of generated double strand breaks with error-prone DNA polymerases and 2) accompanying insertion deletion changes [58] - [60] . In accordance with this mechanism the observed significant differences in distributions of HM* in transcribed and non-tran- scribed DNA strands (Section 3.4; Figure 4) can be explained by similarly selective CTAE-triggered events. Since the comparison of DNA strands does not concern altered nucleotide sequences, inherited but not somatic changes have to be considered, which looks like a contradiction when dealing with somatic mutation changes. Nevertheless, HM WRCH participates also in inherited meiotic hypermutation via a similar though less frequent mechanism than in case of Ig gene hypermutation [56] . Consequently (and in accordance with less frequent incidence of meiotic-mutations), the statistics of HM* occurrence in non-altered DNA strands (see Figure 4) suggests parallel somatic mutation changes via CTAE events in at least some transcribed DNA strands encoding MEP-peptides and containing HM*. In accordance with this parallelism, HM* located in their transcribed DNA appear to be more perilous than others, indicating thus MEP of probably increased hypermutation risk in molecules such as human RhoGEF kinase, MAGEE1, KDR, zinc finger protein 687, rhabdomyo-sarcoma FBO1A, mouse anaplastic lymphoma kinase and zinc finger protein 619 and both reference AR-related NITR molecules of bony fish origin present in Figure 5 or Table WPT2 (see section WP4.3).

In accordance with the theories of ageing, abnormal phosphorylation (or loss of phosphorylation) is assumed as an alteration possibly important during ageing [61] . Abnormal phosphorylation can be among others caused via mutation or hypermutation changes of PPS generating knocked-out, alternatively or weakly reacting PPS. If we assume that such PPS changes frequently occur in different PPS phosphorylated by the same PK, then their change can imitate the loss or decrease of the involved PK activity. In accordance with the given assumption, about eighty percent of MEP appear to be dominant substrates of ATM (section 3.4), which is the reason to consider possible imitation of functional loss of ATM via mutation of these PPS. As well known ATM is involved in regulation of double-strand-break response (cf. Introduction). Functional loss of mutated ATM causes autosomal recessive disease (ataxia-telangiectasia; AT) characterized by median survival 19 - 25 years (a wide range) and death due to cancer and respiratory failure [62] . In addition, AT is also accompanied by neurodegeneration and immunodeficiency or worsened immune response otherwise frequently appearing in elderly people [63] . Consequently, a substantial question arises: Can changes in sequences of PPSIg/MPL-related PPS influence important regulatory or even ageing-related effects of ATM?

Some MPL are not translated in the same reading frame as PPSIg (cf. Section 3.4). The corresponding peptides (M_out) thus can avoid cross-reactivity with Ig epitopes, which mostly contain only "in frame" insertions and deletions [58] [64] . This raises a question of possible usage of at least some mutants of M_out as components of future complex/multicomponent anticancer vaccines. On the other hand, it is a question whether at least some MPL-derived peptides encoded in the same reading frame as PPSIg (M_in) or their mutants can cross- react with rheumatoid autoantibodies. The positive answer could be among others important for protein engineering improving autoantibody specificities with respect to M_in mutants.

5. CONCLUSIONS

Though the databases of empirically proved phosphorylation sites are considerably incomplete, seven MEP contain such database-confirmed sites. The nonrandom main peak of MPL-derived products of phosphorylation-related prediction scores then suggests the phosphorylation of at least some additional MEP (Section 3.3; Figure 6). The significantly prevailing occurrence of HM* and the corresponding W-pairs in non-transcribed DNA strands of MPL indicates certain mutation events in the corresponding gene segments including parallel somatic changes (Section 4.4; Figure 4). In accordance with these parallel changes, MPL segments with HM* present in transcribed DNA strand appear to be most critical (for list see Section 4.4). The occurrence of HM* in MPL meets also three important functional aspects. First of all, most MPL (about 80%) encode peptides identical with or predicted as phosphorylated by ATM molecules, known for their relationship to aging and regulation of DNA double-strand-break response (Sections 3.4 and 4.4). Secondly, cancer-related MRNS are significantly more frequent in humans than in mice (Figure 4; Sections 3.3 and 4.3). Thirdly, feedback similarities of MPL containing W-pairs co-localized mainly with (hypervariable) CDR1 segments of IgV chains forming both

MNSQ (Table 1, Figure 2 and Figure 5, Section 3.4). These three functional aspects moreover concern several existing groups of MPL with extreme properties, e.g. 1) MPL with HM* encoding directly predicted phosphorylated amino acid, 2) CDR1-related MPL containing superior numbers of HM* and 3) MPL/MEP pairs frequently selected in the alternative steps of terminal reselection (Figure 5; Sections 4.3, 4.4 and WP4.2.1-3).

Though the actual carcinogenic effect of mutation changes in phosphorylation sites is known for a long time [65] - [67] , the set of the corresponding investigated sequences is not sufficiently large. In accordance with this fact, we have not yet found experimental confirmation of mutation changes in the seven existing PPS segments (displayed in Figure 5) in the literature. Consequently, these PPS, as well as the displayed predicted segments, represent an inspiration and challenge for the subsequent experimental researches in specialized laboratories and perhaps also subjects interested in certain bioinformatic trends (cf. [68] [69] , Sections 4.2 and 4.4; see also below). In our opinion, future experimental research should comprise 1) comparative sequence studies of ATM- related MPL composing not only DNA of cancer patients (for the corresponding methods see also our papers [70] - [72]) but also DNA of old people and 2) usage of phage displayed libraries or protein engineering in case of the considered immunological relationships ([73] - [75] and Section 4.4). Theoretically based bioinformatic investigation of the immunological relationships could moreover efficiently select the subset of possible candidates for vaccine-related promiscuous epitopes specifically recognized by cytotoxic T cells ([76] - [78] and Section 4.4) even in case of more extended HM*-related set of MEP mutants than is that displayed in Figure 5 (for other corresponding bioinformatic aims and trends see section WP5.8). In addition to the preceding medicinal aspects, an interesting question refers to a possible phylogenic relationship between IgV domains and certain NS encoding peptide segments a) containing several near ATM-related PPS and b) composing for instance immunoglobulin-like archaeal surface layer proteins or related metazoan cell surface proteins [13] [79] .

CITE THIS PAPER

Jaroslav Kubrycht, Karel Sigler, Pavel Souček, Jiří Hudeček (2016) Antibody-Like Phosphorylation Sites in Focus of Statistically Based Bilingual Approach. *Computational Molecular Bioscience*,**06**,1-22.

REFERENCES

1. Kornev, A.P., Haste, N.M., Taylor, S.S. and Ten Eyck, L.F. (2006) Surface Comparison of Active and Inactive Protein Kinases Identifies a Conserved Activation Mechanism. Proceedings of National Academy of Sciences of the United States of America, 103, 17783-17788.
2. Kornev, A.P., Taylor, S.S. and Ten Eyck, L.F. (2008) A Helix Scaffold for the Assembly of Active Protein Kinases. Proceedings of National Academy of Sciences of the United States of America, 105, 14377-14382.
3. Joseph, R.E., Xie, Q. and Andreotti, A.H. (2010) Identification of an Allosteric Signaling Network within Tec Family Kinases. Journal of Molecular Biology, 403, 231-242.
4. Seco, J., Ferrer-Costa, C., Campanera, J.M., Soliva, R. and Barril, X. (2012) Allosteric Regulation of PKCq: Understanding Multistep Phosphorylation and Priming by Ligands in AGC Kinases. Proteins, 80, 269-280.
5. Alberts, B., Johnson, A., Lewis, J., Raff, M., Roberts, K. and Walter, P. (2008) Molecular Biology of the Cell. 5th Edition, Garland Science, New York.
6. Blom, N., Gammeltoft, S. and Brunak, S. (1999) Sequence- and Structure-Based Prediction of Eukaryotic Protein Phosphorylation Sites. Journal of Molecular Biology, 294, 1351-1362.
7. Wong, Y.H., Lee, T.Y., Liang, H.K., Huang, C.M., Wang, T.Y., Yang, Y.H., Chu, C.H., Huang, H.D., Ko, M.T. and Hwang, J.K. (2007) KinasePhos 2.0: A Web Server for Identifying Protein Kinase-Specific Phosphorylation Sites Based on Sequences and Coupling Patterns. Nucleic Acids Research, 35, W588-W594.
8. Kubrycht, J. and Sigler, K. (1997) Animal Membrane Receptors and Adhesive Molecules. Critical Reviews on Biotechnology, 17, 123-147.
9. Kubrycht, J., Borecky, J. and Sigler, K. (2002) Sequence Similarities of Protein Kinase Peptide Substrates and Inhibitors: Comparison of Their Primary Structures with Immunoglobulin Repeats. Folia Microbiologica, 47, 319-358.
10. Kubrycht, J., Borecky, J., Soucek, P. and Jezek, P. (2004) Sequence Similarities of Protein Kinase Substrates and Inhibitors with Immunoglobulins and Model Immunoglobulin Homologue: Cell Adhesion Molecule from the Living Fossil Sponge Geodia Cydonium. Mapping of Coherent Database Similarities and Implications for Evolution of CDR1 and Hypermutation. Folia Microbiologica, 49, 219-246.
11. Kubrycht, J., Sigler, K., Soucek, P. and Hudecek, J. (2013) Structures Composing Protein Domains. Biochimie, 95, 1511-1524.
12. James, L.C., Roversi, P. and Tawfik, D.S. (2003) Antibody Multispecificity Mediated by Conformational Diversity. Science, 299, 1362-1367.
13. Kubrycht, J., Sigler, K., Ruzicka, M., Soucek, P., Borecky, J. and Jezek, J. (2006) Ancient Phylogenetic Beginnings of Immunoglobulin Hypermutation. Journal of Molecular Evolution, 63, 691-706.
14. Muramatsu, M., Kinoshita, K., Fagarasan, S., Yamada, S., Shinkai, Y. and Honjo, T. (2000) Class Switch Recombination and Hypermutation Require Activation-Induced Cytidine Deaminase (AID), a Potential RNA Editing Enzyme. Cell, 102, 553-563.
15. Okazaki, I.M., Hiai, H., Kakazu, N., Yamada, S., Muramatsu, M., Kinoshita, K. and Honjo, T. (2003) Constitutive Expression of AID Leads to Tumorigenesis. The Journal of Experimental Medicine, 197, 1173-1181.
16. Kou, T., Marusawa, H., Kinoshita, K., Endo, Y., Okazaki, I., Ueda, Y., et al. (2006) Expression of Activation-Induced Cytidine Deaminase in Human Hepatocytes during Hepatocarcinogenesis. International Journal of Cancer, 120, 469-476.
17. Endo, Y., Marusawa, H., Kinoshita, K., Morisawa, T., Sakurai, T., Okazaki, I.M., Watashi, K., Shimotohno, K., Honjo, T. and Chiba, T. (2007) Expression of Activation-Induced Cytidine Deaminase in Human Hepatocytes via NF-kappaB Signaling. Oncogene, 26, 5587-5595.

18. Matsumoto, Y., Marusawa, H., Kinoshita, K., Endo, Y., Kou, T., Morisawa, T., Azuma, T., Okazaki, I.M., Honjo, T. and Chiba, T. (2007) Helicobacter Pylori Infection Triggers Aberrant Expression of Activation-Induced Cytidine Deaminase in Gastric Epithelium. Nature in Medicine, 13, 470-476.
19. Gordon, M.S., Kanegai, C.M., Doerr, J.R. and Wall, R. (2003) Somatic Hypermutation of the B Cell Receptor Genes B29 (Igbeta, CD79b) and mb1 (Igalpha, CD79a). Proceedings of National Academy of Sciences of the United States of America, 100, 4126-4131.
20. Duquette, M.L., Pham, P., Goodman, M.F. and Maizels, N. (2005) AID Binds to Transcription-Induced Structures in c-MYC that Map to Regions Associated with Translocation and Hypermutation. Oncogene, 24, 5791-5798.
21. Rogozin, I.B. and Kolchanov, N.A. (1992) Somatic Hypermutagenesis in Immunoglobulin Genes. II. Influence of Neighbouring Base Sequences on Mutagenesis. Biochimica et Biophysica Acta, 1171, 11-18.
22. Wright, B.E., Schmidt, K.H., Davis, N., Hunt, A.T. and Minnick, M.F. (2008) II. Correlations between Secondary Structure Stability and Mutation Frequency during Somatic Hypermutation. Molecular Immunology, 45, 3600-3608.
23. Rogozin, I.B. and Diaz, M. (2004) Cutting Edge: DGYW/WRCH Is a Better Predictor of Mutability at G:C Bases in Ig Hypermutation than the Widely Accepted RGYW/WRCY Motif and Probably Reflects a Two-Step Activation-Induced Cytidine Deaminase-Triggered Process. The Journal of Immunology, 172, 3382-3384.
24. Dorner, T., Foster, S.J., Brezinschek, H.P. and Lipsky, P.E. (1998) Analysis of the Targeting of the Hypermutational Machinery and the Impact of Subsequent Selection on the Distribution of Nucleotide Changes in VHDJH Rearrangements. Immunology Reviews, 162, 161-171.
25. Kubrycht, J. and Sigler, K. (2008) Length of the Hypermutation Motif DGYW/WRCH in the Focus of Statistical Limits. Implications for a Double-Motif or Extended Motif Recognition Models. Journal of Theoretical Biology, 255, 8-15.
26. Beale, R.C.L., Petersen-Mahrt, S.K., Watt, I.N., Harris, R.S., Rada, C. and Neuberger, M.S. (2004) Comparison of the Differential Context-Dependence of DNA Deamination by APOBEC Enzymes: Correlation with Mutation Spectra in Vivo. Journal of Molecular Biology, 337, 585-596.
27. Bishop, K.N., Holmes, R.K., Sheehy, A.M., Davidson, N.O., Cho, S.J. and Malim, M.H. (2004) Cytidine Deamination of Retroviral DNA by Diverse APOBEC Proteins. Current Biology, 14, 1392-1396.
28. Henry, M., Guetard, D., Suspene, R., Rusniok, C., Wain-Hobson, S. and Vartanian, J.P. (2009) Genetic Editing of HBV DNA by Monodomain Human APOBEC3 Cytidine Deaminases and the Recombinant Nature of APOBEC3G. PLoS ONE, 4, e4277.
29. Thielen, B.K., McNevin, J.P., McElrath, M.J., Hunt, B.V., Klein, K.C. and Lingappa, J.R. (2010) Innate Immune Signaling Induces High Levels of TC-Specific Deaminase Activity in Primary Monocyte-Derived Cells through Expression of APOBEC3A Isoforms. The Journal of Biological Chemistry, 285, 27753-27766.
30. Roberts, S.A., Sterling, J., Thompson, C., Harris, S., Mav, D., Shah, R., et al. (2012) Clustered Mutations in Yeast and in Human Cancers Can Arise from Damaged Long Single-Strand DNA Regions. Molecular Cell, 46, 424-435.
31. Nakagawa, K., Taya, Y., Tamai, K. and Yamaizumi, M. (1999) Requirement of ATM in Phosphorylation of the Human p53 Protein at Serine 15 Following DNA Double-Strand Breaks. Molecular and Cellular Biology, 19, 2828-2834.
32. Lavin, M.F. (2007) ATM and the Mre11 Complex Combine to Recognize and Signal DNA Double-Strand Breaks. Oncogene, 26, 7749-7758.
33. Di Domenico, E.G., Romano E., Del Porto, P. and Ascenzioni, F. (2014) Multifunctional Role of ATM/Tel1 Kinase in Genome Stability: From the DNA Damage Response to Telomere Maintenance. BioMed Research International, 2014, Article ID: 787404.
34. Altschul, S.F., Madden, T.L., Schaffer, A.A., Zhang, J., Zhang, Z., Miller, W. and Lipman, D.J. (1997) Gapped BLAST and PSI-BLAST: A New Generation of Protein Database Search Programs. Nucleic Acids Research, 25, 3389-3402.

35. Marchler-Bauer, A., Lu, S., Anderson, J.B., Chitsaz, F., Derbyshire, M.K., DeWeese-Scott, C., et al. (2011) CDD: A Conserved Domain Database for the Functional Annotation of Proteins. Nucleic Acids Research, 39, D225-D229.
36. Thompson, J.D., Higgins, D.G. and Gibson, T.J. (1994) CLUSTAL W: Improving the Sensitivity of Progressive Multiple Sequence Alignment through Sequence Weighting, Position-Specific Gap Penalties and Weight Matrix Choice. Nucleic Acids Research, 22, 4673-4680.
37. Jaroszewski, L., Rychlewski, L., Li, Z., Li, W. and Godzik, A. (2005) FFAS03: A Server for Profile-Profile Sequence Alignments. Nucleic Acids Research, 33, W284-W288.
38. Jaroszewski, L., Li, Z., Cai, X.H., Weber, C. and Godzik, A. (2011) FFAS Server: Novel Features and Applications. Nucleic Acids Research, 39, W38-W44.
39. Rice, P., Longden, I. and Bleasby, A. (2000) EMBOSS: The European Molecular Biology Open Software Suite. Trends in Genetics, 16, 276-277.
40. Diella, F., Cameron, S., Gemünd, C., Linding, R., Via, A., Kuster, B., Sicheritz-Pontén, T., Blom, N. and Gibson, T.J. (2004) Phospho.ELM: A Database of Experimentally Verified Phosphorylation Sites in Eukaryotic Proteins. BMC Bioinformatics, 5, 79.
41. Dinkel, H., Chica, C., Via, A., Gould, C.M., Jensen, L.J., Gibson, T.J. and Diella, F. (2011) Phospho.ELM: A Database of Phosphorylation Sites-Update 2011. Nucleic Acids Research, 39, D261-D267.
42. Gnad, F., Gunawardena, J. and Mann, M. (2011) PHOSIDA 2011: The Posttranslational Modification Database. Nucleic Acids Research, 39, D253-D260.
43. Zvárová, J. (2001) Biomedical Statistics. I. The Fundamentals of Statistics for Biomedical Fields. Karolinum, Prague.
44. Mefford, J. (2012) Bayesian Statistics and Bias Analysis.
45. Lowry, R. (2001) 2x2 Contingency Table.
46. Dembo, A., Karlin, S. and Zeitouni, O. (1994) Limit Distribution of Maximal Non-Aligned Two-Sequence Segmental Score. The Annals of Probability, 22, 2022-2039.
47. Karlin, S. and Altschul, S.F. (1990) Methods for Assessing the Statistical Significance of Molecular Sequence Features by Using General Scoring Schemes. Proceedings of National Academy of Sciences of the United States of America, 87, 2264-2268.
48. Potter, M., Padlan, E. and Rudikoff, S. (1976) Localized Deletion-Insertion Mutations: A Major Factor in the Evolution of Immunoglobulin Structural Variability. The Journal of Immunology, 117, 626-629.
49. Kabat, E.A., Wu, T.T., Perry, H.M., Gottesman, K.S. and Foeller, C. (1991) Sequences of Proteins of Immunological Interest. NIH publication No. 91-3242, Bethesda.
50. Schmid, F.X. (1993) Prolyl Isomerase: Enzymatic Catalysis of Slow Protein-Folding Reactions. Annual Review of Biophysics and Biomolecular Structure, 22, 123-142.
51. Snorek, M. (2002) Neural Networks and Neurocomputers. CVUT Publishing, Prague.
52. Waegeman, W., De Baets, B. and Boullart, L. (2008) Learning Layered Ranking Functions with Structured Support Vector Machines. Neural Networks, 21, 1511-1523.
53. Gao, S., Zhang, N., Duan, G.Y., Yang, Z., Ruan, J.S. and Zhang, T. (2009) Prediction of Function Changes Associated with Single-Point Protein Mutations Using Support Vector Machines (SVMs). Human Mutation, 30, 1161-1166.
54. Rangwala, H., Kauffman, C. and Karypis, G. (2009) SvmPRAT: SVM-Based Protein Residue Annotation Toolkit. BMC Bioinformatics, 10, 439.
55. Ansari, H.R. and Raghava, G.P. (2010) Identification of Conformational B-Cell Epitopes in an Antigen from Its Primary Sequence. Immunome Research, 6, 6.
56. Oprea, M., Cowell, L.G. and Kepler, T.B. (2001) The Targeting of Somatic Hypermutation Closely Resembles That of Meiotic Mutation. The Journal of Immunology, 166, 892-899.
57. Yoder, J.A., Turner, P.M., Wright, P.D., Wittamer, V., Bertrand, J.Y., Traver, D. and Litman, G.W. (2010) Developmental and Tissue-Specific Expression of NITRs. Immuno-genetics, 62, 117-122.

58. Wilson, P., Liu, Y.J., Banchereau, J., Capra, J.D. and Pascual, V. (1998) Amino Acid Insertions and Deletions Contribute to Diversify the Human Ig Repertoire. Immunological Reviews, 162, 143-151.
59. Papavasiliou, F.N. and Schatz, D.G. (2000) Cell-Cycle-Regulated DNA Double-Stranded Breaks in Somatic Hypermutation of Immunoglobulin Genes. Nature, 408, 216-221.
60. Shen, H.M. (2007) Activation-Induced Cytidine Deaminase Acts on Double-Strand Breaks in Vitro. Molecular Immunology, 44, 974-983.
61. Holliday, R. (1995) Understanding Ageing. Cambridge University Press, New York.
62. Crawford, T.O., Skolasky, R.L., Fernandez, R., Rosquist, K.J. and Lederman, H.M. (2006) Survival Probability in Ataxia Telangiectasia. Archives of Disease in Childhood, 91, 610-611.
63. Boohaker, R.J. and Xu, B. (2014) The Versatile Functions of ATM Kinase. Biomedical Journal, 37, 3-9.
64. Ohlin, M. and Borrebaeck, C.A.K. (1998) Insertions and Deletions in Hypervariable Loops of Antibody Heavy Chains Contribute to Molecular Diversity. Molecular Immunology, 35, 233-238.
65. Jeng, Y.M., Wu, M.Z., Mao, T.L., Chang, M.H. and Hsu, H.C. (2000) Somatic Mutations of Beta-Catenin Play a Crucial Role in the Tumorigenesis of Sporadic Hepatoblastoma. Cancer Letters, 152, 45-51.
66. Radivojac, P., Baenziger, P.H., Kann, M.G., Mort, M.E., Hahn, M.W. and Mooney, S.D. (2008) Gain and Loss of Phosphorylation Sites in Human Cancer. Bioinformatics, 24, i241-i247.
67. Hu, Z., Wan, X., Hao, R., Zhang, H., Li, L., Li, L., et al. (2015) Phosphorylation of Mutationally Introduced Tyrosine in the Activation Loop of HER2 Confers Gain-of-Function Activity. PLoS ONE, 10, e0123623.
68. Ndegwa, N., C&ocircté, R.G., Ovelleiro, D., D'Eustachio, P., Hermjakob, H., Vizcaíno, J.A. and Croft, D. (2011) Critical Amino Acid Residues in Proteins: A BioMart Integration of Reactome Protein Annotations with PRIDE Mass Spectrometry Data and COSMIC Somatic Mutations. Database, 2011, bar047.
69. Reimand, J., Wagih, O. and Bader, G.D. (2013) The Mutational Landscape of Phosphorylation Signaling in Cancer. Scientific Reports, 3 2651.
70. Soucek, P., Gut, I., Trneny, M., Skovlund, E., Kristensen, T., B&oslashrresen-Dale, A.-L. and Kristensen, V.N. (2003) Multiplex Single-Tube Screening for Mutations in the Nijmegen Breakage Syndrome (NBS1) Gene in Hodgkin's and Non-Hodgkin's Lymphoma Patients of Slavic Origin. European Journal of Human Genetics, 11, 416-419.
71. Soucek, P., Borovanova, T., Pohlreich, P., Kleibl, Z. and Novotny, J. (2007) Role of Single Nucleotide Polymorphisms and Haplotypes in BRCA1 in Breast Cancer: Czech Case-Control Study. Breast Cancer Research and Treatment, 103, 219-224.
72. Mohelnikova-Duchonova, B., Havranek, O., Hlavata, I., Foretova, L., Kleibl, Z., Pohlreich, P. and Soucek, P. (2010) CHEK2 Gene Alterations in the Forkhead-Associated Domain, 1100delC and del5395 Do Not Modify the Risk of Sporadic Pancreatic Cancer. Cancer Epidemiology, 34, 656-658.
73. Smith, G.P. (1985) Filamentous Fusion Phage: Novel Expression Vectors That Display Cloned Antigens on the Virion Surface. Science, 228, 1315-1317.
74. Schmitz, R., Baumann, G. and Gram, H. (1996) Catalytic Specificity of Phosphotyrosine Kinases Blk, Lyn, c-Src and Syk as Assessed by Phage Display. Journal of Molecular Biology, 260, 664-677.
75. Fack, F., Hugle-Dorr, B., Song, D., Queitsch, I., Petersen, G. and Bautz, E.K.F. (1997) Epitope Mapping by Phage Display: Random Versus Gene-Fragment Libraries. Journal of Immunological Methods, 206, 43-52.
76. Reche, P.A. and Reinherz, E.L. (2005) PEPVAC: A Web Server for Multi-Epitope Vaccine Development Based on the Prediction of Supertypic MHC ligands. Nucleic Acids Research, 33, W138-W141.

77. Shehzadi, A., Rehman, S. and Indrees, M. (2011) Promiscuous Prediction and Conservancy Analysis of CTL Binding Epitopes of HCV 3a Viral Proteome from Punjab Pakistan: An in Silico Approach. Virology Journal, 8, 55.
78. Kubrycht, J., Sigler, K. and Soucek, P. (2012) Virtual Interactomics of Proteins from Biochemical Standpoint. Molecular Biology International, 2012, Article ID: 976385.
79. Jing, H., Takagi, J., Liu, J.H., Lindgren, S., Zhang, R.G., Joachimiak, A., Wang, J.H. and Springer, T.A. (2002) Archaeal Surface Layer Proteins Contain Beta Propeller, PKD, and Beta Helix Domains and Are Related to Metazoan Cell Surface Proteins. Structure, 10, 1453-1464.

CHAPTER 11

Application of Graph Entropy in CRISPR and Repeats Detection in DNA Sequences

Dipendra C. Sengupta, Jharna D. Sengupta

Department of ematics & Computer Science, Elizabeth City State University, Elizabeth City, NC, USA

ABSTRACT

We analyzed DNA sequences using a new measure of entropy. The general aim was to analyze DNA sequences and find interesting sections of a genome using a new formulation of Shannon like entropy. We developed this new measure of entropy for any non-trivial graph or, more broadly, for any square matrix whose non-zero elements represent probabilistic weights assigned to connections or transitions between pairs of vertices. The new measure is called the graph entropy and it quantifies the aggregate indeterminacy effected by the variety of unique walks that exist between each pair of vertices. The new tool is shown to be uniquely capable of revealing CRISPR regions in bacterial genomes and to identify Tandem repeats and Direct repeats of genome. We have done experiment on 26 species and found many tandem repeats and direct repeats (CRISPR for bacteria or archaea). There are several existing separate CRISPR or Tandem finder tools but our entropy can find both of these features if present in genome.

Keywords: CRISPR, Graph Entropy, Tandem Repeats, DNA Sequences

1. INTRODUCTION

Deciphering the enormously long nucleotide sequences that are being uncovered in the human genome is one of the major challenges in our days. Along with serious ethical issues, we encounter a series of tremendously hard scientific problems. These problems mainly arise from the fact that although sequencing techniques are almost completely automatic controlled the analysis of the sequenced data is not. Hence, the major goal of the Human Genome Project is the extraction of biologically and medically relevant information from almost automatically sequenced DNA and RNA molecules. In principle, biochemical methods are able to do this job, but since they are extremely expensive and time consuming, there is a high demand for alternative approaches to extract the information hidden in genome [1] . In this situation, concepts and techniques from information theory turned out to be welcoming tools to handle the problem of extracting valuable information from biosequences such as DNA, RNA, or amino acid chains. The main goal of our work is the presentation of a concept and method derived from information theory that will apply to problems of analysis of DNA.

The motivation for this study is to analyze DNA sequences to determine interesting sections of genome that has repeating features using information theory tool.

In many organisms, the genomic DNA is highly repetitive accounting for close to 5% of the genome size [2] [3] . Repetitive DNA sequences are a major component of eukaryotic genomes and may account for up to 90% of the genome size [4] . The human genome itself has over two-thirds of the sequence consisting of repetitive elements [5] . The identification of repeats has proven to be of significance, as they provide insight into the functional and evolutionary roles of various organisms [6] [7] [8] [9] [10] .

In our study we also focus on a family of repeats known as Clustered Regularly Inter Spaced Palindromic Repeats (CRISPRs) [11] . CRISPRs have attracted a great deal of interest recently in genome editing [12] . CRISPRs have been found only in the genomes of prokaryotes and are composed of short direct repeats currently known to range in sizes from 21 - 47 base pairs. This family of repeats is unique in that they are interspaced by non-repeating sequences of similar size, called spacers. CRISPRs were found in approximately 40% of bacterial genome investigated [13] .

Several software applications are available for identifying various form of repeats in [14] [15] [16] .

2. GRAPH ENTROPY ALGORITHM

A graph is an object that consists of a non-empty set of vertices and another set of edges. Vertices are often called nodes, and edges are referred as connections. The set of edges may be empty, in which case the graph is just a collection of points.

We say that two vertices i and j of a directed graph are connected if there is an edge from i to j or from j and i. Suppose we are given a directed graph with n vertices. We construct an n × n adjacency matrix A associated to it as follows: if there is an edge from vertex i to vertex j, we put 1 as the entry on row i, column j of the matrix A; if there is no edge, we put 0.

If one can walk from vertex i to vertex j along the edges of the graph then we say that there is a path from i to j. If we walked on k edges, then the path has length k. For matrices, we denote by A^k the matrix obtained by multiplying A with itself k times. The entry on row i, column j of A^2corresponds to the number of paths of length 2 from vertex i to vertex j in the graph.

Let us consider a directed graph and a positive integer k. Then the number of directed walks from vertex i to vertex j of length k is the entry on row i and column j of the matrix A^k, where A is the adjacency matrix.

In this section, we will discuss entropy of such adjacency matrix A. Let $\{v_1, v_2, \ldots, v_n\}$ be a set of vertices of a direct graph. Let $A = (a_{ij})$ be the $n \times n$ adjacency matrix with at least one positive element.

Let $\sum_{i,j}^{n} a_{ij} = N$. Let $P = (P_{ij})$ be the matrix such that $P = \frac{A}{N}$.

Hence $\sum_{i,j}^{n} p_{ij} = 1$. p_{ij} is the probability of having a path from vertex v_i to vertex v_j. Adding all elements of each row of P and placing them on the diagonal, we form a diagonal matrix

$$\theta = \begin{pmatrix} \sum_{k=1}^{n} p_{1k} & . & . & . & 0 \\ . & . & . & . & . \\ . & . & . & . & . \\ . & . & . & . & . \\ 0 & . & . & . & \sum_{k=1}^{n} p_{nk} \end{pmatrix} = (\theta_{ij})$$

$1-\theta_{jj} = \left(1-\sum_{k=1}^{n} p_{jk}\right)$ is the probability for a randomly generated path to end at the vertex v_j. Let $Q_{ij}(l)$ be the probability for generating a path of length l that begins at v_i and ends at v_j for any integer l. For example, we have $Q_{ij}(1) = p_{ij}\left(1-\sum_{k=1}^{n} p_{jk}\right)$

and $Q_{ij}(2) = (p_{i1}, p_{i2}, \dots, p_{in}) \begin{pmatrix} p_{1j} \\ . \\ . \\ . \\ p_{nj} \end{pmatrix} \left(1-\sum_{k=1}^{n} p_{jk}\right)$. Let Q_l be the matrix whose ij element is $Q_{ij}(l)$. Then we have $Q_l = P^l(I-\theta)$. Finally, we define the asymptotic walk matrix $\Omega = \left(\Omega_{ij}\right)$ as $\Omega \equiv \sum_{l=1}^{\infty} Q_l$, Where Ω_{ij} is the probability for generating a walk of any length from v_i to v_j.

Note that $\sum_{i,j} \Omega_{ij} = 1.$ $\Omega \equiv \sum_{l=1}^{\infty} Q_l = P(I-\theta) + P^2(I-\theta) + \cdots$

We noticed that the sum of all entrees of the matrix $P^{\lambda}(Q-P)$, for any integer λ, is 0. Since sum of all entrees of P is 1, sum of all entrees of Ω is also1. We therefore define the asymptotic entropy

$$\mathrm{H}(P) = -\sum_{i,j} \Omega_{ij} \log\left(\Omega_{ij}\right)$$

where $\Omega_{ij} \log\left(\Omega_{ij}\right)$ is defined to be 0 for $\Omega_{ij} = 0$. This can also be called the graph entropy of the graph or entropy of the adjacency matrix A. For illustration, Let us consider a short sequence:

ATGCCTGATGCGACGC

Taking 2-letter nodes with one overlap, we can create a graph as following:

$$AT \rightarrow TG \rightarrow GC \rightarrow CC \rightarrow CT \rightarrow TG \rightarrow GA \rightarrow$$
$$AT \rightarrow TG \rightarrow GC \rightarrow CG \rightarrow GA \rightarrow AC \rightarrow CG \rightarrow GC$$

We draw a graph as in the Figure 1.

For our sequence, graph entropy

$$\mathrm{H}(P) = -\sum_{i,j} \Omega_{ij} \log(\Omega_{ij}) \approx 2.9614$$

3. RESULTS

We have downloaded wide range of genome data, eukaryotes (animals, plants, insects, fungus) and prokaryotes (bacteria, archaea) from Gen Bank: ftp://ftp.ncbi.nlm.nih.gov/genomes/.

We have implemented the Graph Entropy Algorithm in MATLAB platform and converted data to MATLAB format. Then we have computed graph entropy using our Graph Entropy Algorithm by scanning the data with a typical sample size of 512 base pairs (bp) and step size of 10 bp taking 3 nodes with 1 overlap. We have drawn graphs of entropy versus genome length of Acidovorax bacteria in Figure 2, Salmonella-Typhi CT18 bacteria in Figure 3, Caldicellulosiruptor Kristianssonii bacteria in Figure 4 and Human Chromosome-21 in Figure 5. We have studied the intervals visually where entropy was low and found some repetitive pattern in the sequence. Once we have a string of repetitive pattern we used MATLAB "strfind" command to find out exact positions of the repetitive patterns. We have included few examples in this paper, only the ones we thought important.

In Figure 2 we looked at the lowest drop of entropy which is at: x: genome length = 871,100, y: entropy = 4.088. We took an interval (871,000, 871,600) around the lowest drop x = 871,100.

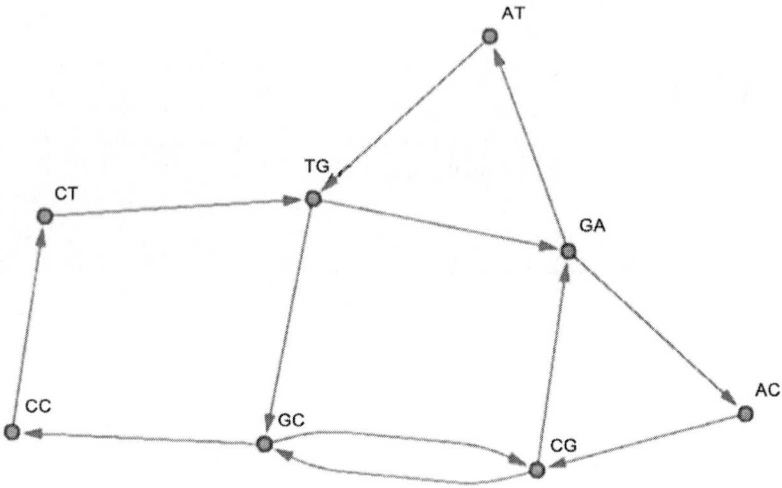

Figure 1. Example of DNA graph.

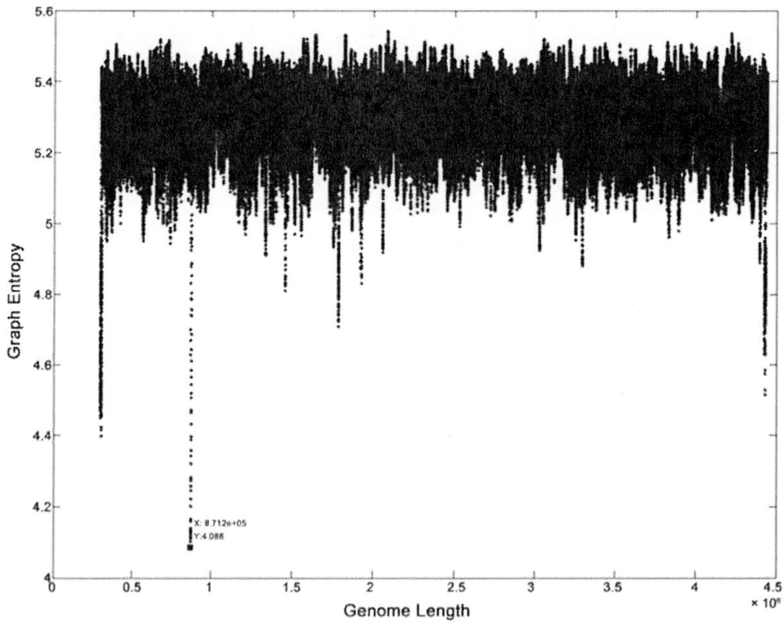

Figure 2. Acidovorax (bacteria) genome length vs entropy.

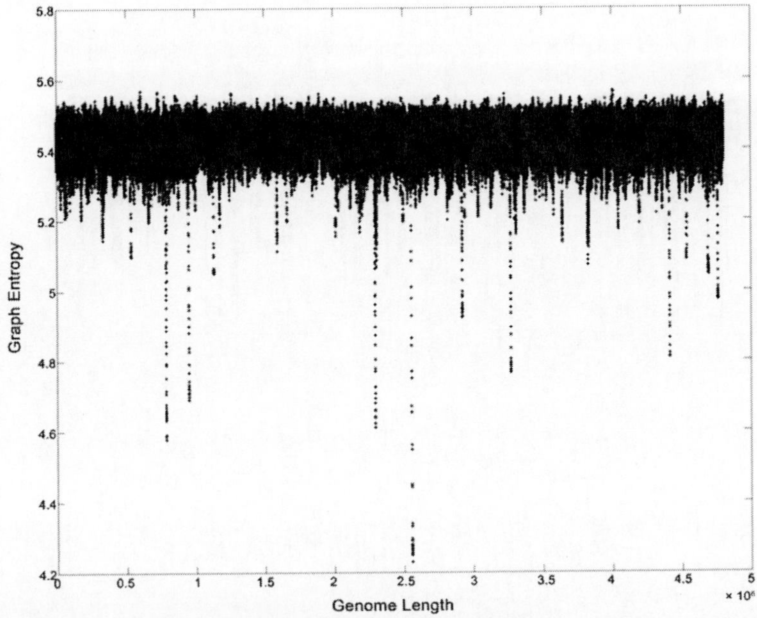

Figure 3. Salmonella-typhi CT18 (bacteria).

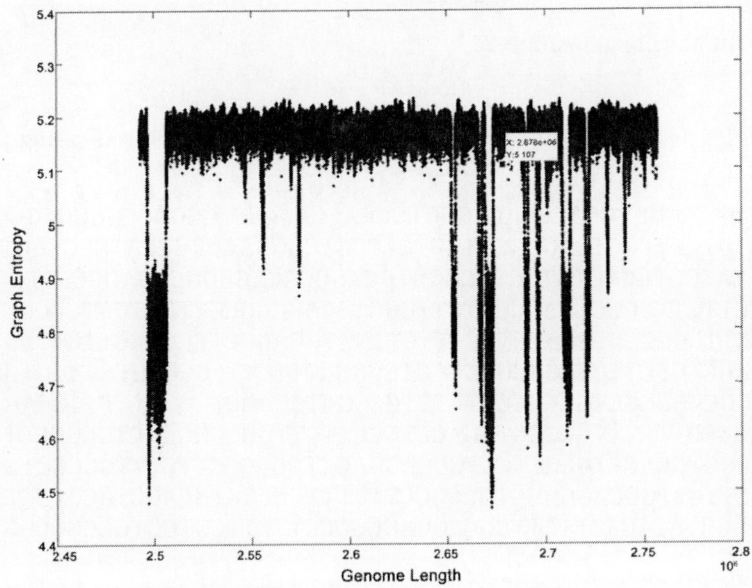

Figure 4. Caldicellulosiruptor kristianssonii (bacteria).

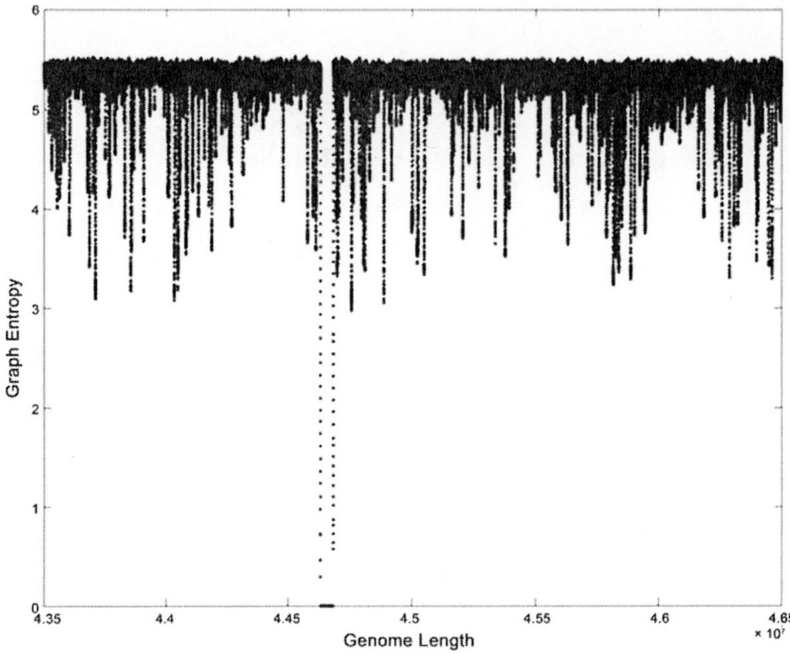

Figure 5. Human chromosome-21.

The following is the sequence in the interval taken. The colored string is repeating.

ATAAAAAAACCCGGTGCATGCACCGGGTGGACCAGCCCCGCGGGCGGGGCGGC
TGGCTGCTGTCGTCGCTCAGGGCTTGGTGCCCGTCGGGAAGGGCCATGCGGCCT
GCGGGTTCAGCGTGGTCTGTGCTGCGGGTGCAGGCGCAGGGGCAGAGGCCTTGG
AGGCCGCCTTTTTCGGGGCAGCCTTCTTCGGTGCAGCGGCCTTGGTCGTGCCGG
TGGCCTTCTTCGCCGGTGCAGCTGCCTTCTTGGTGGAGGCTGCGGCCTTCTTTG
CCGGTGCAGCTGCCTTCTTGGCGGGGGCTGCGGCCTTCTTCGCCGGTGCAGCTG
CCTTCTTGGCGGGGGCTGCGGCCTTCTTTGCCGGTGCAGCTGCCTTCTTGGCAG
GAGCTGCGGCCTTCTTTGCCGGTGCAGCTGCCTTCTTGGCGGGGGCTGCGGCCT
TCTTTGCCGGTGCAGCTGCCTTCTTGGCAGGAGCTGCGGCCTTCTTTGCCGGTG
CAGCTGCCTTCTTGGCAGGAGCTGCGGCCTTCTTTGCCGGTGCAGCTGCCTTCT
TGGCGGGGGCTGCAGCCTTCTTCGCCGGAGCGGCCTTCTTCGTCGTGGCGGCGG
CCTTCTT

Strfind(g,'GCCGGTGCAGCTGCCTTCTTGG') command gave us the following positions of those repeats in the sequence.

871227 871269 871311 871353 871395 871437 871479 871521

The spacers are almost identical. These are tandem repeats.

Similarly in the Figure 3 we looked at the lowest drop of entropy which is at x = 2926000 y = 4.923.

We looked at the DNA sequence in the interval (2926000:2926650) around x = 2926000. The following is the sequence in the interval taken. The colored string is repeating.

AAAAATGCATCCTTCCCGAACGGCAATAGCTGGCACGACGTACGGCTTGATAAT
CAACAGCATATAGACAAGGCGCTGCCAGGGCGGATTGAGCGCCGTAGCCGCGAT
GTAGTGCGGATAATGCTGCCGTTGGTAAAAGAGCTGGCGAAGGCGGAAAAAAC
GTCCTGATATGCTGGTGAAACGTGTTTATCCCCGCTGGCGCGGGGAACACGGAC
AGCAACCCGTGTCGGATATCAGACAGATCGGTTTATCCCCGCTGGCGCGGGGAA
CACACGCGAATCGCCAATCGCCGCCGCGTGAATTGCGGTTTATCCCCGCTGGCG
CGGGGAACACCCACGATGTATGCCGACCGTGATTTTTACCGCCGGTTTATCCCC
GCTGGCGCGGGGAACACAGATACGCCTTTACGTCGCCCTCTTTGGCGCGCGGTT
TATCCCCGCTGGCGCGGGGAACACTAAAACACCGGTTGCGCAACCTCCGCGGGG
ATCGGTTTATCCCCGCTGGCGCGGGGATCGGTTTATCCCCGCTGGCGCGGGGAT
CGGTTTATCCCCGCTGGCGCGGGGAACACTCTAAATCTACCCAATTGAATTTAA
ATACTTTTTTAGCGCACAAAAAACCCACCAACTTTTCCTAATTTTTAAAGATCT
CTAA

We used strfind(g,'CGGTTTATCCCCGCTGGCGCGGGGAACAC') in MatLab and found more repeats outside the interval.

2926243 2926304 2926365 2926426 2926539 2943184 (does not belong to this region). length('CGGTTTATCCCCGCTGGCGCGGGGAACAC')=29 strfind(g,'GTGTTTATCCCCGCTGGCGCGGGGAACAC'): 2926182 strfind(g,'CGGTTTATCCCCGCTGGCGCGGGGATCGG') 2926487 2926513.

Starts: 2926182 Ends: 2926567.

In the interval (2926182, 2926513) we find three strings differing by 2 to 4 letters.

These repeats are called CRISPR. This is only CRISPR so far known for this strain of the bacteria.

Again, we studied the pattern of the DNA sequence of Caldicellulosiruptor Kristianssonii (Bacteria) in intervals around the points of low entropy and found repetitive patterns. In Figure 4, we considered the drop at x = 2,672,000, y = 4.46. Following is the sequence in the interval (2671900, 2672600) around this drop. The repeats are shown in red color.

TATTGCAATTATTGTCCTATGCACAGAGTTTGTAGCCTTCCCGTTGGGGATTGA
AACATAGATTTCATTTCGCAGCCAATAGAGCGGTTTATAGTTTGTAGCCTTCCC
GTTGGGGATTGAAACCTCAATTTCTGTTTCTCTTTTCTCAATTATTCTTGAGTT
TGTAGCCTTCCCGTTGGGGATTGAAACTATAATAGCCCATTCATCAAAAACTTT
TTCATCGAAGTTTGTAGCCTTCCCGTTGGGGATTGAAACTATAATAGCCCATTC
ATCAAAAACTTTTTCATCGAAGTTTGTAGCCTTCCCGTTGGGGATTGAAACCAC

AAAATTATAGTTTGGCGCAATGTAAACACGAACAGTTTGTAGCCTTCCCGTTG
GGGATTGAAACTCTATGTCTTCTTCAAGATACATATCGAGCAGCTTATTGTTTG
TAGCCTTCCCGTTGGGGATTGAAACATACTTTTTTTCTCACGGTCTGTATGGCC
TGTTCAGT

We notice repeats and use Matlab to find the exact locations of that string.

strfind(g,'GTTTGTAGCCTTCCCGTTGGGGATTGAAAC')

Columns 1 through 61

2666352	2666419	2666484	2666551	2666616	2666682	2666748
2666813	2666879	2666947	2667014	2667081	2667147	2667213
2667278	2667344	2667410	2667476	2667544	2667611	2667676
2667741	2667805	2667872	2667939	2670817	2670882	2670949
2671016	2671081	2671147	2671214	2671279	2671345	2671411
2671476	2671544	2671610	2671675	2671740	2671805	2671871
2671936	2672001	2672070	2672135	2672201	2672267	2672333
2672399	2672466	2672534	2672599	2672666	2672733	2672798

2672864 2672930 2672996 2673523 2673590

Length ('GTTTGTAGCCTTCCCGTTGGGGATTGAAAC')=30

61 repeats of length of 30, unique spacers

Starts: 2666352 Ends: 2673620 Period: 65/66/67 Total Length = 7268

These repeats are CRISPR.

In Figure 5, we considered the drop at x = 44010000, y = 4.13 and the interval (44009900, 44010500). Following is the sequence of Human Chromosome-21in that interval. We also found repeats.

CCGTTTATATCCACGCAGGCGTTTCCCCTTACCTGCACCGAGCCTCCATTCCCGT
TTATATCCACGCAGGCGTTTCCCCTTACCTGCACCGAGCCTCCCGCCCCGTTTAC
ATCCACGCAGGCGTTTCCCCTTACCTGCACCGAGCCTCCATTCCCGTTTATATCC
ACGCAGGCGTTTCCCCTTACCTGCACCGAGCCTCCCGCCCCGTTTACATCCACGC
AGGCGTTTCCCCTTACCTGCACCGAGCCTCCATTCCCGTTTATATCCACGCAGGC
GTTTCCCCTTACCTGCACCGAGCCTCCCGCCCCGTTTACATCCACGCAGGCGTTT
CCCCTTACCTGCACCGAGCCTCCATTCCCGTTTATATCCACGCAGGCGTTTCCCC
TTACCTGCACCGAGCCTCCATTCCCGTTTATATCCACGCAGGCGTTTCCCCTTAC
CTGCACCGAGCCTCCATTCCCGTTTATATCCACGCAGGCGTTTCCCCTTACCTGC
ACCGAGCCTCCCGCCCCGTTTATATCCACGCAGGCGTTTCCCCTTACCTGCACCG
GGCCTGCCGCCCCGTTTACATCCACGCATGCGTTTCCCCTTACCTGCACTG

strfind(g,'TTTCCCCTTACCTGCACCGAGCCTCCATTCCCGTTTATATCCACGCAG
GCG')

Columns 1 through 18

44007626 44008952 44009105 44009156 44009258 44009360 44009462 44009513 44009615 44009717 44009819 44009870 44009921 44010023 44010125 44010227 44010278 44010329

The spacers are almost identical with this string except 4 letters (in purple). We also find the spacer string.

Strfind(g,"TTTCCCCTTACCTGCACCGAGCCTCCCGCCCCGTTTACATCCAC GCAGGCG").

Columns 1 through 23

44007575 44007677 44007728 44007779 44007830 44007881 44007983

44008034 44008289 44008493 44008646 44008697 44008799 44008850

44008901 44009207 44009309 44009564 44009666 44009768 44009972

44010074 44010176

This is a repeat of a string without any gap in the region (44007575, 44010329).

DISCUSSION

The importance of identifying repetitive sequences is clear; however, the considerable size of many genomes makes fast and efficient repeat detection very challenging. In this paper, we have presented a new algorithm for finding repeats in DNA sequences. The algorithm is based on our new measure of entropy for any non-trivial graph. In [15] , an algorithm were presented for finding tandem repeats in DNA sequences based on the detection of k-tuple matches. It uses a probabilistic model of tandem repeats and a collection of statistical criteria based on that model. Whereas in [14] and [16] a new tool was introduced for the automatic detection of CRISPR elements in genome. The main advantage of our tool is it will detect both tandem repeats and CRISPR or any other repeats. The main disadvantage of our tool is lack of complete automation and hence it is less efficient compared to the other tools. Our detection technique convert sequences to an alternative representation (namely, graph as it is given in [17]) in an attempt to make analysis more efficient. Future research plans are to modify the presented algorithm so that it is also able to identify repeats efficiently. Our code will be available to the reader upon request through email to one of the authors.

4. CONCLUSIONS

We have studied the following species:

Eukaryotes: Homo sapiens chromosome 19 & 21, Anopheles gambiae, Caenorhabditis elegans, Plasmodium falciparum Saccharomyces cerevisiae.

Prokaryotes: Acidovorax, Ammonifex, Caldicellulosiruptor kristjanssonii, E.Coli, Salmonella Typhi, Listeria Monocyto genes, Bacillus clausii KSM, Chlamydia muridarum Nigg, Cyanobacterium aponinum, Gluconacetobacter diazotrophicus, Haemophilus influenzae R2866, Mycobacterium tuberculosis, Mycoplasma genitalium, Neisseria meningitidis, Streptococcus pneumoniae, Thermosipho africanus, Truepera radiovictrix (Bacteria), A. fulgidus (Archaea).

Viruses: HIV, Hepatitis B. After analyzing the DNA sequence at the points of low entropy for all these species, we conclude that low entropy in a genome graph corresponds to high repeatability in the sequence. These repeats can be classified as CRISPR or Tandem Repeats or something else.

ACKNOWLEDGEMENTS

This paper was written while two authors were Summer Faculty Fellow in SPAWARS YSCEN Atlantic, Charleston, SC funded by Office of Naval Research. Authors are thankful to their mentor for his assistance in the work.

CITE THIS PAPER

Sengupta, D.C. and Sengupta, J.D. (2016) Application of Graph Entropy in CRISPR and Repeats Detection in DNA Sequences. Computational Molecular Bioscience, 6, 41-51.

REFERENCES

1. Grosse, I. (2000) Applications of Statistical Physics and Information Theory to the Analysis of DNA Sequences. Ph.D. Dissertation, Boston University, MA.
2. Ussery, D.W., Binnewies, T.T., Gouveia-Oliveira, R., Jarmer, H. and Hallin, P.F. (2004) Genome Update: DNA Repeats in Bacterial Genomes. Microbilogy, 150, 3519-3521.
3. Hofnung, M. and Shapiro, J. (1999) On Bacterial Repeats. Research in Microbiology (Special November-December Double Issue on Bacterial Repeats), 150, 577-578.

4. Mehrotra, S. and Goyal, V. (2014) Repetitive Sequences in Plant Nuclear DNA: Types, Distribution, Evolution and Function. Genomics Protemics Bioinformatics, 12, 164-171.
5. de Koning, A.P.J., Gu, W., Castoe, T.A., Batzer, M.A. and Pollock, D.D. (2011) Repetitive Elements May Comprise Over Two-Thirds of the Human Genome. PLoS Genet, 7, Article ID: e1002384.
6. Achaz, G., Coissac, E., Netter, P. and Rocha, E.P. (2003) Associations between Inverted Repeats and the Structural Evolution of Bacterial Genomes. Genetics, 164, 1279-1289.
7. Rocha, E.P.C., Danchin, A. and Viari, A. (1999) Functional and Evolutionary Roles of Long Repeats in Prokaryotes. Research in Microbiology, 150, 725-733.
8. Shapiro, J.A. and von Sternberg, R. (2005) Why Repetitive DNA Is Essential to Genome Function. Biological Review, 80, 227-250.
9. van Belkum, A. (1999) Short Sequence Repeats in Microbial Pathogenesis and Evolution. Cellular and Molecular Life Sciences, 56, 729-734.
10. van Belkum, A., Scherer, S., van Alphen, L. and Verbrugh, H. (1998) Short-Sequence DNA Repeats in Prokaryotic Genomes. Microbiology and Molecular Biology Reviews, 62, 275-293.
11. Mojica, F.J., Diez-Villasenor, C., Soria, E. and Juez, G. (2000) Biological Significance of a Family of Regularly Spaced Repeats in the Genomes of Archaea, Bacteria and Mitochondria. Molecular Microbiology, 36, 244-246.
12. Doudna, J.A. and Charpemtier, E. (2014) Genome Editing: The New Frontier of Genome Engineering with CRISPR-Cas 9. Science, 346, 1258096.
13. Jansen, R., Embden, J.D., Gaastra, W. and Schouls, L.M. (2002) Identification of Genes That Are Associated with DNA Repeats in Prokaryotes. Molecular Microbiology, 43, 1565–1575.
14. Bland, C., Ramsey, T.L., Sabree, F., Lowe, M., Brown, K., Kyrpides, N.C. and Hugenholtz, P. (2007) CRISPR Recognition Tool: A Tool for Automatic Detection of Clusetered Regularly Interspaced Palidromic Repeats. BMC Bioinformatics, 8, 209.
15. Benson, G. (1999) Tandem Repeats Finder: A Program to Analyze DNA Sequences. Nucleic Acids Research, 27, 573-580.
16. Grissa, I., Vergnand, G. and Pourcel, C. (2007) CRISPRFinder: A Web Tool to Identify Clustered Regularly Interspaced Short Palindromic Repeats. Nucleic Acids Research, 35, 52-57.
17. Funkhouser, S. (2013) General Formulation for the Entropy of A Graph or Probability Matrix. Unpublished Paper, SPAWAR, SC.

CHAPTER 12

Mean-Field Models for Non-Markovian Epidemics on Networks

Neil Sherborne[1] · Joel C. Miller[2,3,4,5] · Konstantin B. Blyuss[1] · Istvan Z. Kiss[1]

[1] *Department of ematics, School of ematical and Physical Sciences, University of Sussex, Falmer, Brighton BN1 9QH, UK*

[2] *School of ematics, Monash University, Melbourne, VIC, Australia*

[3] *School of Biology, Monash University, Melbourne, VIC, Australia*

[4] *MAXIMA, Monash University, Melbourne, VIC, Australia 5 Institute for Disease Modeling, Bellevue, WA 98005, USA*

ABSTRACT

This paper introduces a novel extension of the edge-based compartmental model to epidemics where the transmission and recovery processes are driven by general independent probability distributions. Edge-based compartmental modelling is just one of many different approaches used to model the spread of an infectious disease on a network; the major result of this paper is the rigorous proof that the edge-based compartmental model and the message passing models are equivalent for general independent transmission and recovery processes. This implies that the new model is exact on the ensemble of configuration model networks of infinite size. For the case of Markovian transmission the message passing model is re-parametrised into a pairwise-like model which is then used to derive many well-known pairwise models for regular networks, or when the infectious period is exponentially distributed or is of a fixed length.

Keywords: Epidemics on networks Non-Markovian transmission and recovery Mean-field models

1 INTRODUCTION

The use of ematical tools to study and understand the spread of infectious diseases is a mature and fruitful area of research. In their 1927 paper Kermack and McKendrick (1927) established the susceptible-infected-recovered (SIR) framework which forms the basis of many models to this day. However, their model assumes that any individual can interact with any other. In reality, in large populations each individual only interacts with a few others, and these connections determine the possible routes of disease transmission. Moreover, studies have found significant heterogeneity in the number of contacts a single individual may have (Pastor-Satorras and Vespignani 2001). The use of graphs or networks to describe these contact patterns represented a major advance in our ability to model more realistic social behaviour. In network-based models individuals are represented by nodes in the network, with edges (or links) encoding the interactions between nodes.

Since the direct analysis of stochastic epidemics on networks is far from trivial, one often relies on deterministic mean-field models that are aimed at approximating some average quantities taken from the stochastic models. Deriving mean-field models can be done in several different ways depending on what one chooses to focus on. For example, considering all nodes and edges in all possible states leads to pairwise models (Keeling 1999; House and Keeling 2011), while considering separately each individual and all possible ways in which it can become infected by its neighbours leads to the message passing (MP) formalism (Karrer and Newman 2010). Focussing on all possible star-like structures, typically defined by a node and all its neighbours, and also taking into account their disease status, yields the so-called effective-degree models (Lindquist et al. 2011). Edge-based compartmental models (EBCM) are based on considering a randomly chosen test node and working out the probability of it staying susceptible, with this probability being then equivalent to the proportion of susceptible nodes in the entire population (Miller et al. 2012). See Danon et al. (2011), Pastor-Satorras et al. (2015) and Kiss et al. (2017) for reviews. All of these models start from the same stochastic model, thus, it is not surprising that some of these models (House and Keeling 2011; Taylor and Kiss 2014; Miller and Kiss 2014; Kiss et al. 2017) are, in fact, equivalent, as we will demonstrate later.

While network models capture contact more accurately, the assumption that the underlying stochastic transmission and recovery processes are memoryless (Keeling and Eames 2005; Volz 2008; House and Keeling 2011) remains restrictive. Of course, memoryless processes are ematically more tractable and relatively simple to analyse when compared to models where the inter-event times are chosen from distributions other than the exponential. However, when compared to data, these assumptions are often violated. For example, diseases can exhibit unique and non-Markovian behaviour in terms of the strength and duration of infection. In this respect, the distribution of the infectious period is usually better approximated by some peaked distribution with a well defined mean, see e.g. Bailey (1954), Gough (1977), Wearing et al. (2005) and references therein.

The MP method does not rely on these assumptions and is able to predict the average behaviour of an epidemic outbreak with general distributions for the transmission times and the duration of infection, although we still require these be independent. Throughout the paper we will denote these distributions as $\tau(a)\tau(\mathrm{a})$ and $q(a)$, where a is the time since the node became infected, known as the age of infection. Once a susceptible node has been exposed to a transmission event, it becomes infected immediately, while the recovery from the disease grants a lifetime immunity. Using these distributions assumes a homogeneous response to disease; whilst this restriction is not always necessary (see e.g. Wilkinson and Sharkey 2014), it is a common simplification in order to obtain a concise model. However, the main focus of this paper is to explore the flexibility of the EBCM in being able to capture epidemics where the infection and recovery processes are described by general independent distributions.

The rest of the paper is organised as follows. In the following section we introduce the MP method (Karrer and Newman 2010) and show how the epidemic model is constructed. We then go on to present the extension of the edge-based compartmental model (Miller et al. 2012) to SIR epidemics with general but independent distributions for time to transmission and duration of the infectious period. This section also contains the main result of the paper, namely, a full rigorous proof that MP model and the EBCM are equivalent, and hence, that the EBCM provides an exact representation of the average stochastic behaviour on the ensemble of infinite Configuration Model (CM) networks (Bender and Canfield 1978; Molloy and Reed 1995, 1998). Section 3 contains a re-parametrisation of the MP model in the special case of Markovian transmission. This proves to be a useful tool in showing how several well-known models can be derived from the MP model or the EBCM when additional assumptions about the network or recovery process are made. In Sect. 4, we compare numerical solutions of the mean-field models to averaged results from explicit stochastic network simulations. The paper concludes with a discussion of main results and possible directions for future research.

2 MODEL SUMMARY

2.1 The message passing (MP) method

In their 2010 paper, Karrer and Newman (2010) introduced the message passing approach to model SIR dynamics on networks. Here, we briefly present the ideas behind their model and its assumptions. Recalling $\tau(a)$τ(a) and $q(a)$ as the densities for transmission and duration of the infectious period one can introduce a new function $f(a)$

$$f(a) = \tau(a) \int_a^{\infty} q(x)\, dx, \tag{1}$$

such that the probability that a node infected at time 0 attempts to transmit the disease to a given neighbour before time t is ∫t0*f*(*a*)*da*∫0tf(a)da, since a neighbour can only transmit the disease if it has not yet recovered. Note that the integration of (1) over all time is the overall probability of an attempted transmission of the disease across a given network edge, commonly known as the *transmissibility* of the disease. This is a quantity which is important in percolation models to determine the epidemic threshold and expected final epidemic size of a major outbreak (Newman 2002; Kenah and Robins 2007).

In order to model the dynamics of disease spread, consider a test node u. This node is placed into a cavity state where it can become infected but is not able to transmit the disease to any of its neighbours. This has no effect on the probability of the node being in any given state (Miller et al. 2012). Now consider a node v which is a neighbour of u; the *message* is the probability that v has *not* attempted to transmit the disease to u by calendar time t, denoted $H^{u \leftarrow v}(t)$. This probability is comprised of two distinct possibilities; the first possibility is that v will make no attempt to transmit the disease before age t, this is given by $1 - \int_0^t f(a)\, da.$ da. This means that even if v is one of the initially infected nodes, it would not attempt to transmit to u. Alternatively, it could be that v will transmit to u at some age $a<t$a<t, but v itself was infected at some time $t_1 > t - a$ and has, therefore, not yet attempted to transmit the disease to its neighbour *u*. This requires *v* to have initially been susceptible (with probability *z*) and to have escaped transmission from each of its neighbours until at least time (*t*−*a*)(t−a). On a tree network with no loops, this is exactly $z \int_0^t f(a) \prod_{w \in \mathcal{N}(v) \backslash u} H^{v \leftarrow w}(t-a)$ da, where N(*v*)N(v) denotes the set of neighbours of *v*. Hence, combining these two gives

$$H^{u \leftarrow v}(t) = 1 - \int_0^t f(a) \left[1 - z \prod_{w \in \mathcal{N}(v) \backslash u} H^{v \leftarrow w}(t-a) \right] da. \quad (2)$$

In principle, one could calculate (2) for all edges (in both directions) to find a full solution for the proportion of the population that is susceptible, infected or removed at any time *t*. For example, the probability that *u* is susceptible is the product of $H^{u \leftarrow w}(t)$ across all neighbours $w \in \mathcal{N}(u)$ multiplied by the probability that it was initially susceptible, *z*. On a single fixed finite tree network, solving (2) for all edges will, in fact, yield the exact solution of the stochastic epidemic (Karrer and Newman 2010). The size of such a system of equations would be twice the number of all edges in the network (since both $H^{u \leftarrow v}(t)$ and $H^{v \leftarrow u}(t)$ would need to be calculated).

Throughout this paper we consider unweighted, bi-directional and static networks constructed according to the configuration model (CM). Every node is assigned a number of neighbours, known as its degree, according to a probability distribution *pk*pk, known as the *degree distribution*, that describes the probability that a randomly chosen node has degree *k*. Let us now focus on an ensemble of CM networks and consider an average message, *H*1H1, instead of considering distinct messages across every edge (Karrer and Newman 2010). For CM networks, as the size of the network tends to infinity, so does the length of the shortest loops, and, therefore, the network becomes locally tree-like. This means that the messages that a node receives from each of its neighbours are independent, and the average message received by the test node *u* is equal to the product of the average message for each neighbour.

The product in (2) is then this *H*1H1 raised to the power of the *excess degree* of the node, its degree excluding the edge which connects it to the test node. The following moment generating functions average this product over the degree distribution

$$G_0(x) := \sum_k p_k x^k, \quad G_1(x) := \frac{1}{\langle k \rangle} \sum_k p_k k x^{k-1},$$
$$G_2(x) := G_1'(x) = \frac{1}{\langle k \rangle} \sum_k p_k k(k-1) x^{k-2}, \tag{3}$$

where $\langle k \rangle = G_0'(1)$ is the mean degree. $G_1(x)$ is the generating function for the *excess degree distribution*, since $k p_k / \langle k \rangle$ describes the probability that a node reached by traversing a randomly selected edge has $(k-1)$ other contacts (Newman 2002); the mean excess degree is given by $G_2'(1)$. The moment generating function $G_2(x)$ will be used to trace the route of infection in later models. Using $G1$G1 to replace the product in (2), the equation for the average message $H_1(t)$ is

$$H_1(t) = 1 - \int_0^t f(a)\,[1 - zG_1(H_1(t-a))]\, da, \tag{4}$$

with $H_1(0) = 1.$

In practice, the trajectory of $H_1(t)$ is found by identifying and then solving a differential or integro-differential equation. For example, the purely Markovian case (i.e. $\tau(a)$ and $q(a)$ are both exponential distributions), with transmission and recovery parameters ββ and γγ, leads to

$$\frac{dH_1}{dt} = \gamma - (\beta + \gamma) H_1(t) + \beta z G_1(H_1(t)),$$

where, again, z is the fraction of the population which was initially susceptible at time t=0t=0(Karrer and Newman 2010). However, the precise form of this equation is not universal, it depends on the particular choice of $\tau(a)$τ(a) and $q(a)$. The proportions of susceptible, infected and recovered individuals at any time t are then given, in terms of the message $H1(t)$H1(t), as

$$\begin{aligned}\langle S\rangle(t) &= zG_0(H_1(t)),\\ \langle R\rangle(t) &= \int_0^t q(a)\,[1-\langle S\rangle(t-a)]\,da, \qquad (5)\\ \langle I\rangle(t) &= 1-\langle S\rangle(t)-\langle R\rangle(t).\end{aligned}$$

The MP model (5) with the average message $H1$H1 is exact when the stochastic epidemic is considered on the ensemble of infinite CM networks (Karrer and Newman 2010). Although this approach is theoretically able to model dynamics for general choices of independent transmission and recovery processes, the need to find a numerically solvable differential equation for $H1$H1 in (4) has restricted the use of MP, and numerical examples are limited, see Karrer and Newman (2010), Wilkinson and Sharkey (2014) for several examples where output from the MP model is compared to results based on simulations.

2.2 EBCM for general transmission and recovery processes

The edge-based compartmental model is an established tool for Markovian dynamics (Miller et al. 2012). We introduce a new extended EBCM which generalises the method to general transmission and recovery processes $\tau(a)$τ(a) and $q(a)$. Again, the EBCM uses the fact that the probability that the test node u remains susceptible is the probability that u escapes transmission from all of its neighbours. This concept is similar to the notion and use of $H1$H1 in MP models. Recovery is modelled using age-structured differential equations. However, the EBCM uses the instantaneous rates of transmission and recovery given by the *hazard functions*rather than the raw densities $\tau(a)$τ(a) and $q(a)$. These are defined as

$$\zeta(a) := \frac{\tau(a)}{\xi_\tau(a)}, \quad \text{and} \quad \rho(a) := \frac{q(a)}{\xi_q(a)}, \qquad (6)$$

where $\xi_\tau(a)$ and $\xi_q(a)$) are the respective *survival functions* (see, e.g., Miller 2011).

$$\xi_\tau(a) = \int_a^\infty \tau(\hat{a})\, d\hat{a} = e^{-\int_0^a \zeta(\hat{a})\, d\hat{a}},$$
$$\xi_q(a) = \int_a^\infty q(\hat{a})\, d\hat{a} = e^{-\int_0^a \rho(\hat{a})\, d\hat{a}}. \tag{7}$$

All of these disease variables and related functions are summarised in Table 1. As before, the contact network is a CM network with degree distribution and generating functions as defined in (3). The basis of the EBCM revolves around finding the probability that a random test node (in a cavity state) is in a susceptible, infected or recovered state at time t. As this test node is chosen at random, these probabilities are equal to the proportions of the population in each state at time t, denoted $S(t)$, $I(t)$ and $R(t)$, respectively.

The first important quantity is $\Theta(t)$, defined in a manner similar to $H1(t)$ in (4) as the probability that the representative test node has not received transmission from a given neighbour by time t. This approach then differs from MP by directly expressing a differential equation for the dynamics of $\Theta\Theta$. The model is known as "edge-based" because it considers the state of the neighbours of the test node; the densities $\Phi_S(t)$, $\Phi_I(t)$ and $\Phi_R(t)$ describe the probability that at time t a random neighbour of the test node is (i) still susceptible, (ii) infected but has not attempted to transmit the disease to the test node, (iii) recovered, and did not transmit to the test node whilst it was infected. The age of infection is, in general, crucial in determining the hazard rates, and so we introduce $i(t, a)$ as the density of infected nodes with the age of infection a. Similarly, $\phi_I(t, a)$ is the density of infected neighbours who have not transmitted to the test node and have age a. Thus, it is clear that $I(t) = \int_0^t i(t, a) da$ and $\Phi_I(t) = \int_0^t \phi_I(t, a)$ da. These variables are summarised in Table 2. We also introduce the Dirac delta function as follows Gel'fand and Shilov (1964)

$$\delta(x) = \begin{cases} +\infty, & x = 0 \\ 0, & x \neq 0 \end{cases}, \qquad \int_{-\infty}^{\infty} \delta(x) dx = 1.$$

Table 1. The variables and functions describing the transmission and recovery processes

Variable	Definition
$\tau(a)$	The density of the transmission process
$q(a)$	The density of the duration of the infectious period
$\xi_\tau(a)$	The *survival function* of the transmission process. The probability that an infected node of age a has not yet attempted to transmit the disease along a given edge: $\int_a^\infty \tau(x)\,dx$
$\xi_q(a)$	The *survival function* of the recovery process. The probability that an infected node reaches at least age a before recovering: $\int_a^\infty q(x)\,dx$
$\zeta(a)$	The *hazard function* of the transmission process. The probability of an edge of age a transmitting in a small interval of time $(a, a+\Delta a)$: $\frac{\tau(a)}{\xi_\tau(a)}$
$\rho(a)$	The *hazard function* of the recovery process. The probability of an infected node of age a recovering in a small interval of time $(a, a+\Delta a)$: $\frac{q(a)}{\xi_q(a)}$
$f(a)$	The probability that, in a small interval, an infectious contact is made by an infected node of age a :$\tau(a)\int_a^\infty q(x)\,dx$
$g(a)$	The probability that, in a small interval, an infectious node of age a recovers, without attempting to transmit the disease to a given neighbour: $q(a)\int_a^\infty \tau(x)\,dx$

Table 2. The list of variables in the EBCM

Variable	Definition
$\Theta(t)$	The probability that the initially susceptible test node has not received a transmission from a random neighbour by time t
$\Phi_S(t)$	The probability that a random neighbour of the test node u is still susceptible
$\Phi_I(t)$	The probability that a random neighbour of the test node u is infected, but has not transmitted to u
$\phi_I(t, a)$	The probability a random neighbour of the test node u to be infected, have not transmitted to u by time t and have age of infection a, $\Phi_I(t) = \int_0^t \phi_I(t, a)\,\mathrm{d}a$
$\Phi_R(t)$	The probability a random neighbour of the test node u has been infected and recovered without transmitting to u
$S(t)$	The density of susceptible nodes
$I(t)$	The density of infected nodes
$i(t, a)$	The density of infected nodes with age of infection a
$R(t)$	The density of recovered nodes
$G_1(x)$	The generating function of the excess degree distribution: $\frac{1}{\langle k\rangle}\sum_{k=0}^\infty p_k k x^{(k-1)}$
$G_2(x)$	The derivative of the generating function of the excess degree distribution: $\frac{1}{\langle k\rangle}\sum_{k=0}^\infty p_k k(k-1) x^{(k-2)}$

Before writing down the new edge-based compartmental model it is worth introducing and explaining the structure of its equations. The message ΘΘ is monotonically decreasing, and it depends on the density and age of infected neighbours, $\phi I(t,a)$ϕI(t,a), and the hazard rate $\zeta(a)$ζ(a). We must consider the possibility of an infected neighbour of any age up to time t transmitting the disease, hence, we have

$$\frac{d\Theta(t)}{dt} = -\int_0^t \zeta(a)\phi_I(t,a)\,da.$$

Given $\Theta(t)$ it follows that $\Phi_S(t) = zG_1(\Theta(t))$. Since the test node is in a cavity state, the probability of a neighbour being susceptible is the probability of it escaping infection from its other contacts averaged over the excess degree distribution.

The rate at which susceptible neighbours are infected is the boundary condition for infected neighbours, i.e.

$$\phi_I(t,0) = -\dot{\Phi}_S(t) = (1-z)\delta(t) + zG_2(\Theta(t))\int_0^t \zeta(a)\phi_I(t,a)\,da,$$

where the first term represents the introduction of the disease at time t=0t=0, and is zero everywhere else. As these neighbours age, they may attempt to transmit the disease, and will eventually recover from it. These events depend on the calendar time and the age of infection, and so we have a von Foerster-type equation

$$\left(\frac{\partial}{\partial t} + \frac{\partial}{\partial a}\right)\phi_I(t,a) = -\left[\zeta(a) + \rho(a)\right]\phi_I(t,a), \qquad 0 < a \le t.$$

The density of nodes in each state depends on ΘΘ and ϕIϕI. By the same logic seen in the MP model, the density of susceptible nodes is $S(t)=zG0(\Theta(t))$S(t)=zG0(Θ(t)). The rate at which susceptible nodes become infected gives the boundary condition of newly infected nodes. Therefore, infected nodes are replenished according to

$$i(t,0) = -\dot{S}(t) = (1-z)\delta(t) + \langle k\rangle zG_1(\Theta(t))\int_0^t \zeta(a)\phi_I(t,a)\,da,$$

where the first term represents the introduction of the disease. As infected nodes age, they recover according to $\rho(a)$ρ(a); these dynamics depend on t and a, and we have a second partial differential equation in the model

$$\left(\frac{\partial}{\partial t} + \frac{\partial}{\partial a}\right) i(t,a) = -\rho(a) i(t,a), \qquad 0 < a \leq t.$$

These equations together form the EBCM for general but independent transmission and recovery processes,

$$\frac{d\Theta(t)}{dt} = -\int_0^t \zeta(a)\phi_I(t,a)\,da, \tag{8a}$$

$$\Phi_S(t) = zG_1(\Theta(t)), \tag{8b}$$

$$\begin{aligned}\phi_I(t,0) &= -\dot{\Phi}_S(t), \\ &= (1-z)\delta(t) + zG_2(\Theta(t))\int_0^t \zeta(a)\phi_I(t,a)\,da,\end{aligned} \tag{8c}$$

$$\left(\frac{\partial}{\partial t} + \frac{\partial}{\partial a}\right)\phi_I(t,a) = -\left[\zeta(a) + \rho(a)\right]\phi_I(t,a), \qquad 0 < a \leq t, \tag{8d}$$

$$\Phi_I(t) = \int_0^t \phi_I(t,a)\,da, \tag{8e}$$

$$\Phi_R(t) = \Theta - \Phi_S - \Phi_I, \tag{8f}$$

$$S(t) = zG_0(\Theta(t)), \tag{8g}$$

$$\begin{aligned}i(t,0) &= -\dot{S}(t), \\ &= (1-z)\delta(t) + \langle k\rangle zG_1(\Theta(t))\int_0^t \zeta(a)\phi_I(t,a)\,da,\end{aligned} \tag{8h}$$

$$\left(\frac{\partial}{\partial t} + \frac{\partial}{\partial a}\right) i(t,a) = -\rho(a) i(t,a), \qquad 0 < a \leq t, \tag{8i}$$

$$I(t) = \int_0^t i(t, a)\, da, \tag{8j}$$

$$R(t) = 1 - S(t) - I(t). \tag{8k}$$

The new edge-based compartmental model (8) offers an alternative way to derive a system of equations that are able to characterise the dynamics of an epidemic outbreak. Although it seems more complex than the MP model, the EBCM is perhaps more intuitive, as many of the variables it involves relate directly to densities of nodes in different states and to the transitions between different states. The EBCM has also proven to be quite versatile and easily extendable to account for different scenarios. For instance, Miller et al. (2012) extended the original EBCM for static networks to dynamic networks where edges are deleted, created or rewired. It may be possible to use similar techniques to extend (8) to model diseases spreading through dynamic networks. To our knowledge, the MP model has so far not been extended beyond static networks, although it may be possible.

As the MP model (5) and the non-Markovian edge-based compartmental model (8) are based on the same underlying stochastic epidemic, it is natural to question how accurate and how similar they are. Karrer and Newman (2010) showed that the MP model (5) is exact on the ensemble of CM networks. Therefore, proving that the EBCM and MP model are equivalent will imply that the EBCM is exact under the same circumstances.

2.3 Model equivalence

We now present and prove the main result of this paper, that the edge-based compartmental and MP model are equivalent for any suitable choices of $\tau(a)$τ(a) and $q(a)$.

Theorem 1

If $\tau(a)$τ(a) and $q(a)$ are independent, integrable probability density functions, then the MP model (5) and the EBCM (8) are equivalent, and will produce identical trajectories for any shared initial conditions.

Proof

The proof consists of showing equivalence for two main elements: the messages for the respective models $H1(t)$ and $\Theta(t)$, and the densities of nodes in each state.

We first show that $H1$H1 and $\Theta\Theta$ satisfy the same evolution equation. To do this, $H1(t)$ (4) is differentiated using Leibniz's rule, which yields

$$\begin{aligned}\frac{dH_1}{dt} &= -f(t)[1 - zG_1(H_1(0))] - \int_0^t f(a)\left[-zG_2(H_1(t-a))\frac{dH_1(t-a)}{dt}\right] da \\ &= -f(t)[1 - zG_1(1)] + \int_0^t f(a)\left[zG_2(H_1(t-a))\frac{dH_1(t-a)}{dt}\right] da \\ &= -f(t)(1 - z) + \int_0^t f(a)\left[zG_2(H_1(t-a))\frac{dH_1(t-a)}{dt}\right] da. \qquad (9)\end{aligned}$$

The dynamics of $\Theta\Theta$ are governed by the following equation

$$\frac{d\Theta}{dt} = -\int_0^t \zeta(a)\phi_I(t,a)\,da.$$

From (8d), one can use the integrating factor exp $\left(\int_0^a [\zeta(\hat{a}) + \rho(\hat{a})]\,d\hat{a}\right)$ to find

$$\phi_I(t,a) = \phi_I(t-a,0)e^{-\int_0^a [\zeta(\hat{a}) + \rho(\hat{a})]\,d\hat{a}} \qquad (10)$$

where

$$\phi_I(t-a,0) = (1-z)\delta(t-a) - zG_2(\Theta(t-a))\frac{d\Theta(t-a)}{dt}.$$

As an alternative, we offer a graphical explanation of (10) in Figure 1.

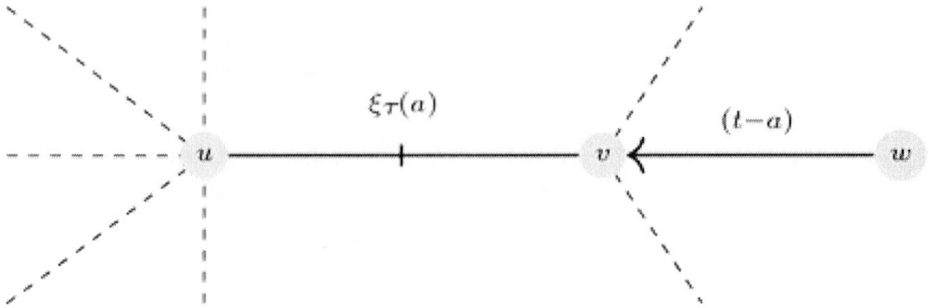

Figure 1. Consider the node labelled u as the test node and thus in a cavity state. For its link with node v to contribute to $\phi_I(t,a)$, it must be the case that v received transmission from some neighbour w at time $(t-a)$. If $t-a=0$, then this is equal to the initial proportion of infected nodes. Otherwise, we take the probability of a transmission event a time ago, which is $\frac{d\Theta(t-a)}{dt}$. For v to have been successfully infected at this time, it must have been susceptible until that point, since two of its neighbours will not have transmitted before this time (u is in a cavity state and w will transmit at $(t-a)$). The probability of this is $zG_2(H_1(t-a))$ for $t>a$, illustrated by the dashed lines. The probability of v not transmitting to u before time t is $\xi_T(a)$ξτ(a). Finally, the neighbour v must still be infected at age a, which is given by the survival function $\xi_q(a)$ξq(a)

Introducing $\hat{f}(a) := \zeta(a)e^{-\int_0^a [\zeta(\hat{a})+\rho(\hat{a})]\,d\hat{a}}$, we have

$$\begin{aligned}\frac{d\Theta}{dt} &= -\int_0^t \hat{f}(a)\left[(1-z)\delta(t-a) - zG_2(\Theta(t-a))\frac{d\Theta(t-a)}{dt}\right]da \\ &= -\hat{f}(t)(1-z) + z\int_0^t \hat{f}(a)G_2(\Theta(t-a))\frac{d\Theta(t-a)}{dt}\,da.\end{aligned} \tag{11}$$

Thus, H_1H1 and ΘΘ have the same dynamics if one can show that $f(a)=\hat{f}(a)$. From the definition of $f(a)$ in (1) and using (7), we obtain

$$f(a) = \zeta(a)\xi_\tau(a)\xi_q(a) = \zeta(a)e^{-\int_0^a [\zeta(\hat{a}) + \rho(\hat{a})]\, d\hat{a}} = \hat{f}(a).$$

Since $H1$H1 and ΘΘ have the same initial condition, this implies that $H1(t)$H1(t) and Θ(t)Θ(t) will exhibit identical dynamics for general but independent transmission and recovery processes. As a direct consequence, this means that the dynamics of susceptibles are the same in both models, since $S(t)$ in (8g) and ⟨S⟩(t)⟨S⟩(t) in (5) differ only in that ΘΘ is replaced by $H1$H1, which are identical and thus interchangeable.

All that remains is to show that the evolution equations for $\langle I\rangle(t)$ in (5) and $\langle I\rangle\langle t\rangle$ in (8j) are identical. From the EBCM we have that

$$I(t) = \int_0^t i(t, a)\, da,$$

which can be differentiated with respect to t

$$\begin{aligned}
\frac{dI}{dt} &= \int_0^t \frac{\partial i(t,a)}{\partial t}\, da + i(t,t) \\
&= -\int_0^t \frac{\partial i(t,a)}{\partial a}\, da - \int_0^t \rho(a) i(t,a)\, da + i(t,t) \\
&= i(t,0) - \int_0^t \rho(a) i(t,a)\, da,
\end{aligned}$$

making use of (8i). The density of infected individuals of age a is the same as the density of infected individuals created at time t–at–a multiplied by the survival probability for the duration of infection, $\xi q(a)$ξq(a), i.e.

$$i(t, a) = \xi_q(a) i(t - a, 0) = \xi_q(a)\dot{S}(t - a).$$

Utilising this substitution and the second identity in (6) gives

$$\frac{dI}{dt} = -\frac{dS}{dt} - \int_0^t q(a)\frac{dS(t-a)}{dt}\,da. \tag{12}$$

From the MP model (5) we can see that

$$\begin{aligned}\frac{d\langle I\rangle}{dt} &= -\frac{d\langle S\rangle}{dt} - \frac{d\langle R\rangle}{dt}\\ &= -\frac{d\langle S\rangle}{dt} - \int_0^t q(a)\frac{d\langle S\rangle(t-a)}{dt}\,da - q(t)(1-z).\end{aligned} \tag{13}$$

This final term describes the rate at which the initially infected individuals recover; this term is implicitly considered in the EBCM in (8h) and so $dI/dt \equiv d\langle I\rangle/dt.$. This completes the proof. □

An implicit analytical relation for the final epidemic size can be found directly from (8). This result corresponds to well-known results based on tools from percolation theory (Newman 2002; Kenah and Robins 2007; Miller 2007), and to the final epidemic size obtained for MP models (Karrer and Newman 2010). It is given in the following corollary.

Corollary 1 The final size of the epidemic $r_\infty = R(\infty)$ *in the EBCM* **(8) with a vanishingly small proportion of infected nodes at time *t*=0t=0 is given by**

$$r_\infty = 1 - G_0(\Theta_\infty),$$

where Θ_∞ solves the equation

$$\Theta_\infty = 1 - \widetilde{T} + \widetilde{T}G_1(\Theta_\infty),$$

and $\widetilde{T} = \int_0^\infty f(a)\,da = \int_0^\infty \zeta(a)\exp(-\int_0^a \left[\zeta(\hat{a}) + \rho(\hat{a})\right]d\hat{a})\,d$ a is the transmissibility of the disease, i.e. the probability that the disease is transmitted along an edge (in isolation) before recovery.

Proof

From (8g), for $z \to 1$ and $t \to \infty$, it immediately follows that

$$r_\infty = 1 - S(\infty) = 1 - G_0(\Theta(\infty)).$$

Furthermore, we have

$$\Theta(\infty) = 1 - \int_0^\infty \int_0^t \zeta(a)\phi_I(t,a)\,da\,dt.$$

Interchanging the order of integration yields

$$\Theta(\infty) = 1 - \int_0^\infty \int_a^\infty \zeta(a)\phi_I(t,a)\,dt\,da.$$

Setting $u=t-a$u=t−a and noting from (10) that $\phi_I(t,a) = \phi_I(u,0)\exp\left(-\int_0^a \left[\zeta(\hat{a}) + \rho(\hat{a})\right] d\hat{a}\right)$ yields

$$\Theta(\infty) = 1 - \int_0^\infty \int_0^\infty \phi_I(u,0)\zeta(a)e^{-\int_0^a[\zeta(\hat{a})+\rho(\hat{a})]\,d\hat{a}}\,du\,da$$

$$\begin{aligned}
&= 1 - \left[\int_0^\infty \phi_I(u,0)du\right]\int_0^\infty \zeta(a)e^{-\int_0^a[\zeta(\hat{a})+\rho(\hat{a})]\,d\hat{a}}\,da \\
&= 1 + \left[\int_0^\infty \dot{\Phi}_S(u)\,du\right]\int_0^\infty \zeta(a)e^{-\int_0^a[\zeta(\hat{a})+\rho(\hat{a})]\,d\hat{a}}\,da \\
&= 1 + (\Phi_S(\infty) - \Phi_S(0))\widetilde{T} \\
&= 1 + \widetilde{T}G_1(\Theta(\infty)) - \widetilde{T}.
\end{aligned}$$

Now that the equivalence between the edge-based compartmental and MP models has been established, we consider the special cases resulting from making extra assumptions about the network (e.g. fully connected and regular) and the infection (e.g. Markovian) and recovery processes (e.g.

Markovian or infectious periods of fixed length). This is motivated by the observation that many earlier epidemic models are based on $\tau(a)$τ(a), $q(a)$ and/or pkpk having the specific forms listed above.

In the following section we aim to produce a model hierarchy by recasting/reducing the EBCM or MP models to earlier models. However, it is not straightforward to see how such earlier models can be derived directly from the EBCM or MP model. This problem can be solved by a re-parametrisation of the MP model in the spirit of pairwise models, and, as a result, one can begin to build a hierarchy of models starting from the most general formulation.

3 MODEL HIERARCHY

Different model families (pairwise, effective degree, MP, EBCM etc.) emerge from different considerations of the same underlying stochastic process. In this section we aim to produce a model hierarchy on CM networks by showing that for specific choices of network topology or recovery process, many well-known models can be derived from the more general MP model. In particular, we will focus on deriving pairwise models (Wilkinson et al. 2017; Volz 2008; Keeling and Grenfell 1997; Kiss et al. 2015). In order to do this, we first present a general re-parametrisation of the MP model (5), and this will act as stepping stone or interpolation between the EBCM/MP and the well-known earlier models. Since all earlier models use a Markovian infection process, the re-parametrisation also uses this assumption.

Pairwise models are based on differential equations for the expected number of nodes in each state. These depend on the number of edges connecting susceptible and infected nodes, and so differential equations are constructed for the expected number of such edges, which themselves depend on the numbers of triples in certain states (e.g. susceptible-susceptible-infected). To break this dependence, a moment closure approximation is commonly used to express the number of triples in terms of pairs and individuals (Keeling 1999).

Recently, Wilkinson and Sharkey (2014) and Wilkinson et al. (2017) have shown that for regular tree networks exact pairwise models can be derived from the MP model when the transmission process is assumed to be Markovian. Here we use similar methods and the notation from Sect. 2.1 to extend this result to the class of CM networks.

Firstly, we define the new variable $\langle SI \rangle(t)$⟨SI⟩(t) as the proportion of edges in the network which connect a susceptible node to an infected one at time t. This can be defined in terms of existing quantities from the MP model. The susceptible node must have been initially susceptible and escaped infection from all other neighbours until time *t*, given by *zG*1(*H*1(*t*))zG1(H1(t)). This must be multiplied by the probability that the remaining neighbour of the susceptible node is infected and has not yet transmitted the disease to this neighbour. To find this probability it is easier to calculate all other possibilities and subtract them from one. These possibilities are: (i) the neighbour is still susceptible, (ii) the neighbour has already transmitted the disease, (iii) the neighbour was infected but has recovered without transmitting the disease. Combining these gives

$$\langle SI\rangle(t) = zG_1(H_1(t))\Big[1 - zG_1(H_1(t)) - \int_0^t \{f(a) + g(a)\}\,[1 - zG_1(H_1(t-a))]\,da\Big], \tag{14a}$$

$$= zG_1(H_1(t))\Big[H_1(t) - zG_1(H_1(t)) - \int_0^t g(a)\,[1 - zG_1(H_1(t-a))]\,da\Big], \tag{14b}$$

where

$$g(a) := q(a)\int_a^\infty \tau(x)\,dx, \tag{15}$$

is the probability of an infected node recovering in the interval (*a*,*a*+Δ*a*)(a,a+Δa) without transmitting to a given neighbour. The corresponding population-level quantity is given by

$$[SI](t) = \langle k\rangle N\langle SI\rangle(t), \tag{16}$$

where *N* denotes the total size of the population.

The transmission process is $\tau(a) = \beta e^{-\beta a}$ which can be substituted into (4). Introducing the change of variable $t' = t - a$, and using the Leibniz rule gives

$$\begin{aligned} \frac{dH_1}{dt} &= -\beta\Bigg[1 - zG_1(H_1(t)) - \int_0^t q(t-t')e^{-\beta(t-t')}\left[1 - zG_1(H_1(t'))\right] da \\ &\quad - \int_0^t \beta e^{-\beta(t-t')}\left(\int_{t-t'}^{\infty} q(x)dx\right)\left[1 - zG_1(H_1(t'))\right] dt'\Bigg] \\ &= -\beta\left[1 - zG_1(H_1(t)) - \int_0^t \{f(a)+g(a)\}\left[1 - zG_1(H_1(t-a))\right] da\right] \\ &= -\beta\frac{\langle SI\rangle(t)}{zG_1(H_1(t))} \end{aligned} \quad (17)$$

$$= -\beta\frac{[SI]}{z\langle k\rangle NG_1(H_1(t))}. \quad (18)$$

For the infected population, using (5) and identities, such as $[S](t) = N\langle S\rangle(t)$, leads to

$$[\dot{I}] = -[\dot{S}] - [\dot{R}] \quad (19)$$

$$= \beta[SI] - \beta\int_0^t q(a)[SI](t-a)da - q(t)N(1-z). \quad (20)$$

The majority of pairwise epidemic models retain an explicit differential equation for the infected population (House and Keeling 2011; Wilkinson et al. 2017). However, we choose to integrate (20) to reduce the number of differential equations that have to be integrated numerically. By noting that $[SI] = \beta\int_0^t [\dot{SI}](t-a)\xi_q(a)\,da$ and $q(a) = -\xi_q'(a)$, '(a), we have

$$[\dot{I}] = \beta\int_0^t \left([\dot{SI}](t-a)\xi_q(a) + \xi_q'(a)[SI](t-a)\right) da - q(t)N(1-z),$$

which is the result of differentiating

$$[I] = \beta \int_0^t [SI](t-a)\xi_q(a)\,da + N(1-z)\xi_q(t). \quad (21)$$

Whilst (20) facilitates easier comparison to existing models, its equivalent representation (21) offers greater computational efficiency.

For the variable ⟨*SI*⟩⟨SI⟩ the calculation is more laborious; working term-by-term from (14b) and using the new relation (18) one obtains

$$\begin{aligned}\langle \dot{SI}\rangle = & -zG_2(H_1)\left(\beta\frac{[SI]}{z\langle k\rangle NG_1(H_1(t))}\right)\left[\cdot\right] \\ & -\beta\frac{[SI]}{\langle k\rangle N} + z\beta[SI]\frac{G_2(H_1(t))}{\langle k\rangle N} - q(t)e^{-\beta t}(1-z)zG_1(H_1(t)) \\ & - z\beta\int_0^t q(a)e^{-\beta a}[SI](t-a)G_2(H_1(t-a))\frac{G_1(H_1(t))}{N\langle k\rangle G_1(H_1(t-a))}\,da,\end{aligned}$$

where [·][·] denotes the large bracket in (14b), and the Leibniz rule has been used again to resolve the integral term. Finally, based on (14), $[\cdot] = \frac{\langle SI\rangle}{zG_1(H_1)}$, which allows us to eliminate [·] and replace it with a term involving ⟨*SI*⟩ Then, multiplying through by ⟨*k*⟩*N*⟨k⟩N one obtains $[SI]^{\cdot}$ in (22a) below.

$$\dot{H}_1 = -\beta\frac{[SI]}{z\langle k\rangle NG_1(H_1)}, \quad (22a)$$

$$\begin{aligned}[\dot{SI}] = & -\beta[SI]^2\frac{G_2(H_1)}{z\langle k\rangle N[G_1(H_1)]^2} - \beta[SI] \\ & + z\beta[SI]G_2(H_1) - q(t)e^{-\beta t}(1-z)zG_1(H_1)\langle k\rangle N \\ & - z\beta\int_0^t q(a)e^{-\beta a}[SI](t-a)G_2(H_1(t-a))\frac{G_1(H_1(t))}{G_1(H_1(t-a))}\,da,\end{aligned} \quad (22b)$$

$$[I] = \beta\int_0^t [SI](t-a)\xi_q(a)\,da + N(1-z)\xi_q(t). \quad (22c)$$

(22c)

At any time *t* the expected number of susceptibles can be found as $[S](t) = zNG_0(H_1(t))$. System (22a) has been derived directly from the MP model, and thus it becomes exact under the same conditions - on the ensemble of CM networks as the network size tends to infinity.

Moreover, retaining the concept of the message, $H1$H1, has meant that system (22a) does not depend on higher order arrangements of nodes (e.g. triples). Therefore, unlike most pairwise models, no further approximations are required to close the model. Similar results have been achieved in the past using heuristic arguments (House and Keeling 2011).

This re-parametrised system (22a) is the first crucial step in being able to move from general to specific models on CM networks, with special focus on unifying various approaches by considering different models from the same perspective.

As one would expect, earlier population-level models were constructed based on some more restrictive assumptions on network and epidemic dynamics. We will show that when these are applied to system/model (22a), earlier models can be easily recovered. The simplifying assumptions refer either to the network [e.g. fully connected or regular (Wilkinson et al. 2017)], or the distribution of the infectious period [e.g. Markovian (Volz 2008), fixed length (Kiss et al. 2015)]. The remainder of this section is devoted to the explicit derivation of the relationships between models as illustrated in Figure 2.

3.1 Degree-regular networks

The first of these reductions concerns the special case of regular networks. For a k-regular (homogeneous) network, all nodes have the same degree, i.e. $k_u = \langle k \rangle = k$, and so the generating functions from (3) simplify to

$$G_0(x) = x^k, \quad G_1(x) = x^{k-1}, \quad \text{and} \quad G_2(x) = (k-1)x^{k-2}.$$

We also introduce two new variables

$$\begin{aligned}
[S](t) &= zNG_0(H_1(t)) = zN[H_1(t)]^k, \\
[SS](t) &= \langle k \rangle N\,(zG_1(H_1(t)))^2 = kN\left(z[H_1(t)]^{k-1}\right)^2,
\end{aligned} \tag{23}$$

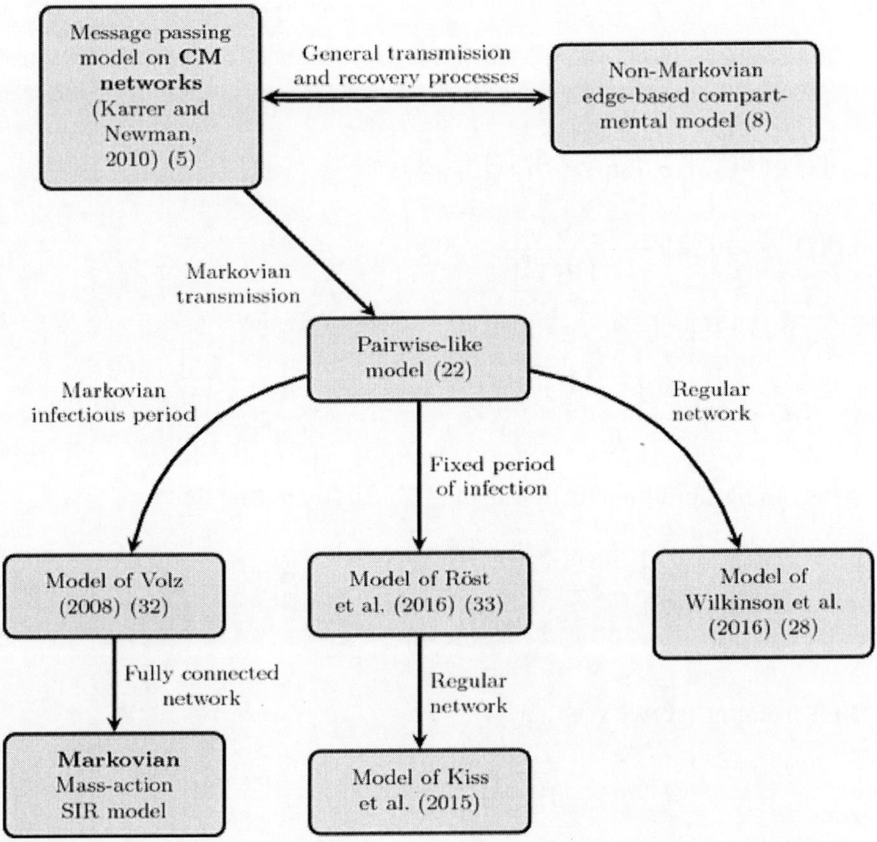

Figure 2. Diagram showing the relationship between the various models discussed in the paper. Labels on each branch state the necessary assumptions in order to reach the model at the end of the *arrow*

that represent the expected number of susceptible individuals and the expected number of edges connecting two susceptible nodes, respectively. $[S](t)$ follows directly from (5), and $[SS](t)$ is defined as the number of edges connecting two nodes who were both initially susceptible at time $t=0$t=0 and have escaped infection from their $(k-1)$(k−1) other neighbours.

Now we return to the system (22a) and the differential equation for $[I]$ (20). Substituting the simpler generating functions yields

$$\dot{H}_1 = -\beta \frac{[SI]}{zkNH_1^{k-1}},$$
$$[\dot{I}] = \beta[SI] - \beta \int_0^t q(a)[SI](t-a)da - q(t)N(1-z),$$
$$[\dot{SI}] = -\beta[SI]^2 \frac{(k-1)H_1^{k-2}}{zkN[H_1^{k-1}]^2} - \beta[SI]$$
$$+ z\beta[SI](k-1)H_1^{k-2} - q(t)e^{-\beta t}(1-z)zH_1^{k-1}kN$$
$$- z\beta \int_0^t q(a)e^{-\beta a}[SI](t-a)(k-1)[H_1(t-a))]^{k-2} \frac{[H_1(t)]^{k-1}}{[H_1(t-a)]^{k-1}} da. \tag{24}$$

This can be simplified further using (23), firstly noting that

$$\frac{[SS]}{[S]} = \frac{Nkz^2H_1^{2(k-1)}}{NzH_1^k} = kzH_1^{k-2}, \tag{25}$$

and, from $H1^{\cdot}$H1˙ we see that

$$\frac{d(H_1^{k-1})}{dt} = -\beta(k-1)\frac{[SI]}{zkNH_1^{k-1}}H_1^{k-2} = -\beta\frac{(k-1)}{k}\frac{[SI]}{[S]}H_1^{k-1}. \tag{26}$$

Solving for H_1^{k-1} using separation of variables gives

$$H_1^{k-1}(t) = \exp\left(-\beta \int_0^t \frac{(k-1)}{k}\frac{[SI](a)}{[S](a)} da\right). \tag{27}$$

The result of this is that the system no longer requires the message $H1$H1, as one can calculate the time derivatives of [S] and [SS] directly from (23). Using the new relations (25) and (27), system (24) can be rewritten to give

$$
\begin{aligned}
[\dot{S}] &= -\beta[SI], \\
[\dot{I}] &= \beta[SI] - \beta\int_0^t q(a)[SI](t-a)da - q(t)N(1-z), \\
[\dot{SS}] &= -2\beta\frac{(k-1)}{k}\frac{[SS][SI]}{[S]}, \\
[\dot{SI}] &= -\beta\frac{(k-1)}{k}\frac{[SI][SI]}{[S]} - \beta[SI] + \beta\frac{(k-1)}{k}\frac{[SS][SI]}{[S]} \\
&\quad - kNq(t)e^{-\beta t}(1-z)z\exp\left(-\beta\int_0^t \frac{(k-1)}{k}\frac{[SI](a)}{[S](a)}da\right) \\
&\quad - \beta\int_0^t q(a)e^{-\beta a}\frac{(k-1)}{k}\frac{[SS](t-a)[SI](t-a)}{[S](t-a)}\mathcal{F}(t)da,
\end{aligned}
\tag{28}
$$

where

$$
\mathcal{F}(t) = \exp\left(-\beta\int_{t-a}^t \frac{(k-1)}{k}\frac{[SI](u)}{[S](u)}du\right). \tag{29}
$$

This is identical to the system proposed by Wilkinson et al. (2017). Recently, Röst et al. (2016) have considered the same problem, namely, an SIR epidemic with Poisson transmission and an arbitrary distribution of the infectious period on a regular network. By constructing an age-structured system of PDEs they were able to reach a very similar, albeit a more compact model. We have, therefore, shown that (22a) extends these recent models by allowing general degree distributions to be modelled.

3.2 Special distributions of the infectious period

As mentioned previously, a popular choice for the duration of infection is to assume an exponential distribution, i.e $q(a) = \gamma e^{-\gamma a}$ for $\gamma > 0$, where $1/\gamma$ is the mean duration of infection. We briefly explain how this assumption simplifies the model and leads to familiar or well-known models. When this choice for $q(a)$ is substituted into (20), we have

$$
[\dot{I}] = \beta[SI] - \gamma\left[\int_0^t e^{-\gamma a}\beta[SI](t-a)da + e^{-\gamma t}N(1-z)\right]. \tag{30}
$$

Note that $e{-}\gamma a$e−γa is the probability of an infected node not recovering before age a, and since the number of infected nodes created a time ago is $\beta[SI](t-a)$ for $a < t$ and $N(1-z)$ for $a = t$, (30) can be rewritten as

$$[\dot{I}] = \beta[SI] - \gamma[I]. \tag{31}$$

A similar result is reached in (22b). In this case the extra terms in the integral describe the probability for the susceptible node of an [*SI*] edge to have survived until age *a* without receiving transmission, either along this edge or from another infected neighbour. Therefore, by the same logic one can replace the final two terms in (22b) with $\gamma[SI]$γ[SI]. This leads to a model, which, although formulated differently, is similar to models of Volz (2008) and House and Keeling (2011), namely,

$$\begin{aligned}
\dot{H}_1 &= -\beta \frac{[SI]}{z\langle k\rangle N G_1(H_1)}, \\
[\dot{I}] &= \beta[SI] - \gamma[I], \\
[\dot{SI}] &= -\beta[SI]^2 \frac{G_2(H_1)}{z\langle k\rangle N[G_1(H_1)]^2} - (\beta+\gamma)[SI] + z\beta[SI]G_2(H_1).
\end{aligned} \tag{32}$$

If one further assumes that the degree is regular, repeating the steps used to derive system (28) recovers the early pairwise model (Keeling and Grenfell 1997).

We examine one final special case, when the duration of infection is a fixed period of time, that σ, so that $q(a) = \delta(a-\sigma)$. This means that as soon as a node is infected at time $t1$t1, it is known that this node will recover at exactly $t_2 = t_1 + \sigma$. Therefore, the integral terms are non-zero only at the point $a{=}\sigma$a=σ. In this case the system of integro-differential equations (22a) simplifies to a delay differential equation model, as stated below

$$
\begin{aligned}
\dot{H}_1 &= -\beta \frac{[SI]}{z\langle k\rangle N G_1(H_1)},\\
[\dot{I}] &= \beta[SI] - \beta[SI](t-\sigma) - \delta(t-\sigma)N(1-z),\\
[\dot{SI}] &= -\beta[SI]^2 \frac{G_2(H_1)}{z\langle k\rangle N[G_1(H_1)]^2} - \beta[SI]\\
&\quad + z\beta[SI]G_2(H_1) - \delta(t-\sigma)e^{-\beta t}(1-z)zG_1(H_1)\langle k\rangle N\\
&\quad - z\beta e^{-\beta\sigma}[SI](t-\sigma)G_2(H_1(t-\sigma))\frac{G_1(H_1(t))}{G_1(H_1(t-\sigma))}.
\end{aligned} \tag{33}
$$

This model generalises the recent work of Kiss et al. (2015) to heterogeneous networks, and once again the original model in that paper can be retrieved when $q(a)$ is chosen to be a delta distribution in (28) (although that original model did not explicitly account for the recovery of initially infected nodes).

Finally, it is worth briefly noting that in the case of a fully connected network, corresponding to a homogeneously well-mixed population we have that [*SI*]=[*S*][*I*][SI]=[S][I], and thus, the Markovian mass-action SIR model, which assumes that the population is unstructured, is recovered. Moreover, with the proper conditions, Wilkinson et al. (2017) proved that the message passing model is equivalent to the mass action model of Kermack and McKendrick (1927) for general transmission and recovery processes.

4 NUMERICAL SIMULATION RESULTS

In order to illustrate the accuracy of (22a), we compare the numerical solution of this model to results of direct stochastic network simulation. A common approach for simulating traditional Markovian models has been to use the Gillespie algorithm (Gillespie 1977). However, as modelling started to move away from the purely Markovian models, novel stochastic simulation methods have been derived (Anderson 2007; Boguná et al. 2014) which provide efficient simulation algorithms that are able to generate true sample paths of the stochastic process.

In this section we take advantage of the fact that in the system (22a) transmission is a Poisson process in order to use an algorithm similar to those described by Barrio et al. (2006). This approach is sometimes known as the rejection method and was proven to be stochastically exact by Anderson (2007). The transmission process is run as in the standard Gillespie algorithm; whenever a node becomes infected, a duration of infection is drawn from the distribution $q(a)$; at each time step the time of next transmission is randomly generated, but if an infected node is scheduled to recover sooner, then the transmission event is rejected, and time is updated to the next recovery time (for full details see Anderson 2007).

In the very early stages of an outbreak stochastic effects dominate the dynamics of the epidemic spread, which means that numerical simulations can often produce results that significantly differ from deterministic predictions. In this situation, methods such as branching process approximations (Heesterbeek 2000) are more appropriate. To ensure that this does not affect our results, we allow every iteration of the algorithm to reach a point where the stochastic effects are no longer a concern, and the infected population behaves deterministically. In practice this is achieved by running each individual realisation of the epidemic from a single initial seed until a specified level of infectivity is reached, at which point time is reset to zero in both the simulation and the mean-field model.

A sufficient number of individual simulations are averaged to ensure that the mean behaviour of the stochastic model is correctly captured and is suitable for comparison with results derived from the deterministic models.

In Figure 3 the results of numerical simulations are shown for three different distributions of the infectious period all having the same mean: a normal distribution, an exponential distribution, and a fixed infectious period. Two important observations can be made. Firstly, the excellent agreement between the average of simulations (markers) and the mean-field model (lines) provides empirical validation of the mean-field model. Secondly, Figure 3 shows marked differences between the epidemics despite the mean of the infectious periods being the same. In particular, the exponential distribution leads to the slowest epidemic growth (and smallest epidemic peak) with the infectious periods of fixed length leading to the fastest growing epidemics (and largest epidemic peak). These results highlight the potential risks of using inaccurate modelling assumptions. The results also suggest that the variance in the duration of the infectious period has a significant effect on the time evolution of the epidemic: a decrease in variance leads to an increase in the initial growth rate (Kiss et al. 2015).

Changes to the degree distribution, transmission and infectious processes will all have an impact on the final epidemic size, as determined by Corollary 1. In the tests shown in Figure 3 only the distribution of the infectious period changes. Whilst this will affect the final epidemic size, it is not possible to tell which of the epidemics is going to produce the largest final epidemic size purely from examining the time evolution of the epidemic in Figure 3. Indeed, it is possible for two epidemics with different time evolutions to have the same final epidemic size. For example, if the degree distribution remains the same, different choices for the distributions of the time to infection and infectious periods can produce identical values for the transmissibility, and the same final epidemic size.

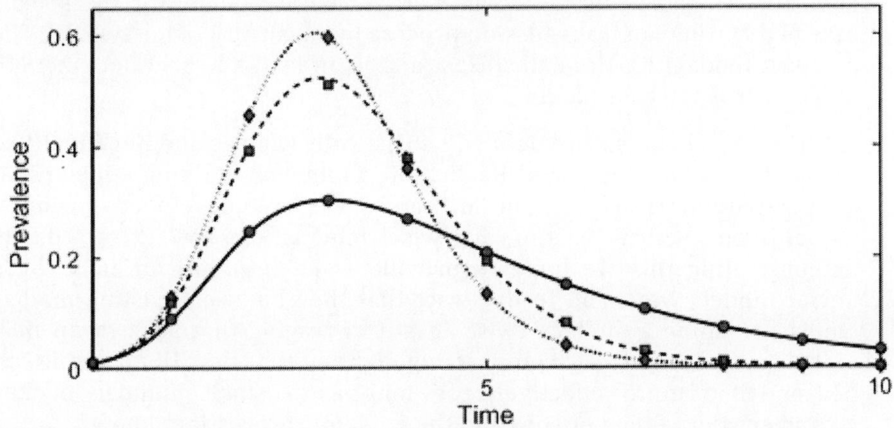

Figure 3 Comparison between system (22a) and the average of numerical simulations. All tests are carried out on randomly generated, truncated scale-free networks of 1000 nodes with exponent 2.5, and the degree bounded between 3 and 60. The transmission rate is set to β=0.3β=0.3 in all cases. Results are plotted where the infectious period is exponentially distributed with parameter γ=0.5γ=0.5 (*solid line, circles*), normally distributed with mean 2 and standard deviation 0.75 (*dashed line, squares*), and a fixed duration σ=2σ=2 (*doted line, diamonds*). The mean duration of the infectious period is equal to 2 for all cases. The differences between the epidemics show that the shape of the distribution of the infectious period can have a significant effect on the dynamics of the epidemic

5 DISCUSSION

In this paper we have reviewed the message passing formalism for SIR epidemics on networks, and introduced a novel extension of the edge-based compartmental model to the case of general but independent transmission and recovery processes. Both of these models are capable of accurately describing the expected dynamics of non-Markovian epidemics on tree networks. The main result of the paper is the complete and rigorous of equivalence between these models, and, as a result, the non-Markovian EBCM (8) is exact on the ensemble of infinite-size CM networks.

Adapting recent methods (Wilkinson and Sharkey 2014; Wilkinson et al. 2017) enabled us to re-parametrise the MP model for the special case of a Markovian transmission process but arbitrary CM networks. The compact model (22a) remains exact and is, in fact, a hybrid between MP and classical pairwise models.

Many pairwise models are defined heuristically (Eames and Keeling 2002; Gross et al. 2006; House and Keeling 2011), but by deriving model (22a) as a re-parametrisation of the MP model we have developed a general model from which existing pairwise models can be extracted. By demonstrating this we hope to provide some intuition for how these newer models work and to illustrate that they build on existing models whilst providing a modern twist. It is encouraging that such mean-field models remain relatively compact, highlighting that the SIR epidemic can be modelled quite effectively, as long as a small number of key characteristics of the network and the epidemic process are known.

The results from our numerical simulations illustrate the dangers of using inaccurate or overly simplistic data to make predictions, in particular, the common assumption of fully Markovian dynamics. The MP and non-Markovian edge-based compartmental models are, therefore, crucial if we are to develop accurate epidemic models on networks.

Numerous extensions of the present work are possible. For example, the implementation of an efficient solver of the novel EBCM is still outstanding. Efficient numerical methods to solve such age-structured models exist, but this was outside the scope of our study. In some sense the novel EBCM is the most complete mean-field model when one considers SIR epidemics on CM networks. This is due to the model being able to handle arbitrary degree distributions, as well as general independent transmission and recovery processes. Additionally, it could be refined to account for dynamic or adaptive contacts. Dynamic networks have already been incorporated in edge-based modelling in the purely Markovian setting (Miller et al. 2012), and it may be possible to extend this to a more general framework to allow for a more unified treatment of models that include the concurrent spread of the disease and link turnover.

ACKNOWLEDGEMENTS

N. Sherborne acknowledges funding for his PhD studies from the EPSRC (Engineering and Physical Sciences Research Council), EP/M506667/1, and the University of Sussex.

REFERENCES

1. Anderson DF (2007) A modified next reaction method for simulating chemical systems with time dependent propensities and delays. J Chem Phys 127(21):214107
2. Bailey NTJ (1954) A statistical method of estimating the periods of incubation and infection of an infectious disease. Nature 174(4420):139–140
3. Barrio M, Burrage K, Leier A, Tian T (2006) Oscillatory regulation of Hes1: discrete stochastic delay modelling and simulation. PLoS Comput Biol 2(9):e117
4. Bender EA, Canfield ER (1978) The asymptotic number of labeled graphs with given degree sequences. J Comb Theory A 24(3):296–307
5. Boguná M, Lafuerza LF, Toral R, Serrano MÁ (2014) Simulating non-Markovian stochastic processes. Phys Rev E 90(4):042108
6. Danon L, Ford AP, House T, Jewell CP, Keeling MJ, Roberts GO, Ross JV, Vernon MC (2011) Networks and the epidemiology of infectious disease. Interdiscip Perspect Infect Dis 2011:284909
7. Eames KT, Keeling MJ (2002) Modeling dynamic and network heterogeneities in the spread of sexually transmitted diseases. Proc Nat Acad Sci 99(20):13330–13335
8. Gel'fand I, Shilov G (1964) Generalized functions, vol 1. Academic Press, New York
9. Gillespie DT (1977) Exact stochastic simulation of coupled chemical reactions. J Phys Chem 81(25):2340–2361
10. Gough K (1977) The estimation of latent and infectious periods. Biometrika 64(3):559–565
11. Gross T, D'Lima CJD, Blasius B (2006) Epidemic dynamics on an adaptive network. Phys Rev Lett 96(20):208701
12. Heesterbeek J (2000) ematical epidemiology of infectious diseases: model building, analysis and interpretation, vol 5. Wiley, New York
13. House T, Keeling MJ (2011) Insights from unifying modern approximations to infections on networks. J R Soc Interface 8(54):67–73
14. Karrer B, Newman ME (2010) Message passing approach for general epidemic models. Phys Rev E 82(1):016101
15. Keeling MJ (1999) The effects of local spatial structure on epidemiological invasions. Proc R Soc B 266(1421):859–867
16. Keeling MJ, Eames KT (2005) Networks and epidemic models. J R Soc Interface 2(4):295–307
17. Keeling MJ, Grenfell B (1997) Disease extinction and community size: modeling the persistence of measles. Science 275(5296):65–67
18. Kenah E, Robins JM (2007) Second look at the spread of epidemics on networks. Phys Rev E 76(3):036113
19. Kermack WO, McKendrick AG (1927) A contribution to the ematical theory of epidemics. Proc R Soc Lond A 115(772):700–721
20. Kiss IZ, Röst G, Vizi Z (2015) Generalization of pairwise models to non-Markovian epidemics on networks. Phys Rev Lett 115:078701
21. Kiss IZ, Miller JC, Simon PL (2017) ematics of epidemics on networks: from exact to approximate models. Springer, New York

22. Lindquist J, Ma J, Van den Driessche P, Willeboordse FH (2011) Effective degree network disease models. J Biol 62(2):143–164
23. Miller JC (2007) Epidemic size and probability in populations with heterogeneous infectivity and susceptibility. Phys Rev E 76:010101
24. Miller RG (2011) Survival analysis, vol 66. Wiley, New York
25. Miller JC, Kiss IZ (2014) Epidemic spread in networks: existing methods and current challenges. Model Nat Phenom 9(2):4–42
26. Miller JC, Slim AC, Volz EM (2012) Edge-based compartmental modelling for infectious disease spread. J R Soc Interface 9(70):890–906
27. Molloy M, Reed B (1995) A critical point for random graphs with a given degree sequence. Random Struct Algorithms 6(2–3):161–180
28. Molloy M, Reed B (1998) The size of the giant component of a random graph with a given degree sequence. Comb Probab Comput 7(03):295–305
29. Newman ME (2002) Spread of epidemic disease on networks. Phys Rev E 66(1):016128
30. Pastor-Satorras R, Vespignani A (2001) Epidemic spreading in scale-free networks. Phys Rev Lett 86(14):3200–3203
31. Pastor-Satorras R, Castellano C, Van Mieghem P, Vespignani A (2015) Epidemic processes in complex networks. Rev Mod Phys 87(3):925
32. Röst G, Vizi Z, Kiss IZ (2016) Pairwise approximation for SIR type network epidemics with non-Markovian recovery.
33. Taylor TJ, Kiss IZ (2014) Interdependency and hierarchy of exact and approximate epidemic models on networks. J Biol 69(1):183–211
34. Volz E (2008) SIR dynamics in random networks with heterogeneous connectivity. J Biol 56(3):293–310
35. Wearing HJ, Rohani P, Keeling MJ (2005) Appropriate models for the management of infectious diseases. PLoS Med 2(7):e174
36. Wilkinson RR, Sharkey KJ (2014) Message passing and moment closure for susceptible-infected-recovered epidemics on finite networks. Phys Rev E 89(2):022808
37. Wilkinson RR, Ball FG, Sharkey KJ (2017) The relationships between message passing, pairwise, Kermack–McKendrick and stochastic SIR epidemic models. J Biol.

CHAPTER 13

A Ematical Model for the Simulation of the Contraction of Burns

Daniël C. Koppenol[1] · Fred J. Vermolen[1] · Gabriela V. Koppenol-Gonzalez[2] · Frank B. Niessen[3] · Paul P. M. van Zuijlen[3,4,5] · Kees Vuik[1]

[1] *Delft Institute of Applied ematics, Delft University of Technology, Delft, The Netherlands*

[2] *Department of Psychology, Education and Child Studies, Erasmus University Rotterdam, Rotterdam, The Netherlands*

[3] *Department of Plastic, Reconstructive and Hand Surgery, MOVE Research Institute, VU University Medical Centre, Amsterdam, The Netherlands*

[4] *Burn Centre, Red Cross Hospital, Beverwijk, The Netherlands*

[5] *Department of Plastic, Reconstructive and Hand Surgery, Red Cross Hospital, Beverwijk, The Netherlands*

ABSTRACT

A continuum hypothesis-based model is developed for the simulation of the contraction of burns in order to gain new insights into which elements of the healing response might have a substantial influence on this process. Tissue is modeled as a neo-Hookean solid. Furthermore, (myo)fibroblasts, collagen molecules, and a generic signaling molecule are selected as model components. An overview of the custom-made numerical algorithm is presented. Subsequently, good agreement is demonstrated with respect to variability in the evolution of the surface area of burns over time between the outcomes of computer simulations and measurements obtained in an experimental study. In the model this variability is caused by varying the values for some of its parameters simultaneously. A factorial design combined with a regression analysis are used to quantify the individual contributions of these parameter value variations to the dispersion in the surface area of healing burns. The analysis shows that almost all variability in the surface area can be explained by variability in the value for the myofibroblast apoptosis rate and, to a lesser extent, the value for the collagen molecule secretion rate. This suggests that most of the variability in the evolution of the surface area of burns over time in the experimental study might be attributed to variability in these two rates. Finally, a probabilistic analysis is used in order to investigate in more detail the effect of variability in the values for the two rates on the healing process. Results of this analysis are presented and discussed.

Keywords Burns Wound contraction Heterogeneous, isotropic, compressible neo-Hookean solid Multiple linear regression analysis Probabilistic analysis Moving-grid finite-element method

1 INTRODUCTION

Burns are a significant global problem; they constitute the second highest incidence of trauma related deaths worldwide and every year nearly half a million citizens of the US require medical treatment as a result of acute thermal injury (Gibran et al. 2013; Goel and Shrivastava 2010). In the US the majority of the injuries are minor and do not require specialized care. However, a small portion of the burn injuries are substantial and as a consequence approximately 3400 persons pass away each year and approximately 40,000 require admission to a hospital or burn center. These numbers have decreased by about 50% over the past two decades due to effective prevention strategies and advances in therapy strategies (Keck et al. 2009). This is an impressive achievement.

Nevertheless, it is in general still very difficult to prevent the development of sequelae such as the formation of permanent shortenings of scar tissue (Hawkins and Finnerty 2012). One of the main causes for this is probably a lack of knowledge about which elements of the wound healing response are the key elements in processes like wound contraction. We think that results obtained with ematical modeling studies can provide new insights into which of these elements are key elements and that these insights might aid in the design of new treatment plans that can reduce the incidence of sequelae.

In this study we develop a ematical model for the simulation of the contraction of burns in order to gain new insights into which elements of the wound healing response might have a substantial influence on the contraction of such wounds. For this end we first show good agreement with respect to the variability in the evolution of the surface area of burns over time between the outcomes of computer simulations obtained in this study and measurements obtained in a previously conducted experimental study. This agreement provides us some confidence about the validity of the model. In the model the variability in the evolution of the surface area of the burns over time is brought about by varying the values for some of its parameters simultaneously. In order to quantify the individual contributions of these parameter value variations to the dispersion in the surface area of healing burns and hence gain insight into which subprocesses have the largest influence on the contraction of such wounds, we subsequently use a factorial design in combination with a multiple linear regression analysis. This analysis shows that almost all variability in the surface area can be explained by variability in the value for the myofibroblast apoptosis rate and, to a lesser extent, the value for the collagen molecule secretion rate. The outcomes of the regression analysis suggest then that most of the variability in the evolution of the surface area of the healing burns over time in the experimental study might be attributed to variability in the apoptosis rate of myofibroblasts and the secretion rate of collagen molecules. We elaborate on this further in the section where we discuss the results obtained with this study.

Before presenting the model, we present in Sect. 2 a general overview of some of the elements involved in the healing of deep burns that cover a substantial surface area. Subsequently, we present the development of the model in Sect. 3. In order to generate the computer simulations a custom-made numerical algorithm had to be developed. An overview of the algorithm is presented in Sect. 4. Thereafter we present in Sect. 5 some details about the factorial design and the regression analysis that have been applied. A probabilistic analysis is also used in this study in order to investigate in more detail the effect of variability in the values for the myofibroblast apoptosis rate and the collagen molecule secretion rate on the healing of burns. More details about this probabilistic analysis are also presented in Sect. 5. Simulation results and the results obtained with the statistical analyses are presented in Sect. 6. Finally, the model and the results are discussed in Sect. 7.

2 BURN WOUND HEALING: AN OVERVIEW

The skin is the largest organ in the human body and is essential for both the maintenance of homeostasis and the protection against potentially damaging environmental factors such as various microorganisms and applied mechanical forces (O'Toole and Mellerio 2010). When the skin is burned, skin tissue is lost rapidly due to tissue necrosis as a consequence of protein denaturation, degradation, and coagulation (Kao and Garner 2000; Keck et al. 2009). The release of local factors such as proteases and toxic oxygen byproducts, tissue edema, and alterations in blood flow may result subsequently in further injury and cell death over time. Taken together these processes cause common local complications such as infections, loss of body heat, and increased water loss.

In addition to the local injury at the site of the burn, severe thermal injury that covers a large portion of the body might also result in acute burn shock (Robins 1990) and a subsequent prolonged period of chronic systemic inflammation, hypermetabolism, and body mass wasting (Farina et al. 2013; Porter et al. 2013). Burn shock is characterized by a collection of acute systemic responses such as fluid and protein movement from the intravascular space into the interstitial space, reduced cardiac output, and hypovolemia. The systemic responses are potentially life-threatening and the delayed responses may have a huge influence on both the quality of the outcome of the wound healing process and the time required to restore the integrity of the skin (Kao and Garner 2000; Keck et al. 2009).

Given the very important functions of the skin and the complications that might arise as a consequence of (thermal) injuries, it is obvious that these injuries need to be repaired quickly and effectively. In order to accomplish this, humans have evolved sophisticated processes for the restoration of the integrity of skin. If the restoration proceeds without complications, then the final outcome of these restoration processes is not normal skin tissue in the case of deep dermal injuries such as severe burns, but whitish, thin, flat scar tissue (Hawkins and Finnerty 2012). This type of tissue is known as normotrophic scar tissue (Verhaegen et al. 2009). Although burns are different from other types of wound due to for example the aforementioned generalized effects on the body, the main processes that lead to the formation of normotrophic scar tissue are the same for all types of wound and are often divided up in three sequential, partially overlapping phases: inflammation, proliferation, and maturation/remodeling (Singer and Clark 1999; Tiwari 2012).

After an initial vasoconstriction of the blood vessels close to the site of injury, this vasoconstriction is reversed quickly and is succeeded by vasodilatation and an increased permeability of the walls of these vessels (Baum and Arpey 2005). Due to these alterations several types of leukocyte such as neutrophils, macrophages, and lymphocytes are able to actively infiltrate the site of injury. This active infiltration marks the beginning of the inflammatory phase (Lawrence 1998; Singer and Clark 1999). The leukocytes clean the wounded area by removing microorganisms and debris through phagocytosis and the release of different types of metalloproteinase (MMP) (Eming et al. 2007). Furthermore, these cells secrete a mixture of signaling molecules (i.e., growth factors and cytokines), which subsequently modulate many of the other processes of the wound healing response (Li et al. 2007).

Soon after the initiation of the inflammatory phase, the proliferative phase of the wound healing cascade starts. The main processes that take place during the proliferative phase are reepithelialization, angiogenesis, fibroplasia, and wound contraction (Baum and Arpey 2005; Singer and Clark 1999). During reepithelialization the presence of an intact epidermis at the site of injury is restored and during angiogenesis a sequence of subprocesses is executed through which the presence of blood vessels in the wounded area is restored. The former process closes the gap while the latter restores a crucial network through which nutrients, cells, oxygen, and other products can be delivered efficiently at the site of injury.

Fibroplasia encompasses the subprocesses that bring about both the restoration of a collagen-rich extracellular matrix (ECM) and the presence of a fibroblast population at the site of injury. Restoration of the presence of a fibroblast population can be accomplished through cell division of fibroblasts present in the injured area, and cell migration of fibroblasts present in the surrounding uninjured tissue into the wounded area. At the site of injury the fibroblasts can differentiate into myofibroblasts (Majno 1979; Tomasek et al. 2002). Taken together this heterogeneous population of (myo)fibroblasts is the main producer of the collagens that constitute the backbone of the newly generated ECM.

Wound contraction is the process that causes the circumferential inward movement of surrounding undamaged skin tissue towards the wounded area (O'Toole and Mellerio 2010). Due to this the surface area of the burn is reduced without the production of new material. Given that wound contraction accounts for up to 40% decrease in the size of the wounded area in the case of deep dermal injuries, wound contraction is an important component of the wound healing response (Li et al. 2007). Myofibroblasts present in the wounded area are the predominant mediators of this contraction process (Tomasek et al. 2002). Due to the presence of actin microfilament bundles within the myofibroblasts, these cells are capable of pulling on the ECM with relatively much force. Given that they are located within the injured area, this pulling at the ECM results in the inward movement of the undamaged skin tissue towards the wounded area.

With the onset of the restoration of a collagen-rich ECM the remodeling of this evolving ECM also starts (Enoch and Leaper 2007). During remodeling the nature of the ECM changes due to a gradual decrease in the cell densities of various cell types, a change in the balances between the production and the breakdown of various constituents of the ECM, and an adjustment in the way that the constituents of the ECM are aligned and interconnected (Monaco and Lawrence 2003). Taken together these alterations result in the formation of normotrophic scar tissue of gradually increasing strength (Hawkins and Finnerty 2012).

3 DEVELOPMENT OF THE EMATICAL MODEL

Given that contraction mainly takes place in the dermal layer of the skin, solely a portion of this layer is incorporated explicitly into the model. The layer is modeled as a continuum. The neighboring dermal structures are incorporated implicitly into the model through mechanical interactions between these structures and the modeled portion of the dermal layer at the individual interfaces.

In order to allow for large strains we model the dermal layer as a heterogeneous, isotropic, compressible neo-Hookean solid (Treloar 1948). With respect to the mechanical component of the model the displacement of the dermal layer (uu) is selected as the primary model variable. In addition, we select the following four constituents of the dermal layer as primary model components: a fibroblast population (*N*), a myofibroblast population (*M*), a generic signaling molecule (*c*), and collagen molecules (ρρ).

We use the general continuum hypothesis-based modeling framework of Tranquillo and Murray (1992) as ematical basis for the model. This framework consists of the following general set of conservation equations:

$$\frac{\partial z_i}{\partial t} + \nabla \cdot [z_i \mathbf{v}] = -\nabla \cdot \mathbf{J}_i + R_i, \tag{1a}$$

$$-\nabla \cdot \sigma = \mathbf{f}. \tag{1b}$$

Equation (1a) is the conservation equation for the cell density/concentration of constituent *i* of the dermal layer and Equation (1b) is the reduced conservation equation for the linear momentum of the dermal layer. Like others, we assume that the inertial forces that work on the dermal layer are negligible (Olsen et al. 1995; Tranquillo and Murray 1992). As a consequence the conservation equation for the linear momentum of the dermal layer reduces to the above force balance equation. Within the above equations, zizi represents the cell density / concentration of constituent *i*, vv represents the displacement velocity of the dermal layer, JiJirepresents the flux associated with constituent *i* per unit area, RiRi represents the chemical kinetics associated with constituent *i*, σσσσ represents the Cauchy stress tensor associated with the dermal layer, and ff represents the total body force working on the dermal layer. Given the chosen primary model variables, we have i∈{N,M,c,ρ}i∈{N,M,c,ρ}. In the remainder of this text, we replace zizi by *i*. Hence, zNzN becomes *N*, zMzM becomes *M*, and so on.

3.1 The cell populations

The random movement of both fibroblasts and myofibroblasts through the dermal layer and the directed movement of both fibroblasts and myofibroblasts up the gradient of the signaling molecule are incorporated into the model. The random movement of the cells is modeled by means of cell density-dependent Fickian diffusion and the directed movement of the cells is modeled by means of a simple model for chemotaxis (Hillen and Painter 2009):

$$\mathbf{J}_N = -D_F F \nabla N + \chi_F N \nabla c, \tag{2}$$

$$\mathbf{J}_M = -D_F F \nabla M + \chi_F M \nabla c, \tag{3}$$

where

$$F = N + M. \tag{4}$$

The parameter DFDF is the cell density-dependent (myo)fibroblast random motility coefficient and χFχF is the chemotactic parameter that depends on both the binding rate and the unbinding rate of the signaling molecule with its receptor and the concentration of this receptor on the cell surface of the (myo)fibroblasts. Certain members of the family of platelet derived growth factors (PDGF) are good examples of molecules that act as a strong attracting stimuli for (myo)fibroblasts (Barrientos et al. 2008).

Furthermore, we incorporate into the model the cell division of fibroblasts and myofibroblasts by using two nearly identical adjusted logistic growth models. The difference between the two growth models is that myofibroblasts solely divide in the presence of the signaling molecule while fibroblasts also divide without the presence of the signaling molecule. The actual rate of cell division of both fibroblasts and myofibroblasts is dependent on the concentration of the signaling molecule. Additionally, we incorporate into the model the cell differentiation of fibroblasts into myofibroblasts under the influence of the signaling molecule. Examples of signaling molecules that can stimulate both the up-regulation of the cell division rate and the cell differentiation rate of fibroblasts into myofibroblasts are certain members of the family of transforming growth factors ββ (TGF-ββ) (Werner and Grose 2003). Finally, we incorporate into the model the removal of (myo)fibroblasts from the dermal layer by means of apoptosis. Taken together we obtain

$$R_N = r_F \left[1 + \frac{r_F^{\max} c}{a_c^I + c}\right][1 - \kappa_F F]N^{1+q} - k_F c N - \delta_N N, \quad (5)$$

$$R_M = r_F \left[\frac{[1 + r_F^{\max}] c}{a_c^I + c}\right][1 - \kappa_F F]M^{1+q} + k_F c N - \delta_M M, \quad (6)$$

where rFrF is the cell division rate, rmaxFrFmax is the maximum factor with which the cell division rate can be enhanced due to the presence of the signaling molecule, aIcacI is the concentration of the signaling molecule that causes the half-maximum enhancement of the cell division rate, κFFκFFrepresents the reduction in the cell division rate due to crowding, *q* is a fixed constant, kFkF is the signaling molecule-dependent cell differentiation rate of fibroblasts into myofibroblasts, δNδN is the apoptosis rate of fibroblasts, and δMδM is the apoptosis rate of myofibroblasts.

3.2 The generic signaling molecule

We assume that both fibroblasts and myofibroblasts release and consume the signaling molecules. The functional form for these processes is based on chemical interactions between the signaling molecules and receptors for these molecules on the cell surfaces of the (myo)fibroblasts. The derivation of the functional form (i.e., the first term on the right hand side of Eq. (8)) can be found in the appendix of the article by Olsen et al. (1995). Furthermore, we incorporate into the model the removal of signaling molecules from the dermal layer through proteolytic breakdown. Finally, we assume that the signaling molecules diffuse through the dermal layer according to linear Fickian diffusion. Taken together this results in

$$\mathbf{J}_c = -D_c \nabla c, \tag{7}$$

$$R_c = k_c \left[\frac{c}{a_c^{II} + c} \right] [N + \eta M] - \delta_c g(F, c, \rho) c, \tag{8}$$

where DcDc is the Fickian diffusion coefficient of the generic signaling molecule, kckc is the maximum net secretion rate of the signaling molecule, ηη is the ratio of myofibroblasts to fibroblasts in the maximum net secretion rate of the signaling molecule and the collagen molecules (see the next subsection), aIIcaII is the concentration of the signaling molecule that causes the half-maximum net secretion rate of the signaling molecule, and δcδc is the proteolytic breakdown rate of the signaling molecules.

The last term of RcRc requires some more explanation. We incorporate into the model the proteolytic breakdown of the signaling molecules by a generic MMP (Mast and Schultz 1996; Van Lint and Libert 2007). It is known that MMPs are involved in the breakdown of collagen-rich fibrils during the remodeling of the ECM and the maintenance of the ECM (Chakraborti et al. 2003; Lindner et al. 2012; Nagase et al. 2006). Furthermore, it is known that (myo)fibroblasts are important producers of MMPs (Lindner et al. 2012) and that the production of MMPs can be reduced in the presence of signaling molecules like TGF-ββ (Overall et al. 1991). Therefore, we assume that the concentration of the generic MMP is a function of the concentration of the collagen molecules, the concentration of the signaling molecule, and the cell density of the (myo)fibroblast population. In this study we take the following relationship:

$$g(F, c, \rho) = \frac{F\rho}{1 + a_c^{III} c}, \tag{9}$$

where $1/[1 + a_c^{III} c]$ represents the inhibition of the synthesis of the generic MMP due to the presence of the signaling molecule.

3.3 The collagen molecules

We assume that secreted collagen molecules are attached to the ECM instantaneously. Hence we assume that no active transport of collagen molecules takes place. Furthermore, we incorporate into the model the production of collagen molecules by both fibroblasts and myofibroblasts. The secretion rate of the molecules is enhanced in the presence of the signaling molecule. Examples of signaling molecules that can upregulate the secretion of collagen molecules by (myo)fibroblasts are certain members of the family of transforming growth factors ββ (TGF-ββ) (Werner and Grose 2003). Finally, we incorporate into the model the proteolytic breakdown of the collagen molecules analogously to the breakdown of the signaling molecules. Taken together we obtain

$$\mathbf{J}_\rho = \mathbf{0}, \tag{10}$$

$$R_\rho = k_\rho \left[1 + \left[\frac{k_\rho^{\max} c}{a_c^{IV} + c}\right]\right][N + \eta M] - \delta_\rho g(F, c, \rho)\rho, \tag{11}$$

where kρkρ is the collagen molecule secretion rate, kmaxρkρmax is the maximum factor with which the secretion rate can be enhanced due to the presence of the signaling molecule, aIVcNacNIV is the concentration of the signaling molecule that causes the half-maximum enhancement of the secretion rate, and δρδρ is the degradation rate of the collagen molecules.

3.4 The force balance

We take the following constitutive stress-strain relation:

$$\sigma = 2C_1 \left[\sqrt{\det(\mathbf{B})}\right]^{-\frac{5}{3}} \left[\mathbf{B} - \frac{1}{3}\mathrm{tr}\,(\mathbf{B})\,\mathbf{I}\right] + 2D_1 \left[\sqrt{\det(\mathbf{B})} - 1\right]\mathbf{I}, \tag{12}$$

$$\mathbf{B} = \left[\mathbf{I} - \nabla\mathbf{u} - [\nabla\mathbf{u}]^{\mathrm{T}} + [\nabla\mathbf{u}]^{\mathrm{T}}\,\nabla\mathbf{u}\right]^{-1}, \tag{13}$$

$$C_1 = \frac{E\rho}{4[1 + \nu]}, \tag{14}$$

$$D_1 = \frac{E\rho}{6[1 - 2\nu]}, \tag{15}$$

where BB is the left Cauchy–Green deformation tensor, II is the second order identity tensor, and νν is Poisson's ratio (Treloar 1948). Like Ramtani (2004) and Ramtani et al. (2002), we assume that the Young's modulus of the tissues is dependent on the concentration of the collagen molecules. In this study we assume that this dependence is linear. Hence we take EρEρ for the Young's modulus, where E is a constant.

Furthermore, we incorporate into the model the forces that are generated by the myofibroblasts due to their pulling on the ECM (Hinz 2007). In this study we model the pulling force as an isotropic stress that is proportional to the product of the cell density of the myofibroblast population and a simple function of the concentration of the collagen molecules (Olsen et al. 1995). No other forces are incorporated into the model. Taken together we obtain

$$\mathbf{f} = \nabla \cdot \left[\xi M \left[\frac{\rho}{R_\rho^2 + \rho^2} \right] \mathbf{I} \right], \quad (16)$$

where ξξ is the generated stress per unit cell density and the inverse of the unit collagen molecule concentration, and RρRρ is a fixed constant.

3.5 The domain of computation

We assume that $\partial u/\partial z = \partial v/\partial z = w = 0$ holds within the modeled portion of the dermal layer, with the *xy*-plane running parallel to the surface of the skin and u=(u,v,w)Tu=(u,v,w)T. Effectively this implies that the plane strain assumption holds within the portion of dermal layer (Lai et al. 1999). Furthermore, we assume that the derivatives of the concentrations and the cell densities of the modeled constituents are equal to zero in the direction perpendicular to the surface of the skin. Hence $\partial i/\partial z = 0$ with $i \in \{N, M, c, \rho\}$. Taken together these assumptions imply that the calculations for obtaining simulations can be performed on an arbitrary, infinitely thin slice of dermal layer oriented parallel to the surface of the skin and that the results from these calculations are valid for every infinitely thin slice of dermal layer oriented parallel to the surface of the skin. For the generation of simulation results we used the computational domain depicted in Figure 1. Using Lagrangian coordinates (X=(X,Y,Z)TX=(X,Y,Z)T), the domain of computation (ΩXΩX) is described ematically by

$$\Omega_X \in \{-12 \leq X \leq 12, \ -12 \leq Y \leq 12, \ Z = 0\}. \quad (17)$$

3.6 The initial conditions and the boundary conditions

The initial conditions give a description of the different cell densities and concentrations at the onset of the proliferative phase of the wound healing cascade. For the generation of simulation results the following general indicator function has been used to describe the shape of the burn:

$$w\left(\mathbf{X}, c^{I}, c^{II}\right) = \begin{cases} 0 & \text{if } \|\mathbf{X}\| < \left[c^{I} - c^{II}\right], \\ \frac{1}{2}\left[1 + \sin\left(\frac{\left[\|\mathbf{X}\| - c^{I}\right]\pi}{2c^{II}}\right)\right] & \text{if } \left|\|\mathbf{X}\| - c^{I}\right| \le c^{II}, \\ 1 & \text{if } \|\mathbf{X}\| > \left[c^{I} + c^{II}\right]. \end{cases} \tag{18}$$

Here w=0w=0 corresponds to completely wounded dermis and w=1w=1 corresponds to unwounded dermis. The values for the parameters cIcI and cIIcII determine respectively the radius of the initial burn and the steepness of the boundary of the wounded area (i.e., the minimum distance between completely wounded dermis and unwounded dermis). We have chosen for a circular shape so that the shape of the burns in the computer simulations is equal to the shape of the burns that were created in the experimental study by Wang et al. (2011). Certain measurements obtained in this experimental study were used for comparison against certain outcomes of the obtained computer simulations in this study (see Sect. 6).

Based on the general function for the shape of the wound we take the following initial conditions for the modeled constituents of the dermal layer:

$$\begin{aligned} N(\mathbf{X}, 0) &= \left[I^{w} + \left[1 - I^{w}\right] w\left(\mathbf{X}, c^{I}, c^{II}\right)\right] \overline{N}, \\ M(\mathbf{X}, 0) &= \overline{M}, \\ c(\mathbf{X}, 0) &= \left[1 - w\left(\mathbf{X}, c^{I}, c^{II}\right)\right] c^{w}, \\ \rho(\mathbf{X}, 0) &= \left[I^{w} + \left[1 - I^{w}\right] w\left(\mathbf{X}, c^{I}, c^{II}\right)\right] \overline{\rho}, \end{aligned} \tag{19}$$

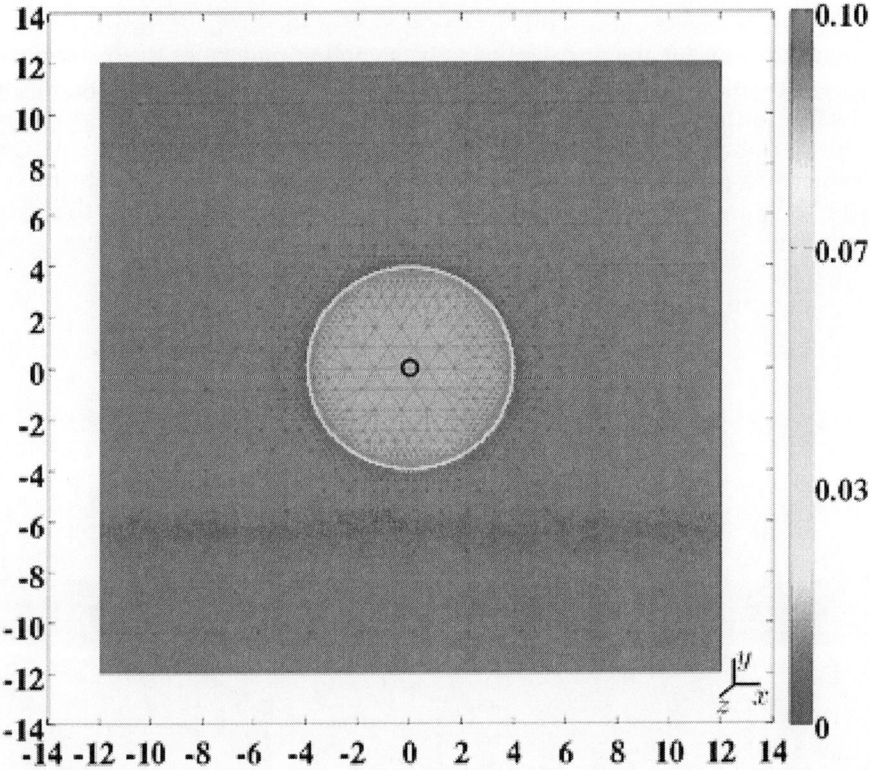

Figure 1. A graphical representation of the domain of computation (the *scale* along both axes is in centimeters). Depicted are the initial shape of the modeled portion of dermal layer and, in color scale, the initial concentration of the collagen molecules (with Iw=10−1Iw=10−1), measured in g/cm3g/cm3. The black circle located more or less at the center of the wound, marks the material point within the dermal layer where the evolution of the concentrations (cell densities) of the different modeled constituents was traced over time for the generation of some of the figures in Sect. 6

where $\overline{N}$, $\overline{M}$, and $\overline{\rho}$ are respectively the equilibrium cell density of the fibroblast population, the equilibrium cell density of the myofibroblast population, and the equilibrium concentration of the collagen molecules, of the unwounded dermis. See Figure 1 for a detailed graphical representation of the initial concentration of the collagen molecules and see the leftmost column of Figure 2 for graphical representations of all initial cell densities and concentrations. Due to early secretion of signaling molecules by for instance leukocytes, signaling molecules are present at the site of injury. The constant cwcw is the maximum of the initial concentration of the signaling molecule in the wounded area. Furthermore, we assume that there are some fibroblasts and collagen molecules present in the wounded area; the parameter IwIw determines the minimum amount of fibroblasts and collagen molecules that are present initially in the wounded area. Finally, we take 3.57≤cI≤3.99 cm3.57≤cI≤3.99 cm and cII=0.10 cmcII=0.10 cm.

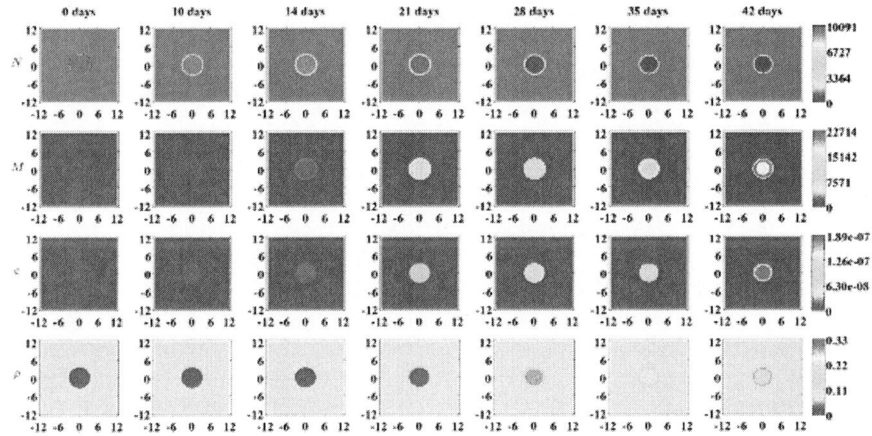

Figure 2. An overview of a simulation with a relatively small burn (cI=3.57 cmcI=3.57 cm), a relatively low minimum initial cell density of fibroblasts and a relatively low minimum initial concentration of collagen molecules (Iw=1×10−1Iw=1×10−1), a relatively low apoptosis rate of myofibroblasts (δM=2×10−2 /dayδM=2×10−2 /day), and an average collagen molecule secretion rate (kρ=1.75×10−8 g/(cells day)kρ=1.75×10−8 g/(cells day)). The *first two rows* show the evolution over time of the cell density of respectively the fibroblast population and the myofibroblast population. The *color scales* represent the cell densities, measured in cells/cm3cells/cm3. The *last two rows*show the evolution over time of the concentration of respectively the signaling molecules and the collagen molecules. The *color scales* represent the concentrations, measured in g/cm3g/cm3. Within the subfigures, the scale along both axes is in centimeters. The values for the remaining parameters are equal to those depicted in Table 1.

With respect to the initial conditions for the displacement of the dermal layer, the following holds. The initial cell density of the myofibroblast population is equal to zero everywhere in the domain of computation. Looking at Eq. (16), this implies $\mathbf{f}(\mathbf{x}, 0) = \mathbf{0}.$ Hence

$$\mathbf{u}(\mathbf{x}, 0) = \mathbf{0} \quad \forall \mathbf{x} \in \Omega_X, \tag{20}$$

where ΩxΩx is the domain of computation in Eulerian coordinates and $\mathbf{x} = (x, y, z)$T are the Eulerian coordinates.

With respect to the boundary conditions for the modeled constituents of the dermal layer, we take the following Dirichlet boundary conditions

$$N = \overline{N}, \quad M = \overline{M}, \quad c = \overline{c}, \tag{21}$$

where c¯c¯ is the equilibrium concentration of the signaling molecules in the unwounded dermis. With respect to the boundary conditions for the mechanical component of the model, we take the following Robin boundary condition

$$\mathbf{n} \cdot \sigma = -s\rho\mathbf{u}, \tag{22}$$

where nn is the unit outward pointing normal vector to the boundary of the domain of computation. This boundary condition implies that the boundaries experience a spring-like force per unit area in the opposite direction of the displacement of the dermal layer that is proportional to the concentration of the collagen molecules and the magnitude of this displacement.

The last column contains the references to the articles that were used for obtaining (estimates of) the values for the parameters. If the range of the value for a parameter was estimated in this study, then this is indicated by the abbreviation TW. If the value for a parameter is a necessary consequence of the values chosen for the other parameters, then this is indicated by the abbreviation NC

Table 1. An overview of the dimensional (ranges of) values for the parameters of the model

Parameter	Value	Dimensions	References
D_F	10^{-7}	$\text{cm}^5/(\text{cells day})$	Sillman et al. (2003)
χ_F	2×10^{-3}	$\text{cm}^5/(\text{g day})$	Murphy et al. (2012)
q	-4.2×10^{-1}	–	NC
r_F	9.24×10^{-1}	$\text{cm}^{3q}/(\text{cells}^q\ \text{day})$	Ghosh et al. (2007)
$r_F^{\max}$	2	–	Strutz et al. (2001)
a_c^I	10^{-8}	g/cm^3	Grotendorst (1992)
κ_F	10^{-6}	cm^3/cells	Van de Berg et al. (1989)
k_F	5.4×10^6	$\text{cm}^3/(\text{g day})$	Desmoulière et al. (1993)
δ_N	2×10^{-2}	/day	Olsen et al. (1995)
δ_M	$(1-6) \times 10^{-2}$	/day	TW
D_c	2.9×10^{-3}	cm^2/day	Murphy et al. (2012)
k_c	4×10^{-13}	$\text{g}/(\text{cells day})$	Olsen et al. (1995)
η	2	–	Rudolph and Vande Berg (1991) & Moulin et al. (1998)
a_c^{II}	10^{-8}	g/cm^3	Olsen et al. (1995)
δ_c	5×10^{-4}	$\text{cm}^6/(\text{cells g day})$	Olsen et al. (1995)
a_c^{III}	2×10^8	cm^3/g	Overall et al. (1991)
k_ρ	$(1.25-2.75) \times 10^{-8}$	$\text{g}/(\text{cells day})$	TW
$k_\rho^{\max}$	10	–	Olsen et al. (1995)
a_c^{IV}	10^{-9}	g/cm^3	Roberts et al. (1986)
δ_ρ	$(1.25-2.75) \times 10^{-6}$	$\text{cm}^6/(\text{cells g day})$	NC
E	1.00×10^2	$(\text{N cm})/\text{g}$	Liang and Boppart (2010)
ν	4.9×10^{-1}	–	Liang and Boppart (2010)
ξ	5×10^{-3}	$(\text{N g})/(\text{cells cm}^2)$	Maskarinec et al. (2009) & Wrobel et al. (2002)
R_ρ	3×10^{-1}	g/cm^3	Olsen et al. (1995)
$\overline{N}$	10^4	cells/cm^3	Olsen et al. (1995)
$\overline{M}$	0	cells/cm^3	Olsen et al. (1995)
$\overline{c}$	0	g/cm^3	NC
$\overline{\rho}$	10^{-1}	g/cm^3	Olsen et al. (1995)
I^w	$(1-2) \times 10^{-1}$	–	TW
c^w	10^{-8}	g/cm^3	Olsen et al. (1995)
s	10^3	N/g	Koppenol et al. (2016)

The last column contains the references to the articles that were used for obtaining (estimates of) the values for the parameters. If the range of the value for a parameter was estimated in this study, then this is indicated by the abbreviation TW. If the value for a parameter is a necessary consequence of the values chosen for the other parameters, then this is indicated by the abbreviation NC

3.7 The (ranges of the) values for the parameters

Table 1 provides an overview of the dimensional (ranges of the) values for the parameters of the model. The majority of these values were either obtained directly from previously conducted studies or estimated from results of previously conducted studies. In addition, we were able to determine the values for three more parameter due to the fact that these values are a necessary consequence of the values chosen for other parameters. These are the equilibrium signaling molecule concentration of the unwounded dermis (c¯c¯), the constant *q*, and the collagen molecule degradation rate (δρδρ). If we $M = 0$ cells/cm^3 in the unwounded dermis, then $c = 0$ g/cm^3 is an attracting equilibrium in the vicinity of $c = 0$ g/cm^3. Hence we take $\overline{c} = 0$ g/cm^3. Furthermore, if $\overline{N} = 10^4$ cells/cm^3, $\overline{M} = 0$ cells/cm^3, and $\overline{c} = 0$ g/cm^3, then $q = [\log(\delta_N) - \log(r_F[1 - \kappa_F \overline{N}])]/\log(\overline{N})$. Finally, if $\overline{M} = 0$ cells/cm^3 and $\overline{c} = 0$ g/cm^3, then $\delta_\rho = k_\rho/\overline{\rho}^2$ cm^6/(cells g day).

For completeness, we want to remark here that there is (some) variability in the values for most of the parameters that were either estimated from results of previously conducted studies or obtained directly from previously conducted studies. However, given the complexity of the model and hence the necessary complexity of the custom-made numerical algorithm, we decided to set the values for these parameters to fixed values within the found ranges during the generation of simulation results in order to prevent the total computation time of the statistical analyses from rising too high.

The ranges of the values for the free parameters δMδM, kρkρ, and IwIw were chosen in such a way that there is good agreement with respect to the variability in the evolution of the surface area of burns over time between the outcomes of computer simulations obtained in this study and measurements obtained in a previously conducted experimental study by Wang et al. (2011) (see Sect. 6). Finally, the range of the value for the radius of the initial burn (i.e., the range of the value for the parameter clcl (see Sect. 3.6)) was chosen in such a way that it matches with the range of the radii of the burns that were created during the experimental study of Wang et al. (2011).

3.8 A qualitative description of the dynamics of the model

Here we give a qualitative description of how the healing of a burn is accomplished in the presented model and how the different subprocesses that have been incorporated into the model, interact with each other. Due to the presence of signaling molecules in the wounded area at the onset of the proliferative phase fibroblasts from the surrounding uninjured tissue are attracted towards the wounded area. Within the wounded area the fibroblasts proliferate and due to the presence of the signaling molecules, the rate of proliferation is enhanced. In addition, the signaling molecules stimulate the cell differentiation of fibroblasts into myofibroblasts. This results in the emergence of a myofibroblast population in the wounded area. While there are signaling molecules present, both fibroblasts and myofibroblasts also secrete signaling molecules which results subsequently in a further enhancement of both the cell differentiation rate and the cell proliferation rate of fibroblasts and myofibroblasts and hence a further growth of the size of the myofibroblast population within the wounded area. The present (myo)fibroblasts in the wounded area produce collagen molecules in order to restore the presence of a collagen-rich ECM and due to the presence of the signaling molecules the rate of production of the collagen molecules is enhanced. Furthermore, the present myofibroblasts in the wounded area pull on the ECM and as a consequence the surface area of the burn is reduced slowly over time.

During the execution of the wound healing processes the concentration of the generic MMP rises slowly in the recovering wounded area and due to this the secretion of signaling molecules becomes smaller than the proteolytic breakdown of these molecules at a certain point in time. This results in a relatively fast disappearance of the signaling molecules from the wounded area and as a consequence of that the rate of production of collagen molecules starts to decline, as does the rate of cell proliferation and cell differentiation. Slowly the cell densities and the concentrations in the recovering wound area will approach the equilibrium cell densities and the equilibrium concentrations of the unwounded area due to, respectively, cell apoptosis and proteolytic breakdown of the molecules, and ultimately the properties of the recovering wounded area will become identical to the properties of the surrounding tissue, which subsequently implies that the size of the surface area of the recovering wound area slowly returns to the surface area of the burn at the onset of the proliferative phase. The decline of the cell density of the myofibroblast population in the dermal layer, the decline of the concentrations of both the signaling molecules and the collagen molecules, the gradual rise of the cell density of the fibroblast population in the wounded area, and the gradual retraction of the recovering wounded area are not (clearly) visible in the results presented in Sect. 6, but these phenomena are properties of the model and are clearly visible in computer simulations where longer healing times are simulated.

4 NUMERICAL ALGORITHM

In this section we present an overview of the custom-made numerical algorithm that had to be developed for the generation of computer simulations. The development of this algorithm was necessary in order to guarantee the positivity of the approximations of the solutions for the constituents of the dermal layer, "catch" the local dynamics of the model, and obtain sufficiently accurate simulations sufficiently fast.

For the kernel of the concrete expression of the algorithm we used MATLAB together with MATLAB's Parallel Computing Toolbox (The Works Inc 2014). Furthermore, we used a slightly adapted version of the mesh generator developed by Persson and Strang (2004), the element resolution refinement / recoarsement tool of the computational fluid dynamics (CFD) software package FEATFLOW2 (Turek 1998), and the scaling and permutation routine HSL_MC64 (HSL 2013) by interfacing these products with the kernel. Finally, we applied the following non-dimensionalisation to the model:

$$\begin{aligned} x &= Lx^*, \quad t = \left[L^2/\left[D_F\overline{N}\right]\right]t^*, \quad \rho = \overline{\rho}\rho^*, \\ N &= \overline{N}N^*, \quad M = \overline{N}M^*, \quad c = c^w c^*, \\ \mathbf{u} &= L\mathbf{u}^*, \quad \mathbf{v} = \left[\left[D_F\overline{N}\right]/L\right]\mathbf{v}^*, \quad \sigma = \left[\left[\xi\overline{N}\right]/\overline{\rho}\right]\sigma^*, \end{aligned} \tag{23}$$

where L=1 cmL=1 cm is the length scale of the model. The variables with the asterisks are the non-dimensionalised variables.

In the following two subsections we present a step-by-step description of the algorithm. Basically the algorithm consists of two parts. The first part of the algorithm is dedicated to the generation of a proper triangulation of the domain of computation and is described in Sect. 4.1. The second part of the algorithm is described in Sect. 4.2 and is dedicated to obtaining an approximation of the solution for the displacement and the modeled constituents of the dermal layer from Eq. (1) after application of the non-dimensionalisation.

4.1 Generation of the initial triangulation of the domain of computation

The first part of the algorithm consists again of two subparts. First, a conforming base triangulation is generated and subsequently the element resolution refinement / recoarsement tool is used to adjust the resolution of the elements of this base triangulation.

In order to create a conforming base triangulation the aforementioned slightly adapted version of the mesh generator developed by Persson and Strang (2004) is used. This results in very high quality triangulations of the domain of computation that consist mainly of equilateral triangles. Most of the triangles that are not equilateral, are located near the left and right boundary of the domain of computation. These latter triangles are nearly equilateral. Using the following measure for the quality of a triangle *ABC*:

$$\alpha(ABC) = 2\sqrt{3}\left[\frac{\|CA \times CB\|}{\|CA\|^2 + \|AB\|^2 + \|BC\|^2}\right], \tag{24}$$

we observed that $\alpha>0.86\alpha>0.86$ for all triangles in the generated triangulations that were used for the generation of the simulation results, where $0\leq\alpha\leq10\leq\alpha\leq1$ (Lo 1989).

For the generation of the simulation results presented in Figs. 2, 3, 4, 5 and 6, we used a base triangulation that consists of triangles with an average initial edge length of 1.85 cm1.85 cm. We repeated the calculations for the generation of the simulation results presented in Figs. 2, 3, 4and 5 two times. The first time we used a base triangulation that consists of triangles with an average initial edge length of $9.24\times10{-}1$ cm$9.24\times10{-}1$ cm and the second time we used a base triangulation that consists of triangles with an average initial edge length of 3.46 cm3.46 cm. We observed that the difference in the simulation results with respect to the outcomes between the different calculations were small.

After generation of the conforming base triangulation the element resolution refinement / recoarsement tool is used to adjust the resolution of the elements of this base triangulation (Möller 2008). First, the L2L2-norm of an estimation of the error of the gradient of the numerical approximation of the indicator function presented in Eq. (18), per element (i.e., ‖e^‖L2(T)‖e^‖L2(T)) is determined (Möller and Kuzmin 2006). Thereafter, the resolution of the elements is adapted in order to adjust the estimated error. These two steps are repeated until either the absolute value of the relative change of the sum of the L2L2-norm of the estimation of the error of the gradient over all elements is smaller than 5%, or the maximum number of allowed for iterations is reached. In this study we set this latter number to ten. The elements are refined and coarsened until the following holds for each element in the triangulation:

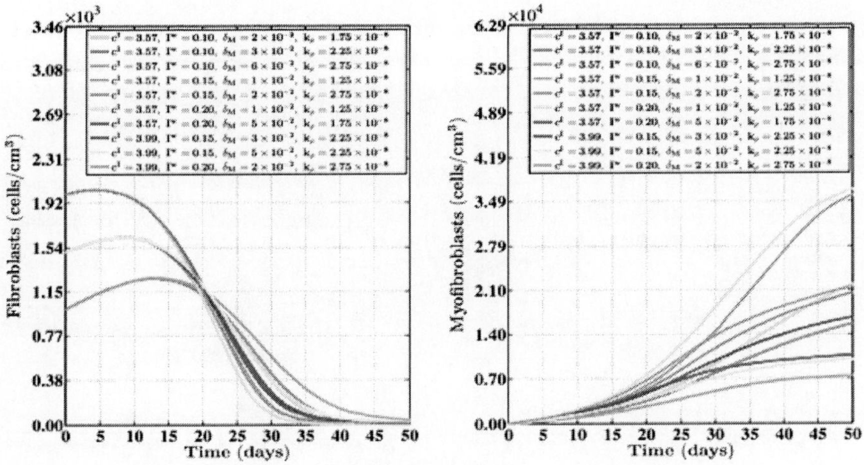

Figure 3. The evolution over time of the cell densities for a random selection of some combinations of the levels of the factors of the factorial design. See Figure 1 for the location where the evolution of the cell densities was traced over time. The values for the remaining parameters are equal to those depicted in Table 1

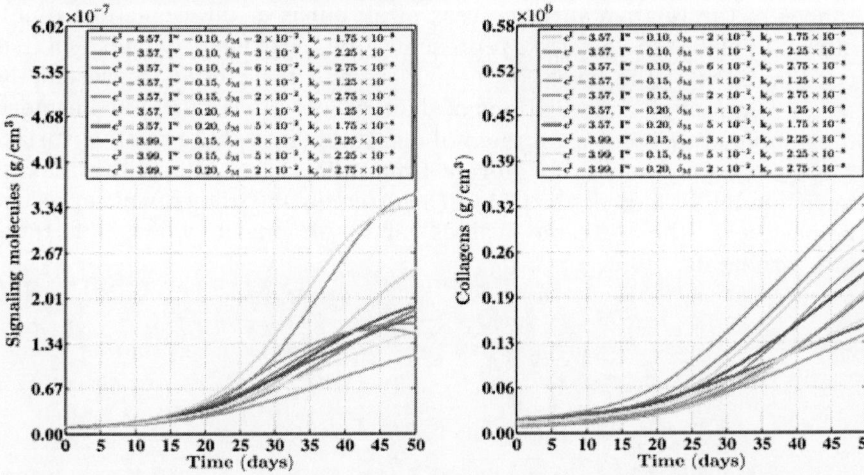

Figure 4. The evolution over time of the concentrations of the signaling molecules and the collagen molecules for a random selection of some combinations of the levels of the factors of the factorial design. See Figure 1 for the location where the evolution of the concentrations was traced over time. The values for the remaining parameters are equal to those depicted in Table 1

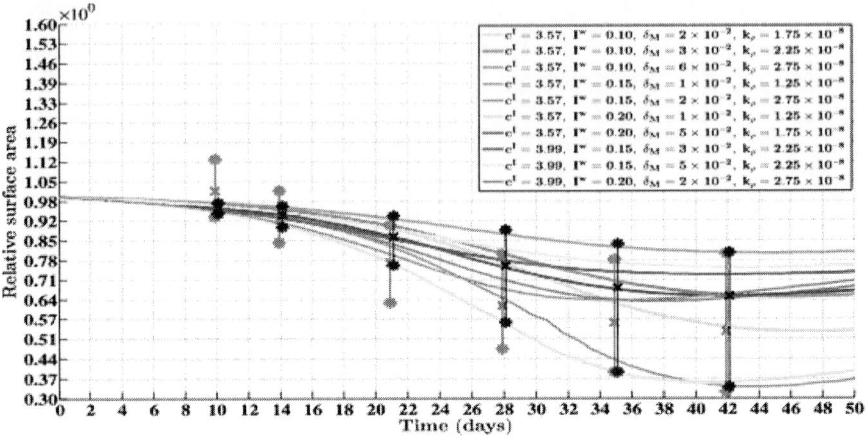

Figure 5. The evolution over time of the relative surface area of burns for a random selection of some combinations of the levels of the factors of the factorial design. The values for the remaining parameters are equal to those depicted in Table 1. The surface area of a burn was compared to its surface area at day 0 at every time point. The *black bars* represent the ranges of the relative surface areas of all burns at different time points after injury. The *asterisks* represent the extreme values of all combinations of the levels of the factors and the crosses represent the mean values of all combinations of the levels of the factors. Furthermore, the *red bars* represent experimental measurements of the relative surface areas of healing burns at different time points after injury in a porcine burn model (Wang et al. 2011). The *asterisks* and the *crosses* represent, respectively, the extreme values and the mean values of these measurements

$$0.04 \leq \left[\frac{|\mathcal{T}_h| \, \|\hat{\mathbf{e}}\|^2_{L_2(T)}}{\sum_{T \in \mathcal{T}_h} \left(\|\sigma_h\|^2_{L_2(T)} + \|\hat{\mathbf{e}}\|^2_{L_2(T)} \right)} \right]^{\frac{1}{2}} \leq 0.2, \tag{25}$$

where ThTh represents the current triangulation $|\mathcal{T}_h|$ denotes the number of elements that constitute the triangulation, and $\|\sigma_h\|_{L_2(T)}$ denotes the L2L2-norm of a low-order estimation of the gradient of the numerical approximation of the indicator function presented in Eq. (18) (Möller and Kuzmin 2006). In this study the resolution of every individual element can be increased at most four times and the size of the individual elements cannot be increased beyond the size they have in the base triangulation.

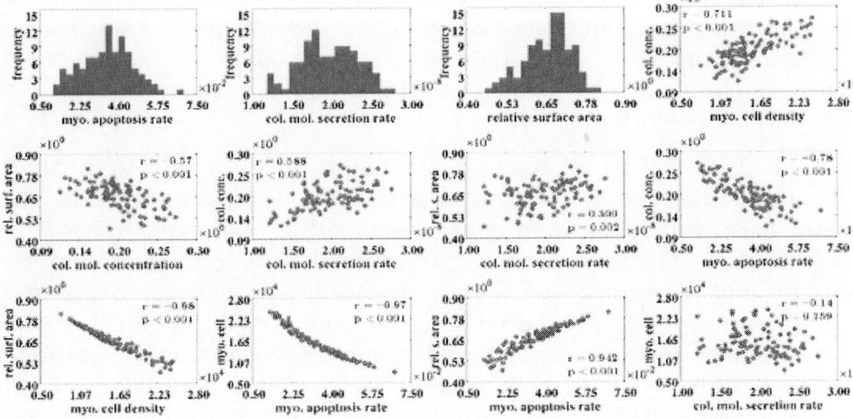

Figure 6. An overview of the results obtained with the probabilistic analysis. The frequency histograms display the drawn samples for the myofibroblast apoptosis rate (δMδM) and the collagen molecule secretion rate (kρkρ), and the obtained distribution of the relative surface area of the healing burns at day 42 compared to their surface areas at day 0. The *scatter plots* provide furthermore graphical representations of the relationships between various properties of the healing burn at day 42. Within the individual subfigures the Pearson correlation coefficients are displayed together with the associated *p* values

4.2 Determination of the approximation of the solution

As was mentioned before, the second part of the algorithm is dedicated to obtaining an approximation of the solution for the displacement and the modeled constituents of the dermal layer from Eq. (1). In order to solve the time-dependent problem the method of lines and the standard fixed-point defect correction method are applied (Van Kan et al. 2014). The two equations of the system are solved in a segregated way. Each time step approximations of the solutions for the modeled constituents of the dermal layer are determined first and subsequently an approximation of the solution for the displacement of the dermal layer is determined. This scheme is iterated until the maximum of the relative 1-norms of the residuals of the approximations is smaller than one, and the maximum of the relative 1-norms of the difference between subsequent approximations per variable is smaller than 5×10−25×10−2. If the fixed-point scheme does not meet the convergence criteria within five iterations, then the scheme is interrupted, the time step is decreased to 85% of its current value, and subsequently the scheme is restarted. The required estimate of the gradient of the solution for the signaling molecule is obtained by using a variational gradient recovery projection scheme (Lyra 1994).

For the discretization of the system of equations, a moving-grid finite-element method is used (Madzvamuse et al. 2003) together with the first-order backward Euler time-integration method. Furthermore, a semi-implicit flux-corrected transport (FCT) limiter developed by Möller et al. (2008), and a source term splitting procedure proposed by Patankar (1980), are applied on the discretized system of equations that describe the dynamics of the modeled constituents of the dermal layer. Taken together these latter two techniques enforce positivity of the approximations of the solutions for the constituents of the dermal layer.

The individual time steps are chosen by using an automatically adaptive time-stepping method with inbuilt local truncation error control (Kavetski et al. 2002). The maximum size of the initial time step is set to 10−510−5 dimensionless units and the upper bound of the size of the time step is set to 10−310−3 dimensionless units. If a time step is accepted, then the subsequent time step is at most 1.25 times the size of the current time step. If a time step is rejected, then the subsequent time step is at least 0.25 times the size of the current time step. For completeness we mention here also that we set the absolute and relative truncation error tolerance to, respectively, 10−210−2 and 5×10−25×10−2 dimensionless units (see the article by Kavetski et al. (2001) for further details on this matter). After obtaining and accepting an approximation for a certain time step, the local extrapolation procedure proposed by Kavetski et al. (2002) is applied to increase the accuracy of the approximation.

Furthermore, the aforementioned element resolution refinement / recoarsement tool is applied every ten time steps to adjust the resolution of the elements of the triangulation. For this end the L2L2-norm of an estimation of the error of the gradient of the numerical approximation of the solution for the concentration of the collagen molecules per element is determined first (Möller and Kuzmin 2006). Thereafter the resolution of the elements is adapted in a fashion identical to the procedure described in Sect. 4.1 in order to adjust the estimated error. For the interpolation of approximations to new nodes in the triangulation, piecewise bivariate Hermite interpolation is used (Feng and Zhang 2013). The required gradients of the approximations at the existing nodes are estimated by using a polynomial preserving gradient recovery scheme (Zhang and Naga 2005).

For the approximation of the individual primary variables of the model functions from the space of triangular finite-elements with linear basis functions were chosen (Quarteroni and Valli 2008). The integrals over the interior of the elements are approximated by a second-order accurate Newton-Cotes quadrature rule and the integrals over the boundaries of the elements are approximated by a second-order accurate Gaussian quadrature rule.

In order to obtain approximate solutions for the resulting linear systems of equations, MATLAB's backslash operator is used (The Works Inc 2014) after using the LU factorization algorithm (Davis and Duff 1997) on scaled and permuted versions of the original linear systems. For the scaling and permutation of the linear systems several inbuilt scaling and permutation algorithms of MATLAB are used (Davis et al. 2004; Duff and Koster 1999) together with the scaling and permutation routine HSL_MC64 (HSL 2013).

5 DETAILS OF THE APPLIED STATISTICAL METHODS

5.1 The factorial design and the regression analysis

In this study we used a simple mixed-level full-factorial design in combination with a multiple linear regression analysis in order to quantify the individual contributions of the variations in the values for the free parameters of the model and the initial radius of the wound, to the dispersion in the surface area of healing burns (Phadke 1989; Tabachnick and Fidell 2007; Taguchi 1987). We chose the relative surface area of the healing burn at day 42 compared to the surface area of the burn at day 0 as the response variable (i.e., the dependent variable). The free parameters of the model and the initial radius of the wound were chosen as factors (i.e., independent variables). These latter parameters were assigned discrete values (i.e., levels) that divide the range of each factor equally. Here we assigned two levels to the radius of the initial burn (cIcI), three levels to the minimum amount of fibroblasts and collagen molecules that are present initially in the wounded area (IwIw), four levels to the collagen molecule secretion rate (kρkρ), and six levels to the myofibroblast apoptosis rate (δMδM). For the analysis all possible combinations of the levels of the factors were examined. Hence in total 144 computer simulations were generated.

The obtained simulation data were analyzed in IBM SPSS by means of the regression analysis (IBM Corp 2011). No interactions or powers of independent variables were included in this analysis. In order to reduce the skewness and kurtosis of the data related to the relative surface area of the healing burn and to improve the normality, linearity, and homoscedasticity of the residuals of the regression analysis, the Rankit rank-based normalization method was used on the data related to the response variable. Furthermore, all data related to the factors of the factorial design were transformed to standard scores (i.e., *z* values).

The surface area of a the burns was determined in two steps. In the first step we determined the displacements of the material points with coordinates $(c^I(\cos(2\pi(i - 1)/40)), c^I(\sin(2\pi(i - 1)/40)), 0)^T$, where $i \in \{1, \ldots, 40\}$. Subsequently, the area of the polygon with vertices located at the displaced material points was computed. This area is approximately equal to the surface area of the healing burn.

5.2 The probabilistic analysis

After having used a regression analysis, a probabilistic analysis was used in order to investigate in more detail the effect of variability in the values for the collagen molecule secretion rate and the myofibroblast apoptosis rate, on both the cell density of the myofibroblasts and the concentration of the collagen molecules at day 42, and the relative surface area of the healing burn at day 42 compared to its surface area at day 0. The collagen molecule secretion rate and the myofibroblast apoptosis rate were chosen because the multiple linear regression analysis demonstrated that varying the value for these two factors has by far the largest impact on the relative surface area of the healing burn (See the presentation of the results of the linear regression analysis in Table 2).

The distribution of the collagen molecule secretion rate (kρkρ) was defined as a Gaussian (normal) distribution with a mean value of 2×10−8 g/(cells day)2×10−8 g/(cells day) and a standard deviation of 3.75×10−9 g/(cells day)3.75×10−9 g/(cells day). The distribution of the myofibroblast apoptosis rate (δMδM) was defined as a Gaussian (normal) distribution with a mean value of 3.50×10−2 /day3.50×10−2 /day and a standard deviation of 1.25×10−2 /day1.25×10−2 /day. The values for the parameters related to the minimum amount of fibroblasts and collagen molecules that are present initially in the wounded area (IwIw) and the radius of the initial burn (cIcI) were set to, respectively 1.5×10−11.5×10−1 and 3.78 cm3.78 cm. The values for all remaining parameters are equal to those depicted in Table 1.

Table 2. An overview of the results obtained with the multiple linear regression analysis.

Variable	β	t score	p value	95% conf. interval		sr^2
				Lower	Upper	
c^I	0.048	2.258	0.026	0.006	0.090	0.002
I^w	−0.073	−3.463	0.001	−0.115	−0.032	0.005
δ_M	0.912	42.978	<0.001	0.869	0.953	0.832
k_ρ	0.313	14.754	<0.001	0.271	0.355	0.098

The regression coefficients (β) are the weights of the linear regression line. The t scores and p values are the associated statistics. The squared semipartial correlations between the dependent variable and the individual independent variables (sr^2) are displayed in the last column of the table

Sampling from the Gaussian distributions was realized by using MATLAB's inbuilt function normrnd (The Works Inc 2014). In this study two samples consisting of 100 observations, were drawn from each distribution. Hence in total 100 computer simulations were generated. We used the Anderson–Darling test to investigate the degree of normality of the samples (Anderson and Darling 1952). Both samples did not show a significant deviation from a normally distributed sample (p>0.5p>0.5 for both samples).

The collagen molecule concentration and the cell density of the myofibroblasts were determined in two steps. In the first step we determined the cell density and the concentration at the material points with coordinates $(0,0,0)^{\mathrm{T}}$, $((c^I/2)(\cos(2\pi(i-1/2)/4)), (c^I/2)(\sin(2\pi(i-1/2)/4)), 0)^{\mathrm{T}}$, and $(c^I(\cos(2\pi(i-1)/4)), c^I(\sin(2\pi(i-1)/4)), 0)^{\mathrm{T}}$, where $i \in \{1,\ldots,4\}$. Subsequently, the averages of these values were calculated and these averages were used for the analysis. See Sect. 5.1 for a description of how the surface area of the healing burns was determined.

6 RESULTS

In this section we present simulation results and the outcomes of the statistical analyses. The discussion about the simulation results and the results obtained with the statistical analyses takes place in Sect. 7.

In order to obtain some insight into the dynamics of the model, we present an overview of a simulation in Figure 2. Furthermore, Figs. 3 and 4 show, respectively, the evolution over time of the different cell densities and the evolution over time of the different concentrations for a random selection of some combinations of factor levels (i.e., some combinations of values for the initial radius of the burn, the minimum initial cell density of fibroblasts and minimum initial concentration of collagen molecules, the apoptosis rate of myofibroblasts, and the collagen molecule secretion rate), at a certain material point within the healing burn. Finally, Figure 5 shows the evolution over time of the relative surface area of the burns for the same random selection of combinations of levels as was used for the creation of Figs. 3 and 4.

Table 2 provides an overview of the outcomes of the multiple linear regression analysis. The multiple correlation (R) is significantly different from zero, $F(4,139)=520,473F(4,139)=520,473$, $p<0.001p<0.001$, with the squared multiple correlation at 0.937. The adjusted squared multiple correlation of 0.936 indicates that more than 93% of the variability in the relative surface area of the healing burn at day 42 compared to the size of the burn at day 0 is predicted by variability in the values for the factors of the factorial design.

Figure 6 shows an overview of the results obtained with the probabilistic analysis. The obtained distribution of the relative surface area of the healing burns at day 42 compared to their surface areas at day 0 is negatively skewed and differs significantly from a Gaussian distribution according to the Anderson–Darling test ($p<0.04p<0.04$) (Anderson and Darling 1952). The mean of the relative sizes of the healing burns at day 42 is 0.657 and the standard deviation of the obtained distribution is 0.073.

7 DISCUSSION

For approximately the last 25 years the ematical modeling of processes involved in the healing of dermal wounds has been an active area of research. Over these years the number of generated models has increased considerably. Surveys of these models such as those compiled in Sherratt and Dallon (2002), Buganza Tepole and Kuhl (2013), and Valero et al. (2015b) indicate that the majority of the models can be placed into one of three categories: continuum hypothesis-based models, discrete cell-based models, and hybrid models. We used the general continuum hypothesis-based modeling framework of Tranquillo and Murray (1992) as ematical basis for the generation of the presented model. Since its initial presentation this framework has been extended and adapted in several modeling studies investigating the impact of the addition and the adaptation of various different components of the wound healing response (Javierre et al. 2009; Murphy et al. 2012; Olsen et al. 1995; Ramtani 2004; Ramtani et al. 2002; Valero et al. 2013, 2014a, b, 2015a; Vermolen and Javierre 2012).

The biochemical components of the model presented in this study are based primarily on the biochemical components of the model developed by Olsen et al. (1995). However, there are also some clear differences between the model presented in this study and the model developed by Olsen et al. For instance, while in the model of Olsen et al. solely the fibroblasts are actively motile, we incorporate in our model the active movement of both fibroblasts and myofibroblasts. Thampatty and Wang (2007) have demonstrated previously that both fibroblasts and myofibroblasts are actively motile. Hence it seems reasonable to incorporate this phenomenon into the model.

In addition, we model the random movement of cells by means of cell density-dependent Fickian diffusion instead of linear Fickian diffusion. All of the models in the above cited modeling studies use linear diffusion to model the random dispersal of cells. As is also mentioned by Olsen et al. (1995), this is the standard representation of random cell movement. However, Hillen and Painter (2009) point out that it is actually far more likely that the random movement of cells depends on the density of these cells. They also show in their study that this type of dependence can show up naturally during the derivation of a Keller-Segel type of system. Hence we assume that the random movement of the cells increases with an increasing cell density (Hillen and Painter 2009; Kowalczyk 2005).

Furthermore, we incorporate into the model the degradation of both signaling molecules and collagen molecules by means of proteolytic cleavage by a generic MMP instead of by means of natural decay or very general enzymatic degradation which are usually used in this type of models. For this end, we use the phenomenological relationship given in Eq. (9) which is based on the deduction presented in the accompanying paragraph.

Finally, with respect to the biochemical components of the model, we want to place some remarks concerning the way that the cell differentiation of fibroblasts into myofibroblasts has been incorporated into the presented model. Similar to Olsen et al. (1995), we assume that the rate of cell differentiation is dependent on the concentration of the signaling molecule with no cell differentiation taking place in the absence of this signaling molecule. Like others such as Javierre et al. (2009), Murphy et al. (2012), and Valero et al. (2014a; b), we are aware of the fact that this cell differentiation can only take place under conditions of sufficient mechanical stiffness. It seems like a good idea then to incorporate this phenomenon also into the model. However, as is clearly pointed out by Van de Water et al. (2013), it is unclear at present what the actual stiffness is that is perceived by fibroblasts. In addition, recent experimental studies have demonstrated that the cell differentiation is also critically dependent on the presence of particular isoforms of fibronectin (Van de Water et al. 2013). Taken together, this implies that the incorporation of the cell differentiation mechanism into a ematical model in a realistic way is basically not possible at present. Hence while being aware of the fact that the cell differentiation process is a very complex process in reality, we had to keep things simple in this modeling study.

With respect to the mechanical component of the presented model, we want to remark the following. Traditionally, the dermis is treated as a linear (visco)elastic solid in mechano-chemical continuum models for (certain aspects of) dermal wound healing (Javierre et al. 2009; Murphy et al. 2012; Olsen et al. 1995; Ramtani 2004; Ramtani et al. 2002; Valero et al. 2014a, b; Vermolen and Javierre 2012). However, skin tissues are actually nonlinear, anisotropic, viscoelastic, and inhomogeneous materials (Fung 1993). Hence the constitutive stress-strain relationships for skin tissues are in reality far more complicated. In order to make the presented model somewhat more realistic while still keeping the model relatively simple, we decided to model the dermal layer as a heterogeneous, isotropic, and compressible neo-Hookean solid (Treloar 1948) where the Young's modulus is linearly dependent on the concentration of the collagen molecules (Ramtani 2004; Ramtani et al. 2002). Although we are not the first to model dermal tissues as hyperelastic materials within the context of dermal wound healing (See for example the studies by Valero et al. (2013; 2015a)), it is not a standard approach. Finally, we want to mention here that, to the best of our knowledge, we are the first to combine an explicit description of the dynamics of a myofibroblast population with the modeling of a dermal tissue as a hyperelastic material.

Ultimately the properties of the recovering wound area will become identical to the properties of the surrounding tissue (i.e., the cell densities and the concentrations in the recovering wound area will approach the equilibrium cell densities and the equilibrium concentrations of the unwounded dermis). Given Eq. (16), this implies that the "body force" in the mechanical force balance will vanish when the solutions related to the constituents of the dermal layer reach their equilibrium solutions. This implies that the displacement field of the dermal layer will become zero when the solutions related to the constituents of the dermal layer reach their equilibrium solutions and therefore the properties of the recovering tissue will indeed become identical to the properties of the surrounding tissue.

Hence it is not possible with the presented model to simulate the often observed permanent deformation of involved tissues and the development of residual stresses within these tissues (Schouten et al. 2012). Recently continuum hypothesis-based models have been formulated that are based on a morphoelastic framework (Bowden et al. 2016; Murphy et al. 2011). With these models it is possible to simulate both the contraction and the growth of involved tissues during the execution of the overall healing process. Due to this combination it is possible with this type of model to simulate the permanent deformation of involved tissues and the development of residual stresses within these tissues. Currently we are working on an extension of the model presented in this study whereby we make use of a morphoelastic framework.

Focusing now on the simulation of the contraction of burns over time we observe the following. The bars in Figure 5 show that the agreement with respect to the variability in the evolution of the surface area of burns over time between the outcomes of the computer simulations obtained in this study and measurements obtained in a previously conducted experimental study by Wang et al. (2011) is quite good. There is in general a reasonably good match between the outcomes of the computer simulations and the measurements obtained in the experimental study with respect to both the ranges of the relative surface areas of the healing burns and the means of these relative surface areas, and this match becomes better as healing progresses. Furthermore, we also observed in the computer simulations that while some healing burns continue to contract until day 50, others stopped to contract at an earlier day and started to retract. This variability is also present in the presented results by Wang et al. (2011).

Taken together these agreements provide us some confidence about the validity of the model. Obviously the number of models with which it is possible to produce the depicted contraction curves is infinite in theory. Therefore, it would have been valuable to validate the presented model against both more experimental data and against experimental data of a different type in order to increase our confidence about the validity of the model. Unfortunately we have not been able to find any other suitable experimental data for the validation of the presented model in the available literature. However, it is a fundamental problem in the field of ematical modeling of dermal wound healing processes to find suitable experimental data for the validation of models. As is pointed out by Bowden et al. (2016), as a consequence of technical difficulties related to obtaining reliable experimental measurement data there are hardly any suitable experimental data available at present for the proper validation of developed ematical models in this field. In our opinion this does not mean that we should refrain from deducing biomedical implications from the results obtained with the presented model. However, we do think that it is very important to be extremely careful when doing so and to keep in mind that these implications are based on results obtained with a ematical model.

Having said that, we look more closely now at the results presented in Figs. 3, 4 and 5. Figures 3and 4 show clearly that varying the values for the free parameters of the model and the radius of the initial burn has in general a huge impact on the evolution over time of the cell densities and concentrations of the modeled constituents of the dermal layer within the wounded area. The dispersion in the cell density of the myofibroblast population at the center of the healing burn becomes quite large over time, and this is also the case for the dispersion in the concentration of the signaling molecules and the collagen molecules. Figure 5 shows furthermore that the evolution over time of the observed dispersion in the relative surface areas of the healing burns from the experimental study by Wang et al. (2011) can be replicated quite well by the model through the variation of the values for radius of the initial burn and the free parameters of the model.

Combined these results provide us the following explanation for the observed dispersion in the relative surface areas of the healing burns in the experimental study by Wang et al. (2011). Variability in the values for the free parameters of the model results in an increasing dispersion in both the cell density of the myofibroblast population and the concentrations of the signaling molecules and collagen molecules over time. Given Eq. (16), this implies that the dispersion in the total generated stress by the myofibroblast population also increases over time, which subsequently results in an increasing dispersion in the relative surface areas of the healing burns over time. Hence the observed variability in the evolution of the surface area of the healing burns over time in the experimental study by Wang et al. (2011) might be attributed to variability in the myofibroblast apoptosis rate, the collagen molecule secretion rate, the minimum amount of fibroblasts and collagen molecules that are present initially in the wounded area, and the radius of the initial burn.

Looking more closely at the outcomes of the multiple linear regression analysis provided in Table 2, it can be observed that all regression coefficients differ significantly from zero. Hence variability in the value for each factor of the factorial design contributes significantly to variability in the relative surface area of a healing burn at day 42. The sizes of the squared semipartial correlations indicate that variability in the value for the myofibroblast apoptosis rate has a very large impact on the relative surface area of the healing burn at day 42. Variability in the value for the collagen molecule secretion rate has a smaller, but still quite substantial impact on the relative surface area of a healing burn at day 42. Variability in the value for the parameter related to the minimum amount of fibroblasts and collagen molecules that are present initially in the wounded area, and variability in the value for the radius of the initial burn have only a relative small impact on the relative surface area of the healing burn.

Taken together the regression analysis suggests that the observed variability in the evolution of the surface area of the healing burns over time in the experimental study by Wang et al. (2011) might be attributed mainly to variability in the myofibroblast apoptosis rate and the collagen molecule secretion rate and to a far lesser extent to variability in the minimum amount of fibroblasts and collagen molecules that are present initially in the wounded area and the radius of the initial burn. Hence if one wishes to adjust the degree of wound contraction during the healing of a burn, then the simulation results and the associated regression analysis presented in this study suggest that this may be accomplished by adjusting the apoptosis rate of myofibroblasts and / or the secretion rate of collagen molecules.

We expect that varying the values for most of the remaining parameters of the model that were not included into the factorial design and for which we could find a range of values in the literature, would probably also contribute significantly to variability in the relative surface area of a healing burn at day 42. We think that varying the values for the proliferation rate of (myo)fibroblasts (rFrF), the generated stress per unit cell density and the inverse of the unit collagen molecule concentration (ξξ), and the constant E related to the Young's modulus within the found ranges of values in particular would contribute to variability in the evolution of the relative surface area of a healing burn over time. Therefore, we would like to add these parameters to the factorial design in order to quantify their contribution to this variability. However, as was mentioned before, given the complexity of the model and hence the necessary complexity of the custom-made numerical algorithm, we basically had to set the individual values for these parameters to fixed values within their respective ranges in order to prevent the total computation time of the statistical analyses from rising too high.

Looking at the results obtained with the probabilistic analysis in Figure 6, we can make a couple of interesting observations. Firstly, there is a good agreement between the regression coefficients for the collagen molecule secretion rate and the myofibroblast apoptosis rate in the linear regression line on the one hand, and the calculated correlation coefficients for the relationship between the relative surface area of a burn at day 42 and the collagen molecule secretion rate and the myofibroblast apoptosis rate on the other hand. Secondly, it is interesting to observe that the obtained distribution of the relative surface areas of the healing burns differs significantly from a Gaussian (normal) distribution, even though the samples for the collagen molecule secretion rate and myofibroblast apoptosis rate were both drawn from a Gaussian distribution, and to observe that this distribution is negatively skewed. Hence relatively speaking, only a small portion of the burns contracts severely while the majority of the burns contract relatively little. This coincides nicely with clinical observations.

Finally, it is interesting to observe that all of the remaining individual correlations displayed in the separate scatter plots make sense intuitively with the exception of the scatter plot that displays the correlation between the collagen molecule secretion rate and the cell density of the myofibroblasts. For example, we would expect positive correlations between the concentration of the collagen molecules and the rate of secretion of these molecules, and between the concentration of the collagen molecules and the cell density of the myofibroblast population. However, taking some of the correlations together shows something peculiar; while the correlation between the collagen molecule secretion rate and the concentration of the collagen molecules, and the correlation between the collagen molecule secretion rate and the relative surface area of a burn at day 42 are both positive, the correlation between the concentration of the collagen molecules and the relative surface area of a burn at day 42 is negative.

In order to explain this phenomenon, we suggest the following. Besides influencing the total amount of generated stress directly through the relationship given in Eq. (16), the concentration of the collagen molecules also influences the amount of generated stress indirectly through the following chain of connections. Given Eq. (9), it is seems reasonable to expect a positive correlation between the concentration of the collagen molecules and the concentration of the generic MMP. Given Eqs. (8) and (6), it also seems reasonable to presume a negative correlation between the concentration of the generic MMP and the concentration of the signaling molecule and a positive correlation between the concentration of the signaling molecule and the cell density of the myofibroblast population. Taken together, these correlations suggest a negative correlation between the collagen molecule secretion rate and the cell density of the myofibroblast population. Although not significantly different from zero, this is exactly what is displayed in the scatter plot that displays the correlation between the collagen molecule secretion rate and the cell density of the myofibroblasts. Figure 6 displays finally a very strong negative correlation between the cell density of the myofibroblast population and the relative surface area of a burn at day 42. We suggest that this chain of connections has a stronger effect on the relative surface area of a burn at day 42 than the direct connection between the concentration of the collagen molecules and the total amount of generated stress. As a consequence the correlation between the concentration of the collagen molecules and the relative surface area of a burn at day 42 is negative while the the correlation between the collagen molecule secretion rate and the relative surface area of a burn at day 42 is positive.

As has been mentioned in the introduction, we think that results obtained with ematical modeling studies can provide new insights into which elements of the wound healing response are key elements in processes like wound contraction, and that these insights might aid in the design of new treatment plans for burns that can reduce the incidence of sequelae. In this study we found that most of the variability in the evolution of the surface area of healing burns over time might be attributed to variability in the apoptosis rate of myofibroblasts primarily and, to a lesser extent, the secretion rate of collagen molecules. Interestingly, this result is in agreement with the suggestion put forward by Desmoulière et al. (1995) that the disruption of apoptosis (i.e., a low apoptosis rate) during wound healing might be a very important factor in the development of pathological scars such as severely contracted scars. Hence, while we keep in mind that we are deducing biomedical implications from results obtained with a ematical model and therefore need to be extremely careful, the results from this study suggest that the degree of wound contraction during the healing of a burn can be adjusted by altering the apoptosis rate of myofibroblasts and / or the secretion rate of collagen molecules. We find this an interesting insight and hope that this insight contributes to the design of better treatment plans for burns.

ACKNOWLEDGMENTS

We would like to thank D. den Ouden and M. Möller for their assistance with the creation of the interface between the element resolution refinement / recoarsement tool of the CFD software package FEATFLOW2 and the kernel of the developed software package.

FUNDING

This study was funded by a grant from the Dutch Burns Foundation (project number WO/12.103).

CONFLICT OF INTEREST

The authors declare that they have no conflict of interest.

REFERENCES

1. Anderson T, Darling D (1952) Asymptotic theory of certain "goodness-of-fit" criteria based on stochastic processes. Ann Stat 23:193–212
2. Barrientos S, Stojadinovic O, Golinko M, Brem H, Tomic-Canic M (2008) Growth factors and cytokines in wound healing. Wound Repair Regen 16:585–601
3. Baum C, Arpey C (2005) Normal cutaneous wound healing: clinical correlation with cellular and molecular events. Dermatol Surg 31:674–686
4. Bowden L, Byrne H, Maini P, Moulton D (2016) A morphoelastic model for dermal wound closure. Biomech Model Mechanobiol 15:663–681
5. Buganza Tepole A, Kuhl E (2013) Systems-based approaches toward wound healing. Pediatr Res 73:553–563
6. Chakraborti S, Mandal M, Das S, Mandal A, Chakraborti T (2003) Regulation of matrix metalloproteinases: an overview. Mol Cell Biochem 253:269–285
7. Davis T, Duff I (1997) An unsymmetric-pattern multifrontal method for sparse LU factorization. SIAM J Matrix Anal Appl 18:140–158
8. Davis T, Gilbert J, Larimore S, Ng E (2004) Algorithm 836: Colamd, a column approximate minimum degree ordering algorithm. ACM Trans Softw 30:377–380
9. Desmoulière A, Geinoz A, Gabbiani F, Gabbiani G (1993) Transforming growth factor-ββ1 induces αα-smooth muscle actin expression in granulation tissue myofibroblasts and in quiescent and growing cultured fibroblasts. J Cell Biol 122:103–111
10. Desmoulière A, Redard M, Darby I, Gabbiani G (1995) Apoptosis mediates the decrease in cellularity during the transition between granulation tissue and scar. Am J Pathol 146:56–66
11. Duff I, Koster J (1999) The design and use of algorithms for permuting large entries to the diagonal of sparse matrices. SIAM J Matrix Anal Appl 20:889–901
12. Eming S, Krieg T, Davidson J (2007) Inflammation in wound repair: molecular and cellular mechanisms. J Investig Dermatol 127:514–525
13. Enoch S, Leaper D (2007) Basic science of wound healing. Surgery 26:31–37
14. Farina J Jr, Rosique M, Rosique R (2013) Curbing inflammation in burn patients. Int J Inflamm 715:645

15. Feng R, Zhang Y (2013) Piecewise bivariate Hermite interpolation for large sets of scattered data. J Appl 2013:1–10
16. Fung Y (1993) Biomechanics, 2nd edn. Springer, New York
17. Ghosh K, Pan Z, Guan E, Ge S, Liu Y, Nakamura T, Ren ZD, Rafailovich M, Clark R (2007) Cell adaptation to a physiologically relevant ecm mimic with different viscoelastic properties. Biomaterials 28:671–679
18. Gibran N, Wiechman S, Meyer W, Edelman L, Fauerbach J, Gibbons L et al (2013) American burn association consensus statements. J Burn Care Res 34:361–365
19. Goel A, Shrivastava P (2010) Post-burn scars and scar contractures. Indian J Plast Surg 43(Suppl S1):63–71
20. Grotendorst G (1992) Chemoattractants and growth factors. In: Cohen I, Diegelmann R, Lindblad W (eds) Wound healing: biochemical and clinical aspects, chap 15, 1st edn. W.B. Saunders, Philadelphia, pp 237–246
21. Hawkins H, Finnerty C (2012) Pathophysiology of the burn scar. In: Herndon DN (ed) Total burn care, chap 46, 4th edn. W.B. Saunders, Philadelphia, pp 507.e3–516.e3
22. Hillen T, Painter K (2009) A user's guide to PDE models for chemotaxis. J Biol 58:183–217
23. Hinz B (2007) Formation and function of the myofibroblast during tissue repair. J Investig Dermatol 127:526–537
24. HSL (2013) A collection of fortran codes for large scale scientific computation. http://www.hsl.rl.ac.uk
25. IBM Corp (2011) IBM SPSS Statistics for Windows. Version 20.0. IBM Corp, Armonk, NY
26. Javierre E, Moreo P, Doblare M, Garcia-Aznar J (2009) Numerical modeling of a mechano-chemical theory for wound contraction analysis. Int J Solids Struct 46:3597–3606
27. Kao C, Garner W (2000) Acute burns. Plast Reconstr Surg 101:2482–2493
28. Kavetski D, Binning P, Sloan S (2001) Adaptive time stepping and error control in a mass conservative numerical solution of the mixed form of richards equation. Adv Water Resour 24:595–605
29. Kavetski D, Binning P, Sloan S (2002) Adaptive backward Euler time stepping with truncation error control for numerical modelling of unsaturated fluid flow. Int J Numer Methods Eng 53:1301–1322
30. Keck M, Herndon D, Kamolz L, Frey M, Jeschke M (2009) Pathophysiology of burns. Wien Med Wochenschr 159:327–336
31. Koppenol D, Vermolen F, Niessen F, Van Zuijlen P, Vuik K (2016) A ematical model for the simulation of the formation and the subsequent regression of hypertrophic scar tissue after dermal wounding. Biomech Model Mechanobiol.
32. Kowalczyk R (2005) Preventing blow-up in a chemotaxis model. J Anal Appl 305:566–588
33. Lai W, Rubin D, Krempl E (1999) Introduction to continuum mechanics, 3rd edn. Butterworth Heinemann Ltd, Waltham, Massachusetts
34. Lawrence W (1998) Physiology of the acute wound. Clin Plast Surg 25:321–340
35. Li J, Chen J, Kirsner R (2007) Pathophysiology of acute wound healing. Clin Dermatol 25:9–18
36. Liang X, Boppart S (2010) Biomedical properties of in vivo human skin from dynamic optical coherence elastography. IEEE Trans Bio-Med Eng 57:953–959
37. Lindner D, Zietsch C, Becher P, Schulze K, Schultheiss HP, Tschöpe C, Westermann D (2012) Differential expression of matrix metalloproteases in human fibroblasts with different origins. Biochem Res Int 2012:1–10
38. Lo S (1989) Generating quadrilateral elements on plane and over curved surfaces. Comput Struct 31:421–426
39. Lyra P (1994) Unstructured grid adaptive algorithms for fluid dynamics and heat conduction. Ph.D. thesis, University of Wales

40. Madzvamuse A, Wathen A, Maini P (2003) A moving grid finite element method applied to a model biological pattern generator. J Comput Phys 190:478–500
41. Majno G (1979) The story of the myofibroblasts. Am J Surg Pathol 3:535–542
42. Maskarinec S, Franck C, Tirrell D, Ravichandran G (2009) Quantifying cellular traction forces in three dimensions. Proc Natl Acad Sci USA 106:22,108–22,113
43. Mast B, Schultz G (1996) Interactions of cytokines, growth factors, and proteases in acute and chronic wounds. Wound Repair Regen 4:411–420
44. Möller M (2008) Adaptive high-resolution finite element schemes. Ph.D. thesis, Dortmund University of Technology
45. Möller M, Kuzmin D (2006) Adaptive mesh refinement for high-resolution finite element schemes. Int J Numer Methods Fluids 52:545–569
46. Möller M, Kuzmin D, Kourounis D (2008) Implicit FEM-FCT algorithms and discrete Newton methods for transient convection problems. Int J Numer Methods Fluids 57:761–792
47. Monaco J, Lawrence W (2003) Acute wound healing: an overview. Clin Plast Surg 30:1–12
48. Moulin V, Castilloux G, Auger F, Garrel D, O'Connor-McCourt M, Germain L (1998) Modulated response to cytokines of human wound healing myofibroblasts compared to dermal fibroblasts. Exp Cell Res 238:283–293
49. Murphy K, Hall C, McCue S, McElwain D (2011) A two-compartment mechanochemical model of the roles of transforming growth factor ββ and tissue tension in dermal wound healing. J Theor Biol 272:145–159
50. Murphy K, Hall C, Maini P, McCue S, McElwain D (2012) A fibrocontractive mechanochemical model of dermal wound closure incorporating realistic growth factor kinetics. Bull Biol 74:1143–1170
51. Nagase H, Visse R, Murphy G (2006) Structure and function of matrix metalloproteinases and TIMPs. Cardiovasc Res 69:562–573
52. Olsen L, Sherratt J, Maini P (1995) A mechanochemical model for adult dermal wound contraction and the permanence of the contracted tissue displacement profile. J Theor Biol 177:113–128
53. O'Toole E, Mellerio J (2010) Wound healing. In: Burns D (ed) Rook's textbook of dermatology, chap 14, 8th edn. Wiley-Blackwell Publishing Ltd, New York, pp 14.1–14.27
54. Overall C, Wrana J, Sodek J (1991) Transcriptional and post-transcriptional regulation of 72-kda gelatinase/type IV collagenase by transforming growth factor-ββ1 in human fibroblasts. J Biol Chem 266:14,064–14,071
55. Patankar S (1980) Numerical heat transfer and fluid flow, 1st edn. McGraw-Hill, New York
56. Persson PO, Strang G (2004) A simple mesh generator in MATLAB. SIAM Rev 46:329–345
57. Phadke M (1989) Quality engineering using robust design, 1st edn. Prentice-Hall, Englewood Cliffs
58. Porter C, Hurren N, Herndon D, Borsheim E (2013) Whole body and skeletal muscle protein turnover in recovery from burns. Int J Burns Trauma 3:9–17
59. Quarteroni A, Valli A (2008) Numerical approximation of partial differential equations, vol 23. Springer Science & Business Media, Berlin
60. Ramtani S (2004) Mechanical modelling of cell/ECM and cell/cell interactions during the contraction of a fibroblast-populated collagen microsphere: theory and model simulation. J Biomech 37:1709–1718
61. Ramtani S, Fernandes-Morin E, Geiger D (2002) Remodeled-matrix contraction by fibroblasts: numerical investigations. Comput Biol Med 32:283–296
62. Roberts A, Sporn M, Assoian R, Smith J, Roche N, Wakefiled L, Heine U, Liotta L, Falanga V, Kehrl J, Fauci A (1986) Transforming growth factor type ββ: rapid induction of fibrosis and angiogenesis in vivo and stimulation of collagen formation in vitro. Proc Natl Acad Sci USA 83:4167–4171
63. Robins E (1990) Burn shock. Crit Care Nurs Clin N Am 2:299–307

64. Rudolph R, Vande Berg J (1991) The myofibroblast in Dupuytren's contracture. Hand Clin 7:683–692
65. Schouten H, Nieuwenhuis M, Van Zuijlen P (2012) A review on static splinting therapy to prevent burn scar contracture: do clinical and experimental data warrant its clinical application? Burns 38:19–25
66. Sherratt J, Dallon J (2002) Theoretical models of wound healing: past successes and future challenges. Comptes Rendus Biol 325:557–564
67. Sillman A, Quang D, Farboud B, Fang K, Nuccitelli R, Isseroff R (2003) Human dermal fibroblasts do not exhibit directional migration on collagen 1 in direct-current electric fields of physiological strength. Exp Dermatol 12:396–402
68. Singer A, Clark R (1999) Cutaneous wound healing. N Engl J Med 341:738–746
69. Strutz F, Zeisberg M, Renziehausin A, Raschke B, Becker V, Van Kooten C, Muller G (2001) TGF-ββ1 induces proliferation in human renal fibroblasts via induction of basic fibroblast growth factor (FGF-2). Kidney Int 59:579–592
70. Tabachnick B, Fidell L (2007) Using multivariate statistics, 5th edn. Pearson Education Inc, Boston
71. Taguchi G (1987) System of experimental design, 1st edn. UNIPUB Kraus International Publications, New York
72. Thampatty B, Wang JC (2007) A new approach to study fibroblast migration. Cell Motil Cytoskelet 64:1–5
73. The Works Inc (2014) MATLAB, Version 8.3.0.532 (R2014a). The Works Inc, Natick, MA
74. Tiwari V (2012) Burn wound: How it differs from other wounds? Indian J Plast Surg 45:364–373
75. Tomasek J, Gabbiani G, Hinz B, Chaponnier C, Brown R (2002) Myofibroblasts and mechano-regulation of connective tissue remodelling. Nat Rev Mol Cell Biol 3:349–363
76. Tranquillo R, Murray J (1992) Continuum model of fibroblast-driven wound contraction: inflammation-mediation. J Theor Biol 158:135–172
77. Treloar L (1948) Stresses and birefringence in rubber subjected to general homogeneous strain. Proc Phys Soc 60:135–144
78. Turek S (1998) FEATFLOW finite element software for the incompressible Navier–Stokes equations: user manual, 1st edn. University of Heidelberg, Heidelberg
79. Valero C, Javierre E, García-Aznar J, Gómez-Benito M (2013) Numerical modelling of the angiogenesis process in wound contraction. Biomech Model Mechanobiol 12:349–360
80. Valero C, Javierre E, García-Aznar J, Gómez-Benito M (2014a) A cell-regulatory mechanism involving feedback between contraction and tissue formation guides wound healing progression. PLoS One 9(e92):774
81. Valero C, Javierre E, García-Aznar J, Gómez-Benito M (2014b) Nonlinear finite element simulations of injuries with free boundaries: application to surgical wounds. Int J Numer Methods Biomed Eng 30:616–633
82. Valero C, Javierre E, García-Aznar J, Gómez-Benito M, Menzel A (2015a) Modeling of anisotropic wound healing. J Mech Phys Solids 79:80–91
83. Valero C, Javierre E, García-Aznar J, Menzel A, Gomez-Benito M (2015b) Challenges in the modeling of wound healing mechanisms in soft biological tissues. Ann Biomed Eng 43:1654–1665
84. Van Kan J, Segal A, Vermolen F (2014) Numerical methods in scientific computing, 2nd edn. Delft Academic Press, Delft, The Netherlands
85. Van Lint P, Libert C (2007) Chemokine and cytokine processing by matrix metalloproteinases and its effect on leukocyte migration and inflammation. J Leukoc Biol 82:1375–1381
86. Van de Berg J, Rudolph R, Poolman W, Disharoon D (1989) Comparative growth dynamics and active concentration between cultured human myofibroblasts from granulating wounds and dermal fibroblasts from normal skin. Lab Investig 61:532–538

87. Van de Water L, Varney S, Tomasek J (2013) Mechanoregulation of the myofibroblast in wound contraction, scarring, and fibrosis: opportunities for new therapeutic intervention. Adv Wound Care 2:122–141
88. Verhaegen P, Van Zuijlen P, Pennings N, Van Marle J, Niessen F, Van der Horst C, Middelkoop E (2009) Differences in collagen architecture between keloid, hypertrophic scar, normotrophic scar, and normal skin: An objective histopathological analysis. Wound Repair Regen 17:649–656
89. Vermolen F, Javierre E (2012) A finite-element model for healing of cutaneous wounds combining contraction, angiogenesis and closure. J Biol 65:967–996
90. Wang XQ, Kravchuk O, Winterford C, Kimble R (2011) The correlation of in vivo burn scar contraction with the level of αα-smooth muscle actin expression. Burns 37:1367–1377
91. Werner S, Grose R (2003) Regulation of wound healing by growth factors and cytokines. Physiol Rev 83:835–870
92. Wrobel L, Fray T, Molloy J, Adams J, Armitage M, Sparrow J (2002) Contractility of single human dermal myofibroblasts and fibroblasts. Cell Motil Cytoskelet 52:82–90
93. Zhang Z, Naga A (2005) A new finite element gradient recovery method: Superconvergence property. SIAM J Sci Comput 26:1192–1213

CHAPTER 14

Mathematical Modelling of Cytokines, MMPs and Fibronectin Fragments in Osteoarthritic Cartilage

Michelle Baker[1], Bindi S. Brook[2], Markus R. Owen[2]

[1] *School of Computer Science, University of Nottingham, Nottingham, UK*

[2] *Centre for ematical Medicine and Biology, School of ematical Sciences, University of Nottingham, Nottingham, UK*

ABSTRACT

Osteoarthritis (OA) is a degenerative disease which causes pain and stiffness in joints. OA progresses through excessive degradation of joint cartilage, eventually leading to significant joint degeneration and loss of function. Cytokines, a group of cell signalling proteins, present in raised concentrations in OA joints, can be classified into pro-inflammatory and anti-inflammatory groups. They mediate cartilage degradation through several mechanisms, primarily the up-regulation of matrix metalloproteinases (MMPs), a group of collagen-degrading enzymes. In this paper we show that the interactions of cytokines within cartilage have a crucial role to play in OA progression and treatment. We develop a four-variable ordinary differential equation model for the interactions between pro- and anti-inflammatory cytokines, MMPs and fibronectin fragments (Fn-fs), a by-product of cartilage degradation and up-regulator of cytokines. We show that the model has four classes of dynamic behaviour: homoeostasis, bistable inflammation, tristable inflammation and persistent inflammation. We show that positive and negative feedbacks controlling cytokine production rates can determine either a pre-disposition to OA or initiation of OA. Further, we show that manipulation of cytokine, MMP and Fn-fs levels can be used to treat OA, but we suggest that multiple treatment targets may be essential to halt or slow disease progression.

Keywords: Osteoarthritis Cytokine Modelling Simulation Non-linear dynamics

1 INTRODUCTION

Osteoarthritis (OA) is a degenerative disease of the joints and is a leading cause of disability worldwide. Characterised by pain and stiffness in joints, OA leads to a significant loss of mobility. OA is strongly correlated with increasing age, with estimates that OA affects as many as 80% of those over 75 years of age (Arden and Cooper 2005). Other major risk factors include genetic predisposition, obesity, and injury or trauma to the joint (Felson et al. 2000). Recent research has also implicated inflammatory processes in the pathogenesis and progression of OA (Hedbom and Huselmann 2002; Scanzello and Goldring 2012).

The main disease mechanism in OA is the degeneration of the cartilage lining the joint. This allows the bone ends to rub together and eventually the whole joint is compromised and breaks down. Onset of OA varies between individuals and can be fast, known as acute onset OA. Typically, however, the onset will be gradual, with sporadic periods of symptoms followed by asymptomatic periods for several years as the condition progresses.

Despite a desperate need for disease-modifying treatments for OA, currently none have progressed beyond clinical trials (Roubille et al. 2015), and only pain-relieving treatments are available. There is a need for research into effective treatment in early OA to slow the progression of the disease.

Initiation of OA is the loss of the homoeostatic balance between extracellular matrix (ECM) synthesis and degradation (Sandell and Aigner 2001). Cartilage breakdown is accelerated through raised pro-inflammatory cytokine levels, stimulating production of matrix metalloproteinases (MMPs), which degrade the tissue (Martel-Pelletier et al. 2008). This degradation includes fibronectin breakdown and waste fibronectin fragments (Fn-fs) act as an irritant stimulating further pro-inflammatory cytokine response (Martel-Pelletier 2004). Pro-inflammatory cytokines and Fn-fs also induce production of anti-inflammatory cytokines offsetting some of this activity.

To date there are few published ematical models of cytokine interactions specifically in OA but models which consider cartilage lesions, cytokine networks and inflammation may be relevant. A model of articular cartilage lesion formation and recovery (Wang et al. 2015; Graham et al. 2012), showed that in healthy tissue the balance of pro- and anti-inflammatory signalling is the determining factor in returning the system to homoeostasis after a blunt force injury. The model we present here also considers this balance but focuses on behaviour when the balance is lost. A model of acute systemic inflammation as a result of pathogen infection was presented by Kumar et al. (2004). The model identified five possible outcomes of infection some of which seem similar to those present in OA, particularly healthy response and recurrent inflammation. A general model of inflammation was proposed by Herald (2010) considering macrophage-mediated inflammation. The model showed that even small cytokine-dependent inflammatory responses to infection may become chronic rather than being resolved.

Two key models of cytokine signalling may be relevant to OA. One is a model of the dynamics of IL-1, TNF-αα and IL-10 in monocytes (Seymour and Henderson 2001). The model showed different types of behaviour dependent upon the parameter values, including uncontrolled production of IL-1, stable equilibria and stable limit cycles which the authors linked to observed behaviour in RA and Septic Shock. A related model by Jit et al. (2005) looked in more depth at pro-inflammatory TNF-αα and modelled the effects of anti TNF-αα drugs in the inflamed synovial joint. From the model results the authors suggested that cytokine levels in RA were usually in equilibrium and anti TNF-αα forced a shift from a disease equilibrium to a healthy equilibrium. However, they suggested that SIRS was a non-equilibrium condition and as such could not be moved to a healthy equilibrium state. A model of epithelium homoeostasis by Domínguez-Hüttinger et al. (2013) shows the importance of positive and negative feedback in homoeostatic regulation of inflammation.

We have previously published a two-variable ODE model analysing the interactions of cytokines (Baker et al. 2013) classified into two groups, pro- and anti-inflammatory. We showed that the feedback mechanisms gave rise to monostable and bistable behaviour, which we were able to associate with disease processes within the RA synovium. We suggested possible mechanisms of disease progression in RA as well as implications for treatment strategies. In that previous model we considered cytokine dynamics in isolation. However, we know that in OA other factors such as mechanical damage can lead to OA initiation and progression. Given that anti-cytokine therapies have so far proved unsuccessful in OA we suggest that not only the cytokine network but its relationship to other factors in OA, such as the waste products of joint damage, may be crucial in perpetuating the feedback networks leading to disease. For this reason, we extend this previous model to include MMPs and Fn-fs and investigate the interactions between these variables.

In the following section, we describe the model development, non-dimensionalisation and analysis of the steady states of the system. We also look at parameter sensitivity. In Sect. 3 we conduct a bifurcation analysis of the model, which shows the range of behaviours in the model which we classify based on the nature and stability of the steady states. In Sect. 4, we look at possible treatment strategies for the categories of behaviour observed in the bifurcation analysis. Finally, in Sect. 5, we discuss the main findings of the model and their implications.

2 MODEL EQUATIONS AND STEADY STATE ANALYSIS

Our four model variables are pro-inflammatory cytokines (p), anti-inflammatory cytokines (a), MMPs (m) and Fn-fs (f). The network of interactions between these variables is shown in Figure 1and is based on the joint and cytokine biology as described in Wojdasiewicz et al. (2014), Fernandes et al. (2002), Goldring (2000b) and Westacott and Sharif (1996).

Pro-inflammatory cytokines are normally present at low levels in the cartilage as they play a role in mediating the normal turnover of the ECM (Poole 1993), which involves remodelling the cartilage ECM at a very slow rate to maintain tissue integrity. Production of pro-inflammatory cytokine is up-regulated in response to trauma or infection, as part of the inflammatory response and repair mechanism. This response is usually kept in homoeostatic balance by anti-inflammatory cytokines, which act both to inhibit the synthesis of pro-inflammatory cytokines and also to block pro-inflammatory cytokine receptors (Opal and DePalo 2000). To model these processes we assume that production of pro-inflammatory cytokines, p, is dependent on itself, anti-inflammatory cytokines, a, and Fn-fs, f,

$$\frac{\mathrm{d}p}{\mathrm{d}t} = \left(c_0 + c_1\frac{p^n}{c_2{}^n + p^n} + c_3\frac{f^n}{c_4{}^n + f^n}\right)\left(\frac{c_5{}^n}{c_5{}^n + a^n}\right) - d_p p. \quad (1)$$

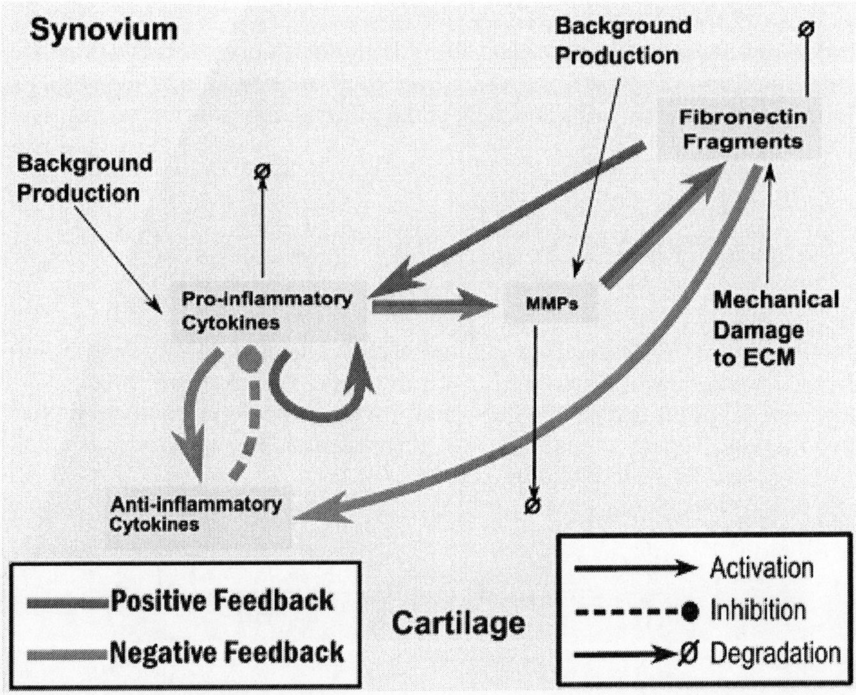

Figure 1. A simplified network of cytokine interactions within articular cartilage. Cytokines are classed as either pro- or anti-inflammatory. MMPs directly degrade the ECM leading to the release of fibronectin fragments (Fn-fs). Fn-fs are an irritant leading to an increased cytokine response

We assume that production rates will be limited biologically and therefore model functions of p, a and f as saturating Hill functions with Hill coefficients, n. For simplicity in Eq. 1 and the following we assume equal Hill coefficients in all regulatory Hill functions. In the discussion we consider the implications of unequal Hill coefficients. We also assume that pro-inflammatory cytokines will degrade naturally at rate dpdp. We choose to make the stimulatory terms additive since these two pathways are biochemically distinct and activate different cell receptors. This means that even if there is no cartilage degradation there may still be a large cytokine response due to an increase in p. Since anti-inflammatory cytokines reduce production and effectiveness of pro-inflammatory cytokines regardless of the source we apply the anti-inflammatory inhibition term to all the source terms.

Anti-inflammatory cytokine production is stimulated by both pro-inflammatory cytokines and fibronectin fragments. The dynamics of anti-inflammatory cytokines, a, therefore includes source terms (saturating Hill functions of p and f) representing their up-regulation by both pro-inflammatory cytokines and Fn-fs, and a natural degradation term,

$$\frac{\mathrm{d}a}{\mathrm{d}t} = c_6\frac{p^n}{c_7{}^n + p^n} + c_8\frac{f^n}{c_9{}^n + f^n} - d_a a. \tag{2}$$

MMPs mediate ECM degradation and the synthesis of MMPs is stimulated by pro-inflammatory cytokines (Vincenti and Brinckerhoff 2002). MMPs are also found at low levels in normal cartilage so we assume some basal production. Together with natural degradation, the dynamics of MMPs (m) are therefore modelled by

$$\frac{\mathrm{d}m}{\mathrm{d}t} = c_{10} + c_{11}\frac{p^n}{c_{12}{}^n + p^n} - d_m m. \tag{3}$$

We model Fn-fs production as a result of ECM degradation at a rate proportional to the MMP level, m, with an additional constant rate c14c14 representing mechanical damage. We also include natural degradation, giving

$$\frac{\mathrm{d}f}{\mathrm{d}t} = c_{13}m + c_{14} - d_f f. \tag{4}$$

We non-dimensionalise the model using the scalings:

$$p = c_2\tilde{p} \quad a = c_5\tilde{a} \quad m = \frac{\tilde{m}c_4 d_a}{c_{13}} \quad f = c_4\tilde{f} \quad t = \frac{\tilde{t}}{d_a}$$

where the tilde denotes dimensionless quantities. Time is scaled with the degradation of anti-inflammatory cytokines, which we expect to be in the order of minutes. Dropping the tildes for convenience gives the dimensionless model:

$$\frac{\mathrm{d}p}{\mathrm{d}t} = \left(P_{bp} + P_{pp}\frac{p^n}{1+p^n} + P_{fp}\frac{f^n}{1+f^n}\right)\left(\frac{1}{1+a^n}\right) - \gamma_p p \tag{5}$$

$$\frac{\mathrm{d}a}{\mathrm{d}t} = A_{pp}\frac{p^n}{{A_{ph}}^n + p^n} + A_{fp}\frac{f^n}{A_{fh}^n + f^n} - a \tag{6}$$

$$\frac{\mathrm{d}m}{\mathrm{d}t} = M_{bp} + M_{pp}\frac{p^n}{{M_{ph}}^n + p^n} - \gamma_m m \tag{7}$$

$$\frac{\mathrm{d}f}{\mathrm{d}t} = m + F_{dam} - \gamma_f f \tag{8}$$

where,

$$P_{bp} = \frac{c_0}{c_2 d_a}, \quad P_{pp} = \frac{c_1}{c_2 d_a}, \quad P_{fp} = \frac{c_3}{c_2 d_a}, \quad A_{pp} = \frac{c_6}{c_5 d_a},$$
$$A_{ph} = \frac{c_7}{c_2}, \quad A_{fp} = \frac{c_8}{c_5 d_a}, \quad A_{fh} = \frac{c_9}{c_4}, \quad M_{bp} = \frac{c_{10}c_{11}}{c_4 {d_a}^2},$$
$$M_{pp} = \frac{c_{11}c_{13}}{c_4 {d_a}^2}, \quad M_{ph} = \frac{c_{12}}{c_2}, \quad F_{dam} = \frac{c_{14}}{c_4 d_a}, \quad \gamma_p = \frac{d_p}{d_a},$$
$$\gamma_m = \frac{d_m}{d_a}, \quad \gamma_f = \frac{d_f}{d_a}.$$

The meaning of each of these new parameters is summarised in Table 1.

Insight into the nature of the steady states of this system can be gained from the nullclines, which are hypersurfaces given by

$$\dot{p} = 0 \Longleftrightarrow a = \left(\frac{1}{\gamma_p p}\left(P_{bp} + P_{pp}\frac{p^n}{1+p^n} + P_{fp}\frac{f^n}{1+f^n}\right) - 1\right)^{\frac{1}{n}}, \tag{9}$$

$$\dot{a} = 0 \Longleftrightarrow a = A_{pp}\frac{p^n}{A_{ph}^n + p^n} + A_{fp}\frac{f^n}{A_{fh}^n + f^n}, \tag{10}$$

$$\dot{m} = 0 \Longleftrightarrow m = \frac{M_{bp}}{\gamma_m} + \frac{M_{pp}}{\gamma_m}\frac{p^n}{M_{ph}^n + p^n}, \tag{11}$$

Table 1. The parameters in the system (5)–(8), their interpretation and the reference values used throughout the paper

Parameter	Description	Value
P_{bp}	Background pro-inflammatory production	0.01
P_{pp}	Pro-inflammatory cytokine driven pro-inflammatory cytokine production	10
P_{fp}	Fibronectin fragment driven pro-inflammatory cytokine production	10
A_{pp}	Pro-inflammatory cytokine driven anti-inflammatory cytokine production	10
A_{ph}	Pro-inflammatory cytokine concentration at which pro-inflammatory cytokine driven anti-inflammatory cytokine production is half maximal	1
A_{fp}	Fibronectin fragment driven anti-inflammatory cytokine production	10
A_{fh}	Concentration at which Fibronectin fragment driven anti-inflammatory cytokine production is half maximal	1
M_{bp}	Background MMP production	0.01
M_{pp}	Pro-inflammatory cytokine driven MMP production	10
M_{ph}	Pro-inflammatory cytokine concentration at which MMP production is half maximal	1
F_{dam}	Mechanical damage parameter	0
γ_p	Relative rate of clearance of pro-inflammatory cytokine to anti-inflammatory cytokine	1
γ_m	Relative rate of clearance of MMP to anti-inflammatory cytokine	1
γ_f	Relative rate of clearance of fibronectin fragments to anti-inflammatory cytokine	1
n	Hill coefficients of regulatory Hill functions	2

$$\dot{f} = 0 \iff f = \frac{m + F_{dam}}{\gamma_f}. \tag{12}$$

The steady states of the model are the points where all the nullclines intersect. Substituting Eqs. 11 and 12, $m = N_m(p)$ and $f = N_f(N_m(p))$ into 9 and 10 reduces the problem to two simultaneous equations for *a* in terms of *p*:

$$a = N_p(p) = \left(\frac{h(p)}{\gamma_p p} - 1\right)^{\frac{1}{n}}, \tag{13}$$

where

$$h(p) = P_{bp} + P_{pp}\frac{p^n}{1+p^n} + P_{fp}\frac{\left(F_{dam}\gamma_m + M_{bp} + M_{pp}\frac{p^n}{M_{ph}^n+p^n}\right)^n}{(\gamma_f\gamma_m)^n + \left(F_{dam}\gamma_m + M_{bp} + M_{pp}\frac{p^n}{M_{ph}^n+p^n}\right)^n}, \tag{14}$$

and

$$a = N_a(p) = A_{pp}\frac{p^n}{A_{ph}^n + p^n} + A_{fp}\frac{\left(F_{dam}\gamma_m + M_{bp} + M_{pp}\frac{p^n}{M_{ph}^n+p^n}\right)^n}{(\gamma_f\gamma_m)^n A_{fh}^n + \left(F_{dam}\gamma_m + M_{bp} + M_{pp}\frac{p^n}{M_{ph}^n+p^n}\right)^n}. \tag{15}$$

The intersections of $N_p(p)$ and $N_a(p)$, given by Eqs. (13) and (15), give the steady states, although analytical solutions are in general not tractable. The forms of Eqs. (13) and (15) mean that they will always intersect at least once, so the system will always have at least one steady state. With a Hill coefficient of n=1n=1, Eq. (15) increases monotonically and Eq. (13) decreases monotonically, giving only one possible steady state. For n>2n>2, the nullclines take a similar form as with n == 2, although the parameter values differ and the steeper gradients allow for the possibility of additional steady states, in smaller regions of parameter space. For this reason we focus on the case n=2n=2 as in our previous work, Baker et al. (2013), through the rest of this analysis.

The first term in Eq. (15) is a Hill function of *p* and the second is a Hill function of *p* embedded within another Hill function. This allows $N_a(p)$ to take two qualitative forms, either a sigmoidal shape, if A_{ph} and A_{fh} are close in value (Figure 2a), or a double sigmoidal shape if A_{ph} and A_{fh} are sufficiently different (Figure 2b).

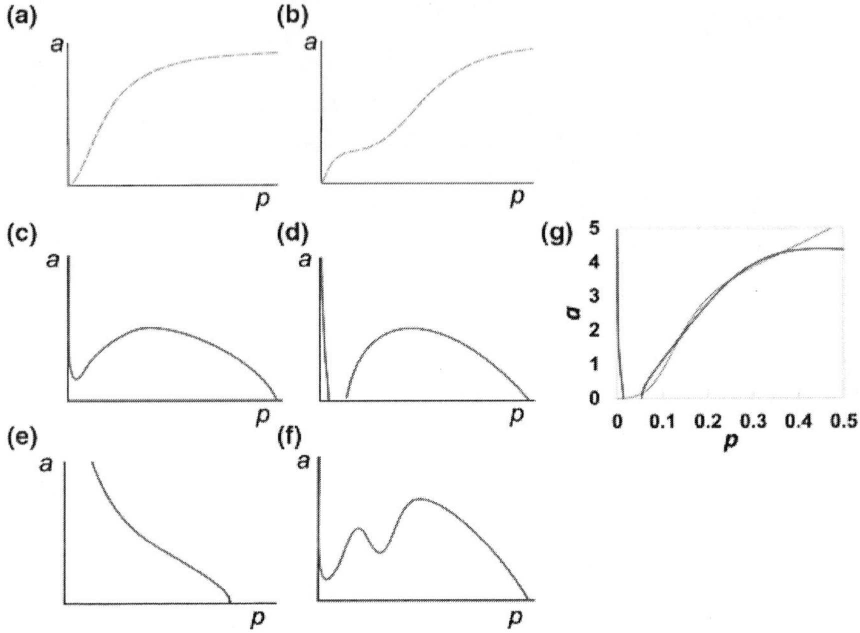

Figure 2. Possible forms of the Np(p)Np(p) and Na(p)Na(p), whose intersections define steady states of the OA-model [Eqs. (5)–(8)]. **A, B** Schematic representation showing how Na(p)Na(p) can be sigmoidal (**A**) or double-sigmoidal (**B**). **C–F**Np(p)Np(p) can take several forms that may meet the *p*-axis one (**C, E, F**) or three (**D**) times. **G** A specific example showing both Np(p)Np(p) and Na(p)Na(p) which cross five times, using the parameters $P_{bp} = M_{bp} = 0.01,\ P_{pp} = P_{fp} = A_{pp} = M_{pp} = 10,\ A_{fp} = 3.2,\ A_{ph} = 3.2,\ A_{ph} = M_{ph} = 1,\ A_{fh} = 0.2,\ \gamma_p = \gamma_f = \gamma_m = 1$ and $F_{dam} = 0$

Equation (13) is only valid when $h(p) \geq \gamma_p p$. Since $h(0) > 0$ (for positive basal production), as $\lim_{p\to\infty} \frac{h(p)}{\gamma_p p}$ and $h(p)$ is continuous, $N_a(p)$ always meets the *p*-axis for a large enough value of *p*.

The function *h*(*p*), similar to Eq. (15), can have either a sigmoidal or double sigmoidal shape. The double sigmoidal shape is possible when the terms involving PppPpp and PfpPfp are sufficiently different. This leads to several possible shapes of $a = N_p(p) = \sqrt{\frac{h(p)}{\gamma_p p} - 1}.$ These are illustrated qualitatively in Figure 2(c–f).

It is clearly possible for these curves to also intersect three times. We have found a maximum of five (see Sect. 3.5) with the Hill coefficients of n=2n=2 but could conceivably have more. Two or four steady states will only occur when the curves meet tangentially. Since this only occurs at bifurcations we will focus on cases with one, three or five steady states.

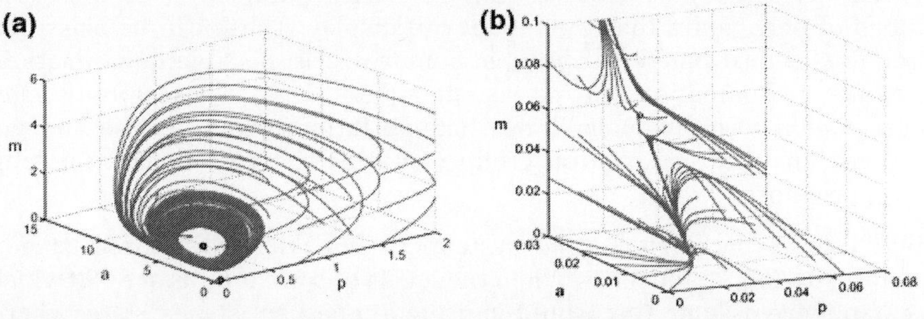

Figure 3. For the reference parameter set (Table 1), the OA model has a stable quiescent state and a stable inflamed limit cycle. The figure shows a projection of the phase space for the reference parameter set (Table 1), showing trajectories for the cartilage model [Eqs. (5)–(8)] for various regularly spaced initial conditions in (p, a, m) space. **A** Shows all three steady states whilst, **B** focuses on the behaviour around the quiescent steady state. The *black circles* show the position of unstable fixed points and the *red dot* shows the stable fixed point. The trajectories either move to the stable fixed point or the stable limit cycle which surrounds an unstable fixed point. The unstable steady state influences the path taken by trajectories (colour figure online)

Due to the difficulties of obtaining measurements of cytokine levels and rates both in vivo and in vitro, there is little data concerning the parameters in the model. We therefore chose an illustrative initial parameter set, summarised in Table 1, in which similar processes have similar rates and concentration dependence, in the absence of data suggesting otherwise. The relative thresholds in Hill functions (A_{ph}, A_{fh}, M_{ph}) are all set to one, equal to the dimensionless thresholds for pro-inflammatory cytokine regulation in Eq. 5. Similarly, the relative degradation rates $(\gamma_p, \gamma_m, \gamma_f)$ are equal to the dimensionless degradation of a. We set all the cytokine and MMP production parameters $(P_{pp}, P_{fp}, A_{pp}, A_{fp}, M_{pp})$ to 10 and background rates (P_{bp}, M_{bp}) to 0.01 in order that regulated production is relevant and not dominated by background rates. These choices avoid unnecessary bias in the network. In the following sections we consider parameter variations about this initial set in order to uncover the relevant biologically plausible behaviours to be expected from the system. Variations reasonably close to this reference parameter set show the range of behaviours that the model can display. These can be classified into four robust behaviour groups: homoeostasis, bistable inflammation, tristable inflammation and persistent inflammation. These groups may better align to individuals clinical manifestation of the disease and we propose that the behavioural group would be the determinant for long term prognosis.

With these parameters the system is bistable with three steady states: S0,S1S0,S1 and S2S2. When the levels of *p* are low, as at S0S0,which is stable, we assume this would indicate a quiescent steady state, where cells are producing only the basal level of *p* necessary for homoeostatic matrix turnover. The stable limit cycle around the unstable state S2S2 is likely to indicate an inflamed state due to high and fluctuating levels of *p* and *f*. A projection of the phase-space for this system is shown in Figure 3a, b, with the latter showing the steady states in more detail. The diagrams suggest that the stable inflamed limit cycle has a large basin of attraction whilst the quiescent steady state has only a small basin of attraction.

A sensitivity analysis of small parameter changes around the reference parameter set gives some insight into the behaviour of the system. Figure 4 shows sensitivity of the parameters to changes of up to $\pm 30\%$ as measured by three features: the concentration of *p* at the steady state, the amplitude and the period of any limit cycles. We use a one-at-a-time sensitivity analysis to measure the sensitivity gain for each feature according to the sensitivity function,

$$S_k^{\phi} = \frac{\delta\phi/\phi}{\delta k/k}, \tag{16}$$

where ϕϕ is the feature being measured and *k* is the parameter being changed. We repeat this for 1000 iterations choosing a uniformly distributed random percentage change each time within the range [−30%, 30%]. Parameters related to anti-inflammatory cytokine production $(A_{pp}, A_{ph}, A_{fp}, A_{fh})$ and clearance rate parameters

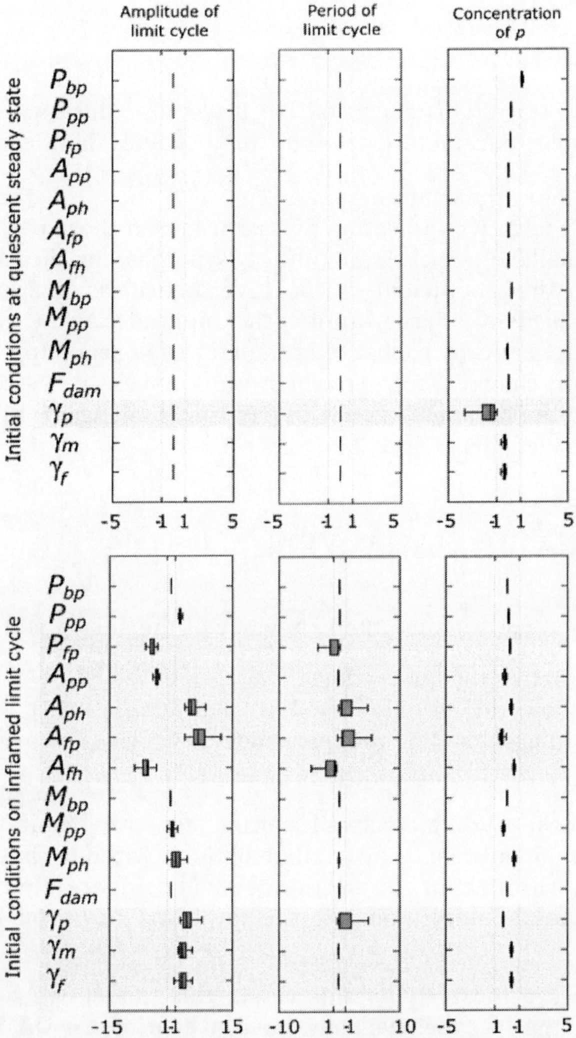

Figure 4. Parameters related to anti-inflammatory cytokine production are particularly sensitive. Box-plots of the sensitivity of steady state pro-inflammatory cytokine (p) and the period and amplitude of limit cycles, to variations in model parameters. The *top row* shows sensitivity at a quiescent state whilst the *bottom row* shows sensitivity at an inflamed state. *Boxes* show the interquartile range (IQR) of the relative sensitivity coefficient as the parameters are varied randomly from a uniform distribution within the range [−30%, 30%] (n == 1000). A value of 1 is representative of an equal change in the feature for a given change in the parameter

$(\gamma_p, \gamma_m, \gamma_f)$ are consistently the most sensitive when the system is at the inflamed state. This suggests that if we alter these parameters from the reference parameter set we may have significantly different behaviour. For the parameters P_{fp}, A_{pp} and A_{fh} the sensitivity coefficient is negative indicating that increases in these parameter values lead to a smaller limit cycle amplitude. Whereas, for the other sensitive parameters the coefficient is positive hence an increase in these parameters leads to a larger limit cycle amplitude. At the quiescent state the system is generally robust to parameter changes up to ±30% with the exception of changes to γpγp which can move the system from the quiescent to the inflamed state with parameter changes within this range. We will consider this in Sect. 3.6.

3 BIFURCATION ANALYSIS

In this section we examine the behaviour of the steady states as we vary the parameters. In the limit when $M_{bp} = 0$ and $M_{pp} = 0,$, this model reduces to the cytokine-only model of Baker et al. (2013). Figure 5 shows bifurcation diagrams for the parameters which are present in both models, with $M_{bp} = 0$ and $M_{pp} = 0.$. Here we can see how changes to the cytokine-related parameters can result in transitions between homoeostasis and inflammation without any mechanical component. Throughout this section we will compare the current model with the cytokine-only model to see the effect of fibronectin-mediated feedback.

Since there are fourteen parameters in this model, we focus on those parameters that we believe may have a bearing on OA initiation and progression. Pro-inflammatory cytokine production (P_{bp}, P_{pp}, P_{fp}), MMP production (M_{bp}, M_{pp}), Fn-fs production (FdamFdam) and Fn-fs clearance (γfγf) have all been implicated in disease progression and are considered potential treatment targets (Pearle et al. 2005; Homandberg et al. 1998; Martel-Pelletier et al. 2008). The remaining parameters, those relating to anti-inflammatory production, cytokine and MMP clearance and half maximal parameters, are discussed more briefly, with figures in the "Appendix", since we are not considering these as possible treatment targets.

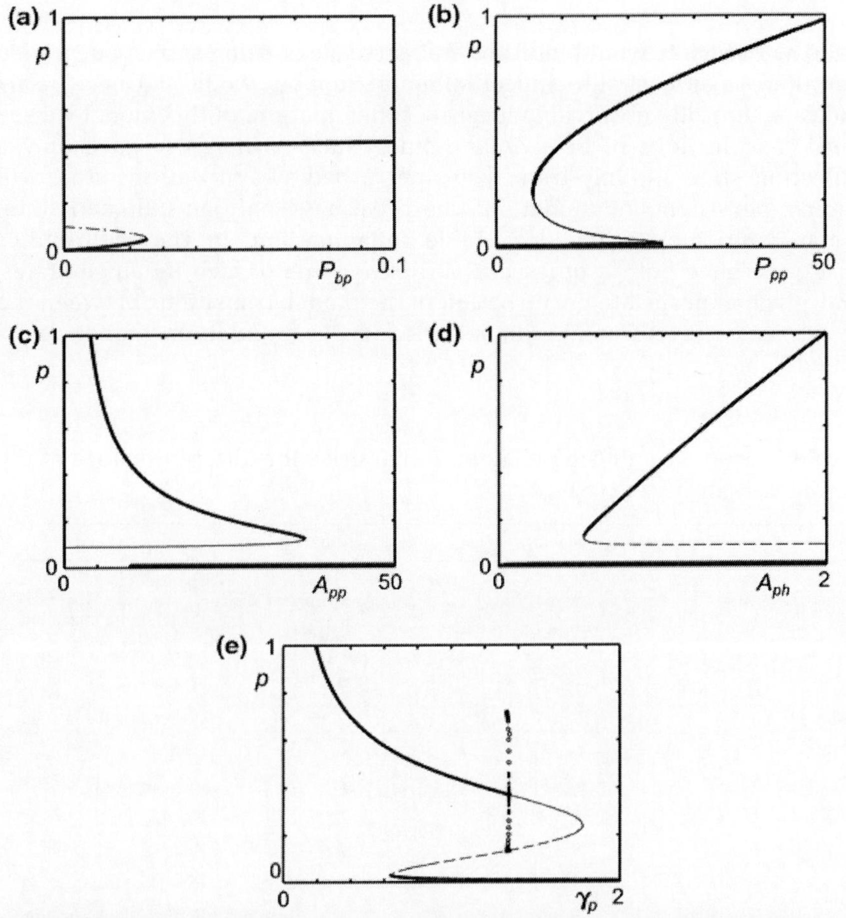

Figure 5. Bifurcation diagrams of the cytokine only model differentiate cytokine-driven from fibronectin-driven effects. Bifurcation diagrams of the cytokine-only model, (Baker et al. 2013), using the reference parameter values used in this paper showing the transitions from monostable to bistable. This model does not have any fibronectin involvement so comparison of **A**, **B** with Figs. 6a, b and **C**, **D**, **E** with 10a, c, e respectively, shows the effect of the fibronectin fragment driven feedback

Henceforth, we use the level of p as a measure of whether an individual is in a quiescent or inflamed state. A quiescent state, with low p, would indicate that either (i) an individual is healthy and shows no signs of OA or (ii) an individual has OA but the disease is not actively progressing at that time (i.e. the individual is not in a period of *flare up*). An inflamed state, with high p, would indicate that cartilage is being actively degraded and there is at least sub-clinical inflammation, but would not necessarily indicate clinically observable levels of inflammation. In this model we see four classifications of behaviour: homoeostasis, where we have only a quiescent state and any trajectories perturbed away from this state will return; persistent inflammation where we have only an inflamed state; bistable inflammation and tristable inflammation. In the multi-stable cases, we have both a quiescent and either one or two inflamed states, and, given appropriate perturbation in the model, transitions between the states can occur. A summary of behaviour types found in the model can be found Table 2.

Table 2. Summary of the behaviours that arise for different values of the parameters in system (5)–(8)

Name	No. steady states	Stability of steady states	Limit cycles	Type
Ai	1	S_0^S	–	Homoeostasis
Aii	1	S_0^S	–	Persistent inflammation
Aiii	1	S_0^U	L_1^S	Persistent inflammation
Bi	3	S_0^S, S_1^U, S_2^U	–	Homoeostasis
Bii	3	S_0^U, S_1^U, S_2^S	–	Persistent inflammation
Biii	3	S_0^S, S_1^U, S_2^S	–	Bistable
Ci	3	S_0^S, S_1^U, S_2^U	L_1^S	Bistable
Cii	3	S_0^U, S_1^U, S_2^S	L_1^S	Bistable
Ciii	3	S_0^S, S_1^U, S_2^S	L_1^S	Bistable
Civ	3	S_0^S, S_1^U, S_2^S	L_1^U, L_2^S	Bistable
Cv	3	S_0^U, S_1^U, S_2^U	L_1^S	Persistent inflammation
Di	3	S_0^S, S_1^U, S_2^S	L_1^U, L_2^S	Bistable
Dii	3	S_0^S, S_1^U, S_2^U	L_1^S, L_2^U	Bistable
Diii	3	S_0^U, S_1^U, S_2^S	L_1^U, L_2^S	Bistable
Ei	5	$S_0^S, S_1^U, S_2^S, S_3^U, S_4^S$	–	Tristable
Eii	5	$S_0^S, S_1^U, S_2^U, S_3^U, S_4^S$	–	Bistable
Fi	5	$S_0^S, S_1^U, S_2^U, S_3^U, S_4^S$	L_1^S	Tristable
Fii	5	$S_0^S, S_1^U, S_2^S, S_3^U, S_4^S$	L_1^U	Tristable

The abbreviation S means Stable and U means Unstable, indicating the stability of the steady state or limit cycle

3.1 Loss of the quiescent steady state

Since the quiescent steady state of the system occurs only when all the variables are small we can simplify the system [Eqs. (5)–(8)] to analyse the loss of this steady state. To find the leading order terms of Eq. (5) around the reference parameter set we rescale all the parameters using the small parameter ϵϵ where ϵ∼0.01ϵ∼0.01 such that all the hatted parameters are O(1). If *a* is small we lose the anti-inflammatory inhibition term since it is approximately 1. Additionally, if *f* is small Eq. (8) is a quadratic in *p*, hence Eq. (5) in the limit when all the variables are small, becomes,

$$\epsilon\frac{\mathrm{d}\widehat{p}}{\mathrm{d}t} = \epsilon\widehat{P_{bp}} + \frac{\epsilon^2}{\sqrt{\epsilon}}\widehat{P_{pp}}\widehat{p}^2 + \frac{\epsilon^3}{\sqrt{\epsilon}}\frac{\widehat{P_{fp}M_{pp}}^2}{\widehat{\gamma_m}^2\widehat{\gamma_f}^2\widehat{M_{ph}}^4}\widehat{p}^4 + 2\epsilon^2\frac{\widehat{P_{fp}M_{bp}M_{pp}}}{\widehat{\gamma_m}^2\widehat{\gamma_f}^2\widehat{M_{ph}}^2}\widehat{p}^2$$
$$+ 2\epsilon^2\frac{\widehat{P_{fp}M_{pp}F_{dam}}}{\widehat{\gamma_m\gamma_f}^2\widehat{M_{ph}}^2}\widehat{p}^2 - \epsilon\widehat{\gamma_p}\widehat{p} + \frac{\epsilon^2}{\sqrt{\epsilon}}\frac{\widehat{M_{bp}}^2\widehat{P_{fp}}}{\widehat{\gamma_m}^2\widehat{\gamma_f}^2}$$
$$+ \frac{2\epsilon^2}{\sqrt{\epsilon}}\frac{\widehat{P_{fp}M_{bp}F_{dam}}}{\widehat{\gamma_m\gamma_f}^2} + \frac{\epsilon^2}{\sqrt{\epsilon}}\frac{\widehat{P_{fp}F_{dam}}^2}{\widehat{\gamma_f}^2} = 0, \tag{17}$$

where,

$$p = \epsilon\widehat{p}, \qquad f = \epsilon\widehat{h}, \qquad P_{bp} = \epsilon\widehat{P_{bp}},$$
$$P_{pp} = \frac{\widehat{P_{pp}}}{\sqrt{\epsilon}}, \qquad P_{fp} = \frac{\widehat{P_{fp}}}{\sqrt{\epsilon}}, \qquad M_{bp} = \epsilon\widehat{M_{bp}},$$
$$M_{pp} = \frac{\widehat{M_{pp}}}{\sqrt{\epsilon}}, \qquad M_{ph} = \widehat{Mph}, \qquad F_{dam} = \epsilon\widehat{F_{dam}},$$
$$\gamma_p = \widehat{\gamma_p}, \qquad \gamma_m = \widehat{\gamma_m}, \qquad \gamma_f = \widehat{\gamma_f} = 0.$$

Taking leading order terms (O(ϵϵ)) and next order terms $(O(\epsilon^2/\sqrt{\epsilon}))$ we have,

$$\frac{\mathrm{d}p}{\mathrm{d}t} = \widehat{P_{pp}}\widehat{p}^2 - \widehat{\gamma_p}\widehat{p} + \widehat{P_{bp}} + \frac{\widehat{M_{bp}}^2\widehat{P_{fp}}}{\widehat{\gamma_m}^2\widehat{\gamma_f}^2} + \frac{\widehat{2P_{fp}M_{bp}F_{dam}}}{\widehat{\gamma_m\gamma_f}^2}$$
$$+ \frac{\widehat{P_{fp}F_{dam}^2}}{\widehat{\gamma_f}^2} = 0 \tag{18}$$

The discriminant of (18) determines where we switch from two roots to zero roots, which represents the loss of the small amplitude steady state. Hence we will lose the quiescent state if

$$\widehat{\gamma_p} < \widehat{P_{pp}}\left(\widehat{P_{bp}} + \frac{\widehat{M_{bp}}^2\widehat{P_{fp}}}{\widehat{\gamma_m}^2\widehat{\gamma_f}^2} + \frac{2\widehat{P_{fp}}\widehat{M_{bp}}\widehat{F_{dam}}}{\widehat{\gamma_m}\widehat{\gamma_f}^2} + \frac{\widehat{P_{fp}}\widehat{F_{dam}^2}}{\widehat{\gamma_f}^2}\right). \tag{19}$$

This shows that reducing clearance rates (γmγm or γfγf) or increasing production rates (Pbp,Pfp,MbpPbp,Pfp,Mbp or MppMpp) is likely to lead to inflammation. Mechanical damage is a key driver in moving the system towards inflammation, since the parameter FdamFdam appears in two terms. In the absence of mechanical damage, cytokine- and fragment-driven feedback are key to driving the system to inflammation. Referring back to the network diagram, Figure 1, this indicates that where we do not have mechanical damage, the cytokine-only feedback loops will dominate system behaviour and drive the move from quiescence. However, if we have mechanical damage, the MMP- and fibronectin-driven feedback loops become much more important and can become disease drivers even where the cytokine feedback parameters are at basal levels.

Decreases in γpγp the left-hand side of the inequality [Eq. (19)] push the system towards inflammation. However, the clearance rate of *p* is determined by the half-life of the pro-inflammatory cytokines, which is unlikely to vary either with time or between individuals, so is unlikely to be a driver of inflammation in most cases.

3.2 Pro-inflammatory cytokine production parameters

The parameters P_{bp}, P_{pp} and P_{fp} .govern the production of pro-inflammatory cytokines. These cytokines are raised in OA and this has been implicated in disease progression (Hedbom and Huselmann 2002). The mechanism by which these raised levels occur is unclear but could be the result of higher than normal production rates of pro-inflammatory cytokines. If any of these three parameters are sufficiently high, bistability is lost via a fold bifurcation, and there is a single steady state (Figure 6).

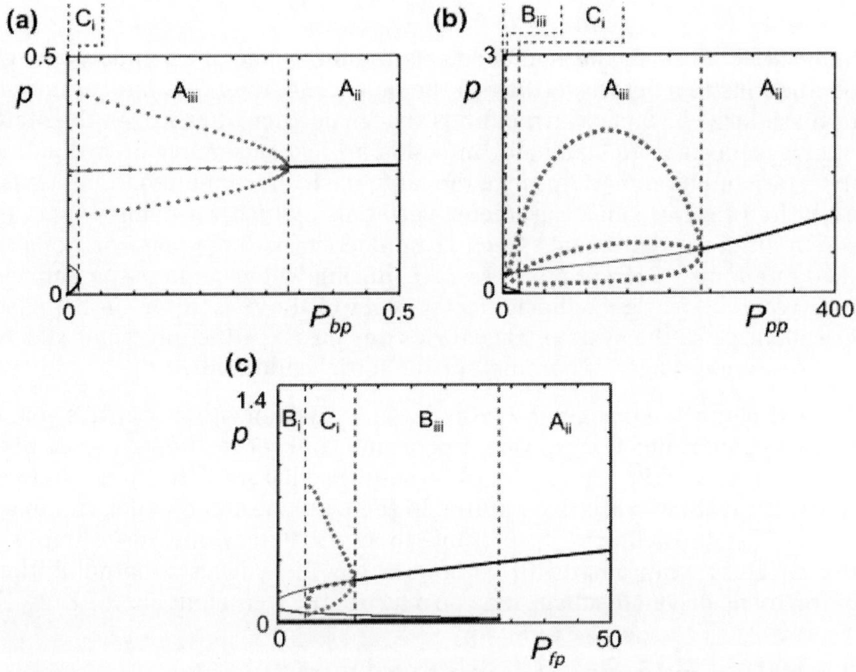

Figure 6. Increases in pro-inflammatory cytokine production lead to persistent inflammation. Bifurcation plots of the pro-inflammatory cytokine level (p) against pro-inflammatory cytokine production parameters **A**PbpPbp, **B**PppPpp and **C**PfpPfp. The *blue dotted lines* represent the minimum, maximum and average values of the limit cycles. The *vertical black dashed lines* denote the transition between different behaviours, which are labelled and summarised in Table 2 (colour figure online)

For intermediate PbpPbp and PppPpp values, the loss of bistability is followed by a single stable limit cycle (Figure 6a, b), representing an oscillatory inflamed state. Then at higher levels of these parameters this is lost, via a Hopf bifurcation, leaving a single stable steady state. For increases in PfpPfp, the Hopf bifurcation is encountered before the fold bifurcation, giving rise to two stable steady states for some values of PfpPfp. For low values of PfpPfp, we can have a single quiescent steady state representing homoeostasis. As the limit cycle collides with S1S1 at a homoclinic bifurcation it leaves only one stable and two unstable steady states. Examination of the phase space suggests that, as for the reference parameter set, the basin of attraction of the inflamed state is large in the bistable region and remains large even as we move towards the homoclinic. Even in the homoeostasis region, trajectories undergo large fluctuations in p before settling to the quiescent state. One interpretation of these large basins of attraction is that major deviations from the state of quiescence due to trauma or infection are likely to move an individual to a state of inflammation, since the large basins for inflammation persist over the range of small parameter variations, which we might expect to see in different individuals. Even in homoeostasis large perturbations to the system take a long time to resolve, during which some tissue damage may accumulate. This behaviour may point to the reason for OA being so prevalent since the system trajectories deviate from the quiescent steady state for wide ranges of parameters and initial conditions.

Comparison of Figure 5a, b (cytokine-only dynamics) with Figure 6a, b shows similar behaviour. One important difference, however, is that both PbpPbp and PppPpp in Figure 6 show oscillatory behaviour whereas the inflamed states are fixed points in the cytokine-only model. If a fixed inflamed state is clinically preferable to an oscillatory one, for example if this was less symptomatic for patients, the model suggests that inhibition of fragment-driven feedback may be a justifiable treatment aim.

3.3 MMP production parameters

The parameters MbpMbp and MppMpp determine the maximum rates of MMP production. Since MMP levels are known to be raised in OA these parameters are of great interest. Figure 7 shows that, for variations in these parameters about the reference parameter set, there are no regions of homoeostasis. At high levels of MMP production, with either high MbpMbp or MppMpp, there is a region of persistent inflammation with a stable steady state. In both cases, at lower production levels, including the reference parameter set, a fold bifurcation leads to bistability with the introduction of stable and unstable steady states, providing the possibility of moving to a quiescent state. For MppMpp (Figure 7b), within the bistable region there is a region of oscillatory inflammation due to two Hopf points.

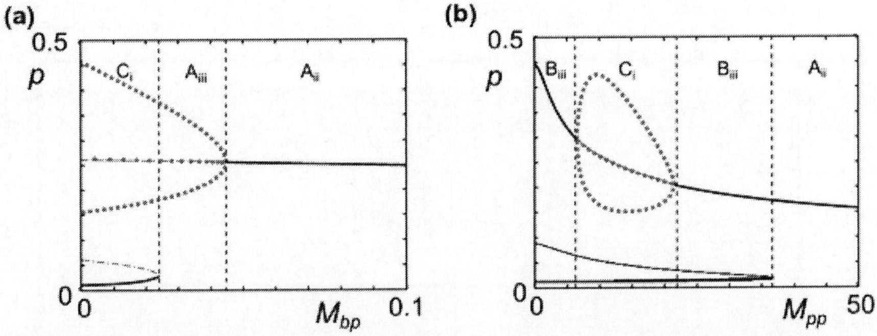

Figure 7. High rates of MMP production lead to persistent inflammation. Bifurcation diagrams of pro-inflammatory cytokine level (p) against the MMP production parameters MbpMbp and MppMpp. The *blue dotted lines* represent the minimum, maximum and average values of the limit cycles. The *vertical black dashed lines* denote the transition between different behaviours, which are labelled and summarised in Table 2

For this parameter set, for an individual in an oscillatory inflamed state, an increase in MMP production pushes the system to a steady inflamed state in which p is lower than the average of the limit cycle in the oscillatory states. However, the level of f is lower at smaller values of MppMppso the lower levels of p do not necessarily imply less cartilage degradation. This counter-intuitive result arises from the balance in the positive and negative feedback pathways. MppMpp is part of both pathways and for the reference parameter set the negative feedback is dominant. If we make the negative feedback weaker, by reducing the value of AfpAfp, the positive pathway dominates, and the value of p at the steady state increases with higher values of MppMpp as illustrated in Figure 8.

3.4 Fibronectin fragment related parameters

Fibronectin fragments have been shown to exacerbate cartilage degradation in OA (Homandberg 1999). Fibronectin levels in the model are directly influenced by a source parameter, FdamFdam, and a sink parameter, γfγf.

The bifurcation plot for γfγf, the clearance rate of Fn-fs (Figure 9a), shows persistent inflammation for low γfγf and bistability for higher γfγf. The bistable region is divided into an oscillatory inflamed state and a steady inflamed state. The value of p at the steady state lies within the range of

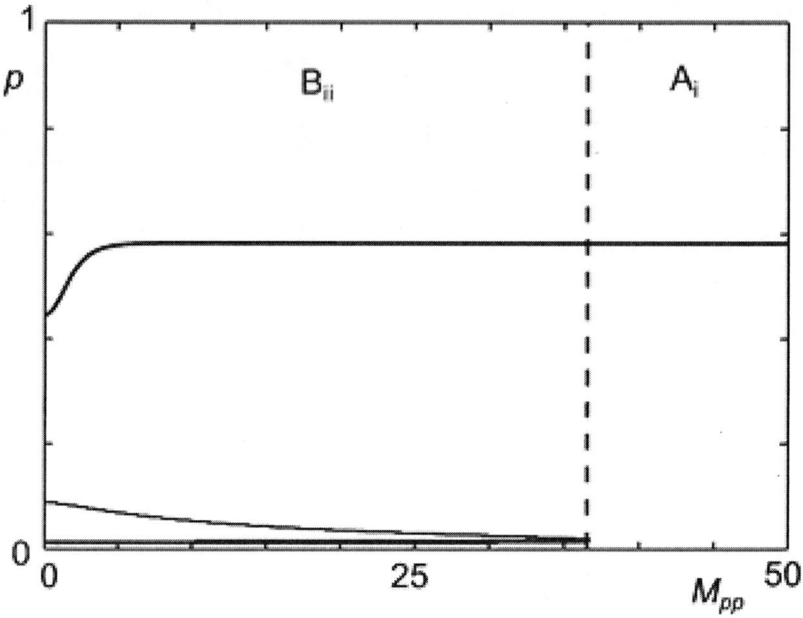

Figure 8 .The MMP production rate, MppMpp, is involved in both negative and positive feedback. Hence, its bifurcation behaviour is dependent on the parameter set. Reduced fibronectin driven anti-inflammatory cytokine production allows the positive feedback pathway to dominate, hence increases in MppMpp lead to higher values of *p* at the steady state. Bifurcation diagram of pro-inflammatory cytokine level (*p*) against the MMP production parameter MppMpp. The value of AfpAfp is reduced from 10 (seen in Figure 7b) to 2 whilst all other parameters are the same as in the reference set. The *vertical black dashed lines* denote the transition between different behaviours, which are labelled and summarised in Table 2

the oscillatory behaviour so it is not clear which behaviour type would be most destructive to the cartilage. These bifurcations suggest that increasing Fn-fs clearance in an individual could be beneficial, possibly improving the outcome of treatment to reduce cytokine levels since increasing γfγf can move the system from persistent inflammation to bistability. This is of particular interest since research has suggested that Fn-fs clearance may be affected by mechanical loading such as exercise (Evans and Quinn 2006; Zhang and Szeri 2005; O'Hara et al. 1990). Additionally, γfγf principally represents removal of Fn-fs from the ECM via diffusion, which is likely to be reduced by the changes in the joint that are seen in OA, e.g. increased water content and immobility.

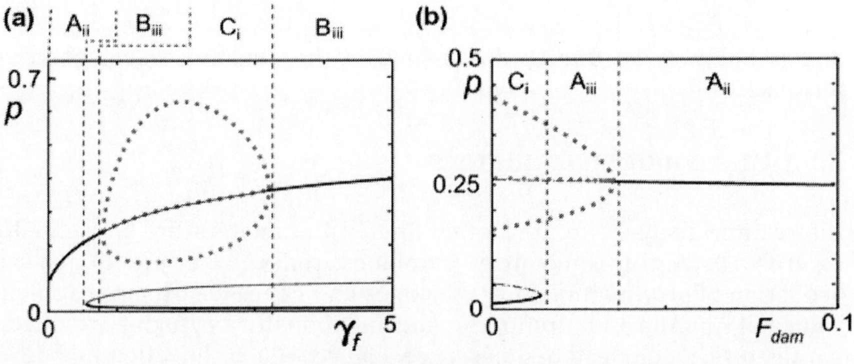

Figure 9. A Increased rates of fibronectin clearance, γfγf, can move the system from persistent inflammation, **B** increased damage, FdamFdam, has the opposite effect. Bifurcation plots of the pro-inflammatory cytokine level (*p*) against the mechanical damage and clearance parameters, γfγf and FdamFdam. The *blue dotted lines* represent the minimum, maximum and average values of the limit cycles. The *dashed lines* denote the transition between different behaviours, which are labelled and summarised in Table 2 (colour figure online)

Mechanical damage is considered to be a major risk factor in OA and the first stage of repair after mechanical damage to the cartilage is necrosis of the damaged tissue which leads to increased concentrations of Fn-fs. We represent mechanical damage to the cartilage by an increase in the Fn-fs production rate, FdamFdam. Since FdamFdam was zero in the reference parameter set, there is bistability at Fdam=0Fdam=0 as before (Figure 9b), which we might associate with the situation in normal cartilage. This behaviour has been observed in experimental work by Homandberg et al. (1997), who found that in normal cartilage tissue IL-6 levels became raised after treatment with high Fn-fs, and remained high over a period of 28 days. In contrast, for lower concentrations of Fn-fs the levels of IL-6 remained close to the control levels, suggesting a bistable system. As FdamFdam increases, a fold bifurcation removes the lower two states leaving only persistent inflammation. This has not yet been shown experimentally. However repeating the experiments of Homandberg et al. (1997) with OA cartilage tissue may provide evidence of a behavioural change, and cytokine levels would be raised.

For the reference parameter set damage leads to sustained inflammation. However, as FdamFdamincreases we see that the level of *p* at the inflamed state reduces as does the level of *f*. These reduced levels still represent inflammation but may indicate slower disease progression. That an increase in FdamFdam could slow disease progression is counter-intuitive and is likely to be due to dominance of the negative fragment-driven feedback.

3.5 Other model parameters

Bifurcation analyses for the other model parameters are presented in Figure 10. Anti-inflammatory cytokines reduce the production and activation of pro-inflammatory cytokines and hence we expect that higher values of AppApp and AfpAfp, the anti-inflammatory cytokine production parameters, would lead to quiescence. Figure 10a, b show that high levels of either parameter result in the loss of the inflamed state and a return to homoeostasis. This suggests that increases in anti-inflammatory cytokine production could be beneficial and lead to reduced or even halted cartilage degradation. Where there is an oscillatory inflamed state (Figure 10b) decreases to anti-inflammatory production, move the system to a fixed inflammatory state at a lower value of p than the maximum of the inflamed limit cycle which may be beneficial depending on the relationship between cytokine level and cartilage destruction. Comparing Figure 10a with the cytokine-only diagram (Figure 5c), the bistable region occurs over a much wider range when we have fragment-driven dynamics. In some cases this could mean that inhibition of fragment-driven feedback could move the system from inflammation to quiescence.

The parameters AphAph and AfhAfh are the concentrations of p and f at which the anti-inflammatory cytokine production terms are half maximal. As such they represent the sensitivity of the anti-inflammatory cytokine response to pro-inflammatory cytokine and Fn-fs stimulation. When either AphAph or AfhAfh is small, the anti-inflammatory cytokine response is maximal at low inputs, and we only have a low single steady state, indicative of homoeostasis (Figure 10c, d). At higher values of either AphAph or AfhAfh, larger concentrations of p or f are required for anti-inflammatory cytokine production, and there is a fold bifurcation.

MphMph governs the sensitivity of MMP production to activation by pro-inflammatory cytokine. When MphMph is small there is a single stable steady state with a high value of p indicative of persistent inflammation (Figure 10e). For higher values of MphMph we move to bistability through a fold bifurcation, allowing the possibility of a move to a quiescent state. The inflamed state is oscillatory for some values of MphMph due to two Hopf bifurcations.

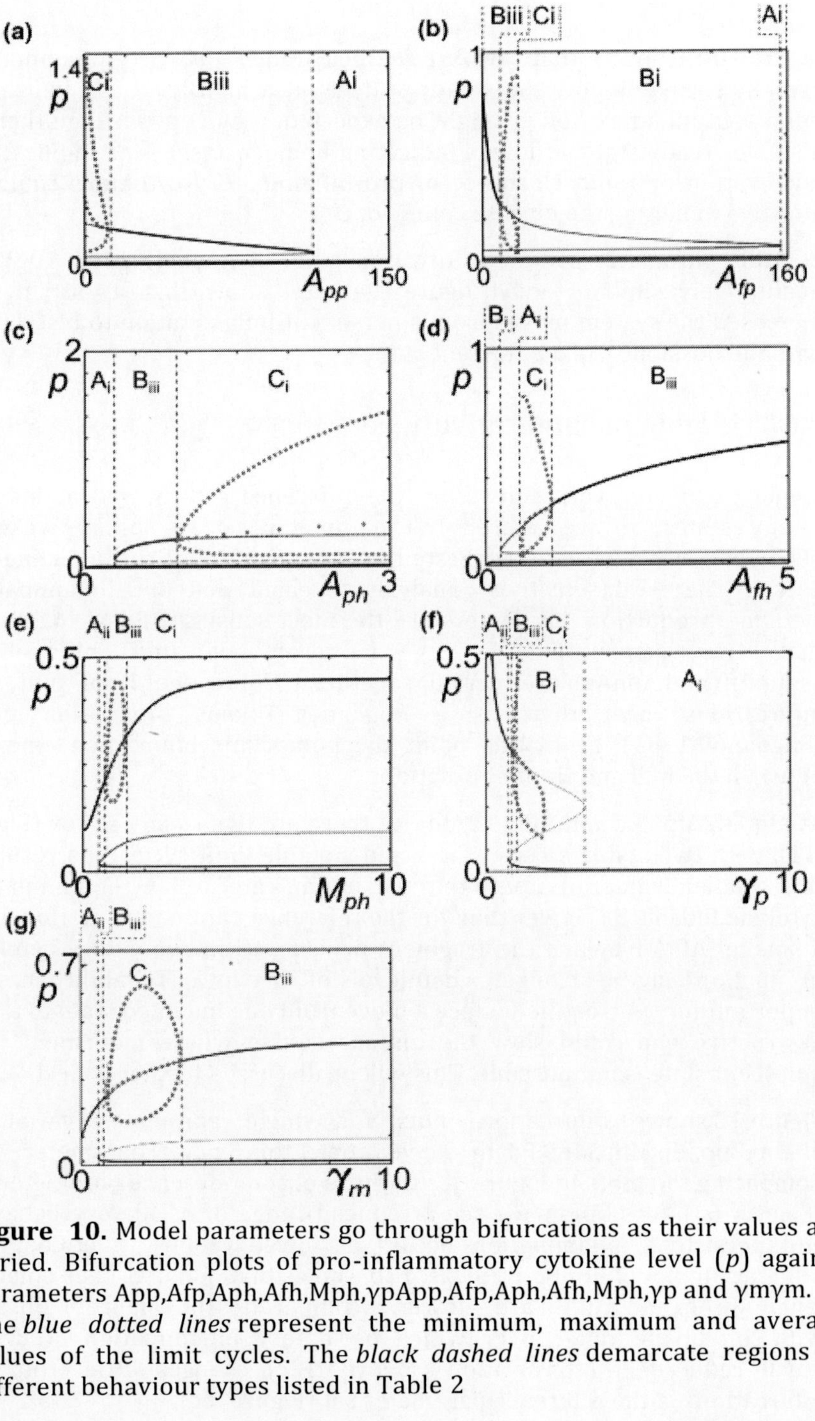

Figure 10. Model parameters go through bifurcations as their values are varied. Bifurcation plots of pro-inflammatory cytokine level (p) against parameters App,Afp,Aph,Afh,Mph,γpApp,Afp,Aph,Afh,Mph,γp and γmγm. The *blue dotted lines* represent the minimum, maximum and average values of the limit cycles. The *black dashed lines* demarcate regions of different behaviour types listed in Table 2

Figure 10f shows that when the clearance of pro-inflammatory cytokines, γpγp, is low the steady state is at a high level of *p* indicating persistent inflammation as might be expected. When γpγp is high there is a stable steady state, at low *p*, indicating homoeostasis. This implies that inactivation or rapid clearance of pro-inflammatory cytokines could be effective in halting the disease course of OA.

For this parameter set the bifurcation plots of γmγm (Figure 10g) are qualitatively similar to γfγf(Figure 9a) and show that as as γmγm is increased the system moves from a persistent inflammation to bistability, with the possibility of a quiescent state.

3.6 Tristable parameter values

Single parameter variations from the reference set showed at most 3 steady states, so we altered the parameter set to see if five-state behaviour emerged as is suggested by our analysis of the nullclines in Sect. 2. The local sensitivity analysis revealed that anti-inflammatory cytokine production parameters are the most sensitive. Figure 11 shows that five states emerge as a result of fold bifurcations as AfpAfp and AphAph are varied together. Where the Hopf and fold bifurcations meet there is a Bogdanov–Takens bifurcation point (Bogdanov 1981). From this point the homoclinic bifurcation emerges between the fold and Hopf bifurcations.

At Afp=3.2Afp=3.2 and Afh=0.2Afh=0.2 there are five steady states (Figure 11), with two stable steady states and a stable limit cycle. As a result of the smaller value of AfpAfp the maximum rate of anti-inflammatory cytokine feedback is lower than for the reference parameter set. However, a smaller AfhAfh means the fragment-driven anti-inflammatory cytokine production may be stronger when levels of *f* are low. Tristability may be important for OA treatment since a move from one inflamed state to a less destructive one could slow the disease course where movement to a quiescent state is not possible. This will be discussed further in Sect. 4.

Figure 12 shows bifurcation plots for single parameter variations of Pbp,Ppp,Pfp,Mbp,Mpp,Fdam and γf rom this new parameter set. Comparing the plots in Figure 12 to those of the reference parameter set (Figure 6a), for PbpPbp we see a reduced range of oscillatory behavior and persistent inflammation occurs at lower values of PbpPbp. The bifurcation plot of PppPpp, Figure 12b, shows that homoeostasis emerges when AfpAfp and AfhAfh are lower, compared to Figure 6b. As with PbpPbp the range over which oscillatory inflammation occurs is much reduced. PfpPfp variation (Figure 12c) changes little from the bifurcations in the reference parameter set (Figure 6c).

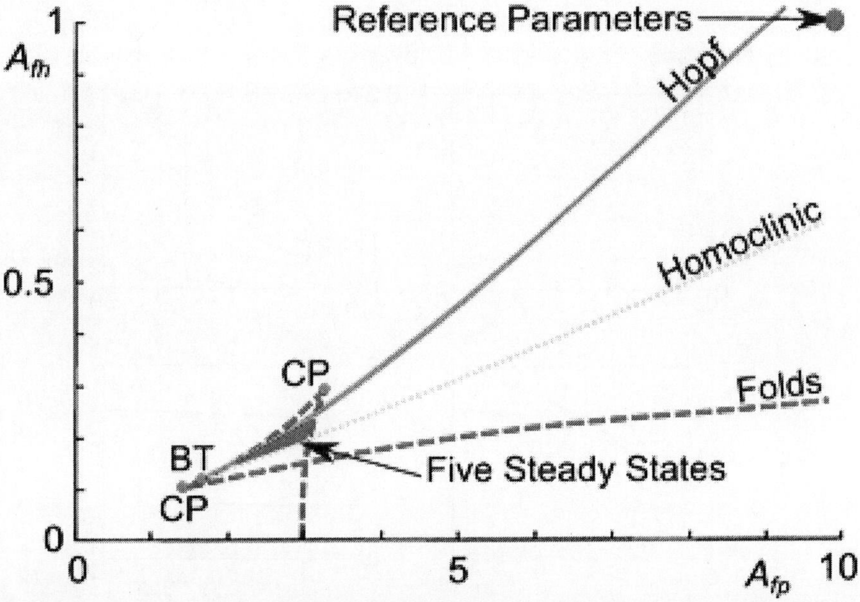

Figure 11. A small region with five steady states emerges as a result of Hopf and fold bifurcations. Two parameter bifurcation diagram showing curves of the Hopf and fold bifurcations in AfpAfp–AfhAfh space, for the reference parameter set. The diagram shows that if we reduce the value of both AfpAfp and AfhAfh from the values of the reference set there is a region where there are five steady states as a result of fold and Hopf bifurcations (shown as dashed *blue* and *solid green lines* respectively). A homoclinic bifurcation is shown in *dotted black*, which emerges from a Bogdanov–Takens bifurcation (labelled BT). Cusp points are labelled CP (colour figure online)

Bistability persists at high values of MbpMbp when AfpAfp and AfhAfh are lower (Figure 12d), however the levels of *p* in both states are high and suggest differing intensities of inflammation rather than quiescence and inflammation. The plot for MppMpp (Figure 12e) compared to that of the reference parameter set (Figure 7b), shows additional fold bifurcations in the upper branch. Additionally the limit cycle branches now collide with unstable branches at homoclinic bifurcation points. This has a significant effect as there is now a region of homoeostasis between two regions of bistability. This is also seen for. $\gamma_f \cdot F_{dam}$ (Fig. 12g), like M_{bp}, (Figure 12g), like MbpMbp, has a new region of bistability between two inflamed states.

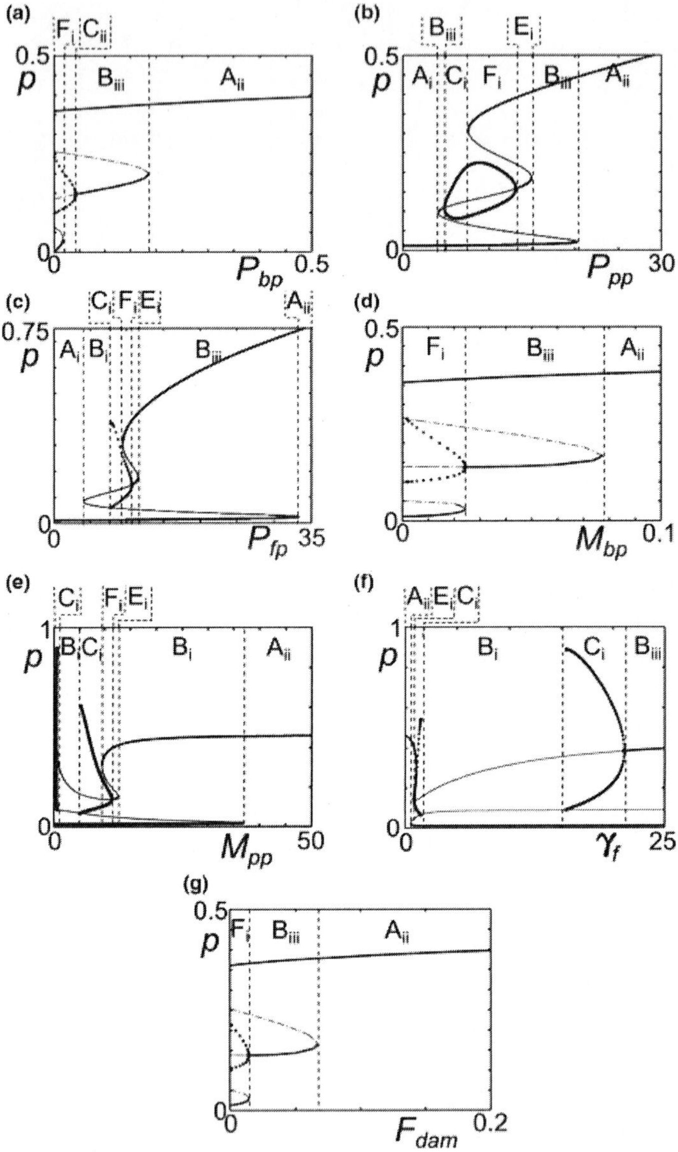

Figure 12. For some parameter ranges, one parameter variations show mono-, bi- and tri-stability. Bifurcation diagrams for Eqs. (5)–(8) in a parameter region with five steady states. Pro-inflammatory cytokine level plotted against **a** P_{bp}, **b** P_{pp}, **c** P_{fp}, **d** M_{bp}, **e** M_{pp}, **f** γ_f and **g** F_{dam}. \. The *dashed lines* denote the transition between different behaviours, which are labelled and described in Table 2. The reference parameters, see Table 1 have been used except Afp=3.2Afp=3.2 and Afh=0.2Afh=0.2

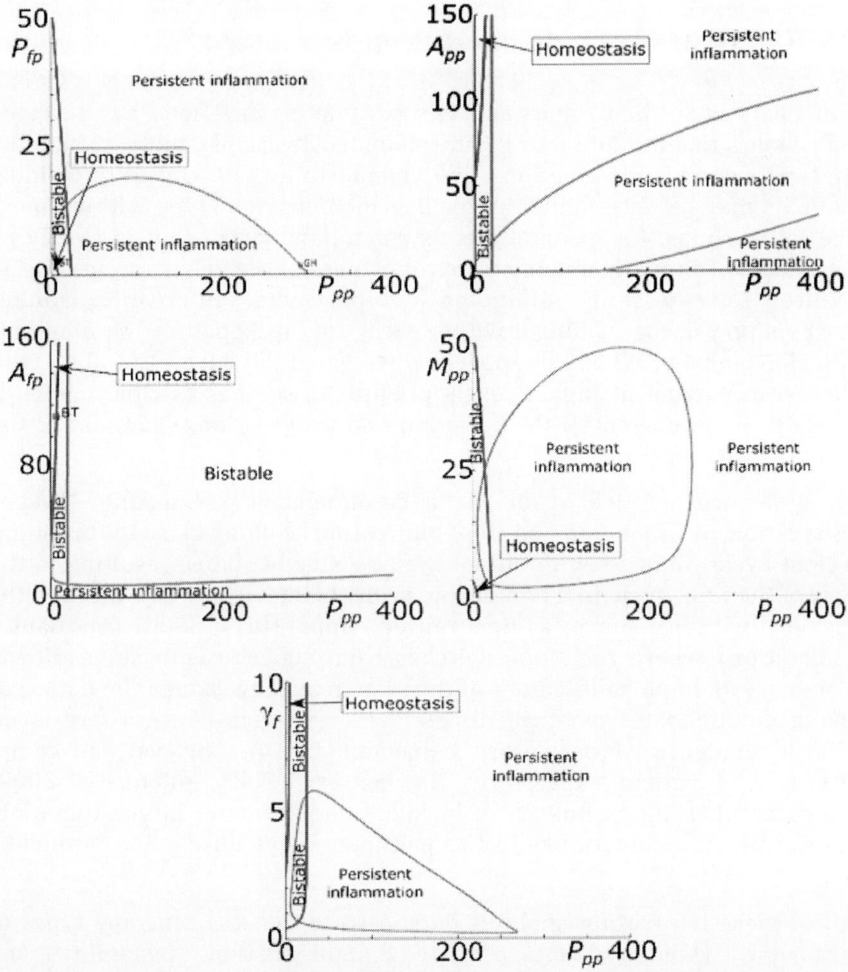

Figure 13. Two parameter bifurcation analysis showing robustness of the model [Eqs. (5)–(8)] behaviour to parameter variations. Two parameter bifurcation diagrams showing PppPppagainst P_{pp} against P_{fp}, A_{pp}, A_{fp}, M_{pp} and γfγf. Fold bifurcations are shown as *blue lines* and Hopf bifurcations as *green lines.* Areas of homoeostasis, bistability and persistent inflammation are indicated. The reference parameters in Table 1 have been used (colour figure online)

3.7 Robustness of behaviour groups

Our analysis so far of one- and two-parameter variations has revealed behaviours that fall into four groups: homoeostasis, bistability, tristability and persistent inflammation. The manifestation of OA in individual patients may be determined by the specific behaviour type within one of the four groups, for example persistent inflammation that is steady or oscillatory. To further demonstrate robustness of the behaviour groups, in Figure 13, we show additional two-parameter bifurcation studies for PppPpp varying along with each of Pfp,App,Afp, MppPfp,App, Afp,Mpp and γfγf. When PppPpp is low we generally have bistability, with disease occurring at higher levels of PppPpp, so that this parameter is likely to be important in OA initiation and progression,regardless of the other parameters.

As in Figure 11, we again see a co-dimension 2 Bogdanov–Takens bifurcation, at which fold and Hopf bifurcations collide. Close to this point, a limit cycle connects with one of the two steady states, resulting in the loss of the limit cycle to a homoclinic orbit (Izhikevich 2006). Often in the vicinity we also have a Generalised Hopf (GH) point (or Bautin bifurcation) where the Hopf switches from subcritical to supercritical. These are an important feature of the system as they change the nature of persistent disease from fixed to oscillatory or vice versa. There is no clinical evidence of oscillatory behaviour in OA, however, anecdotal evidence of patient experience (Hawker et al. 2008; Allen et al. 2009) suggests this is a possibility. We include two parameter bifurcation plots for the other combination of the six parameters in Figure 13 in "Appendix 2".

All of these bifurcation analyses demonstrate that the four key types of behaviour (homoeostasis, persistent inflammation, bistability and tristability) are robust core features of the system. In the next section we will consider possible treatment strategies for each of these behaviour groups.

4 TREATMENT STRATEGIES

Clinical trials of disease modifying drugs for OA have not shown a slowing down of disease progression, as measured by pain, inflammation and joint space, or have had unexpected complications. Here we consider two of the main treatments that have reached clinical trials, anti-cytokine drugs and MMP inhibitors, as well as a further theoretical treatment possibility Fn-fs inhibition. Anti-cytokine therapy, licensed for use in RA, inhibits either the production or functioning of pro-inflammatory cytokines, usually TNF-αα. We model this as an instantaneous reduction in the level of pro-inflammatory cytokine, p. Additionally we consider an anti-cytokine therapy which instantaneously increases the level of anti-inflammatory cytokine, a. This type of treatment is licensed for RA but trials for OA have been unsuccessful (Chevalier et al. 2009). We model the effect of MMP inhibitors as an instantaneous reduction of m. In addition to these three types of therapies we also consider the possibility of Fn-fs as a target for OA treatment and model this as a reduction in the level of f. Here we consider how persistent inflammation, bistable and tristable behaviours may respond to these treatments.

4.1 Treatment for bistable cases

In the bistable case (as in the reference parameter set we consider here) it is theoretically possible for an individual in an inflamed state to be moved to a state of quiescence, and this should be the aim for disease-modifying treatment, to achieve the best clinical outcome.

In this model, for the reference parameter set, we tried single doses of anti-cytokine therapy, MMP inhibition and Fn-fs inhibition. In each case we simulated the effect of applying the largest possible dose by reducing the level to zero, but none of these treatments moved the system to quiescence (Figure 14a), since the system was not moved outside the basin of attraction of inflammation. This result is in line with data from clinical trials of anti-cytokine and MMP inhibition treatments, that have shown no long term benefit in single dose therapy (Qvist et al. 2008).

We have found that a combined treatment strategy can bring the system to a state of quiescence. Over several simulations we varied the timing and reduced the magnitude of the doses used, until we found the smallest dose of each treatment that would move the system from inflammation to quiescence when all three treatments were combined (Figure 14b). In this case the magnitudes of the variable reductions are $0.2(p)$, $0.5(m)$ and $0.4(f)$.

Timing of the dose is of crucial importance, particularly for this parameter set since the inflamed state is oscillatory. If a dose is given at the wrong point in the limit cycle then it may not be large enough to move out of the basin of attraction of the inflamed state and may result in a period of increased amplitude oscillations as it moves back to the inflamed state (Figure 14b). This type of behaviour could have large implications both for clinical trial results and treatment regimens for drugs taken to market.

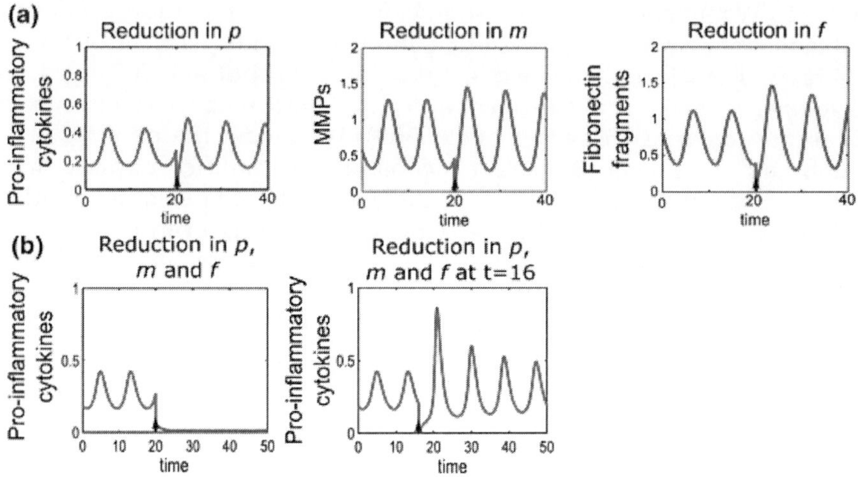

Figure 14. Monotherapy treatments are unable to move the system to quiescence, but combined treatments can be effective. **A** Time course simulations of single treatments where the system displays bistable behaviour. At t=0t=0 the system is at the disease limit cycle. A single dose of anti-cytokine (reduction in p), MMP inhibition (reduction in m) or Fn-fs inhibition (reduction in f) treatment was simulated at t=20t=20. The dose size given in each case was the maximum possible (i.e. an instantaneous decrease to zero of each of the variables). **B** Time course simulations of combined treatments where we have bistable behaviour in the system. A single combined dose of anti-cytokine, MMP inhibition and Fn-fs inhibition treatment was simulated at t=20t=20 and t=16t=16. The dose size is the minimum dose size (see text) that moves the system to health [0.2(p), 0.5(m) and 0.4(f)]. The diagrams show that dose timing as well as dose size is important. The reference parameter set was used for these simulations

We find that multiple doses of treatment given over time can also reduce the system to quiescence and allow smaller individual doses to be given. Figure 15 shows a series of six doses, given ten time units apart, which moves the system to quiescence; the magnitude of each dose is 0.1(p), 0.2(m) and 0.1(f). Compared to the single dose strategy this represents a large reduction in dosage at any particular time. This may be beneficial if there are side effects associated with the drugs, although overall, the total amount of drugs given would be higher in this case. As with the single dose, the timing and size of the dose is important as well as the total number of doses.

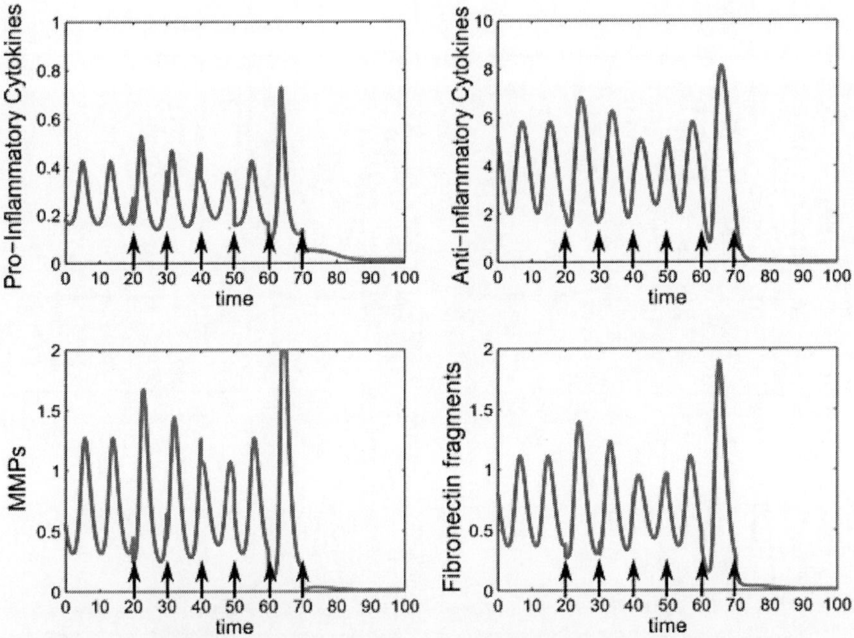

Figure 15. Multiple doses of combined therapy allow the individual dose magnitude to be reduced. Time course simulations of multiple combined treatments where we have bistable behaviour in the system. At t=0t=0 the system is at the inflamed limit cycle. Six combined doses of anti-cytokine, MMP inhibition and fibronectin fragment inhibition treatment are simulated starting at t=20t=20, with a dose interval of ten time units. The dose magnitude for each of the six doses is $0.1(p)$, $0.2(m)$ and $0.1(f)$. The reference parameter set was used for these simulations

Anti-inflammatory cytokines are not currently used in anti-cytokine therapy as they have had poor joint response in clinical trials. We found that a single dose of anti-inflammatory cytokines was able to bring the system to a quiescent state from the inflamed state. Figure 16 shows that a dose of 40 units moves the system to quiescence, when given at t == 20. This dose is the lowest that will bring the system to quiescence. However, this dose is an order of magnitude greater than the anti-inflammatory cytokine level at the inflamed state, so may not be clinically feasible.

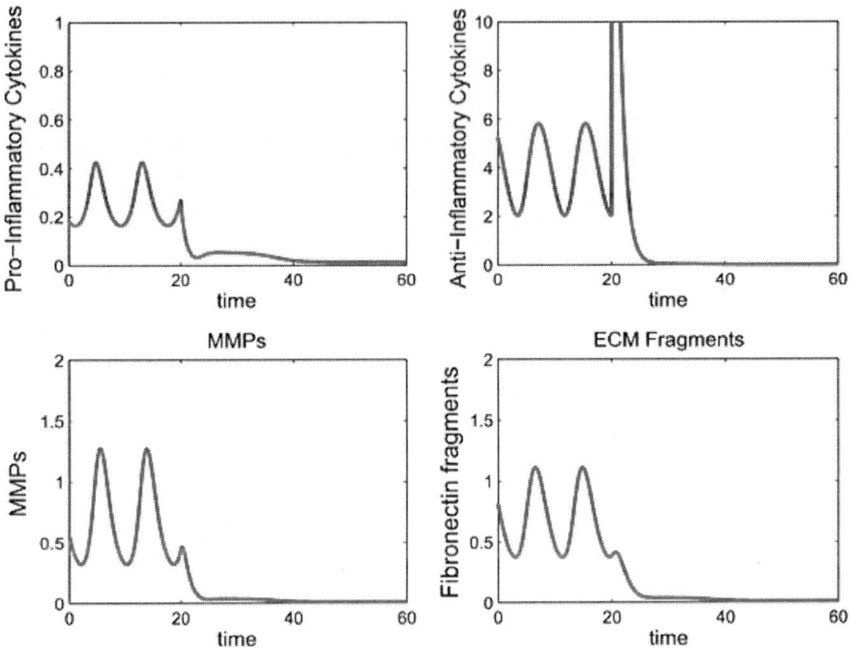

Figure 16. A high dose of anti-inflammatory cytokines can bring the bistable system to the quiescent state. Time course simulations of single doses of anti-inflammatory cytokines where we have bistable behaviour in the system. At t=0t=0 the system is at the inflamed limit cycle. A dose of 40 units of a is given at t == 20 bringing the system to quiescence. The reference parameter set (Table 1) was used for these simulations

Dose timing is not trivial; the optimal timing is not at the highest point of p of the limit cycle as might be expected, but at the point where the system is closest to the basin of attraction of the quiescent state. This point may vary between individuals so individually tailored treatment plans may be necessary for most effective treatment.

We investigated application of multiple doses of anti-inflammatory cytokines in order to reduce the dose size necessary. By giving 3 doses at intervals of 14 time units starting at t == 20 we were able to bring the system to quiescence with a dose size of 20, reduced from 40 in the single dose case at this time (Figure 17). Again timing of the initial dose,the dose interval and number of doses are of crucial importance.

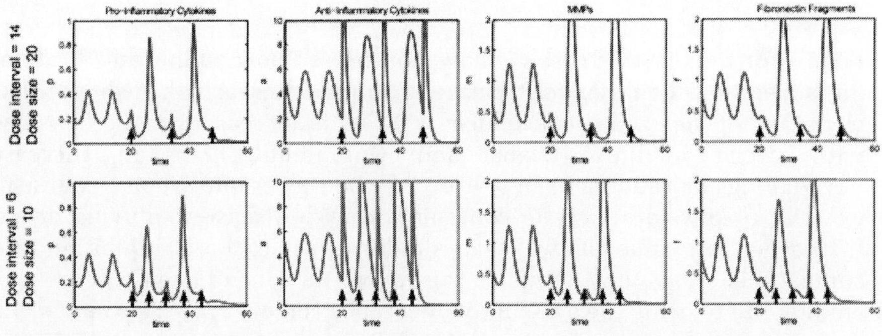

Figure 17. Multiple smaller doses of anti-inflammatory cytokine can move the system to quiescence but dose timing, interval and size are all crucial to treatment outcome. Time course simulations of multiple doses of anti-inflammatory cytokines where we have bistable behaviour in the system. At t=0t=0 the system is at the disease limit cycle. In the *top row*three doses of 20 units of *a* are given as indicated by the *black arrows.* In the *bottom row* five doses of 10 units of *a* are given as indicated by *black arrows*. The reference parameter set (Table 1) was used for these simulations

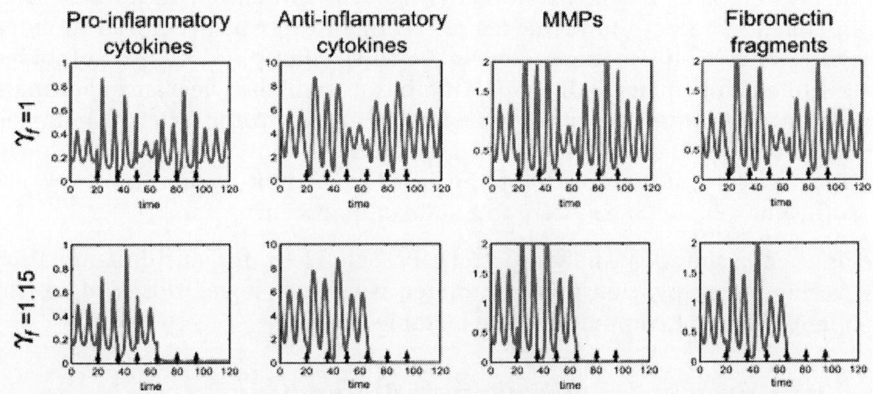

Figure 18. An increased rate of Fn-fs clearance can replace Fn-fs inhibition therapy in combined treatment regimes. Time course simulations of multiple combined treatments where the system displays bistable behaviour. The *first row* shows the system with the reference parameter set, whilst the *second row* shows the same parameters except that γfγf is increased by 15%. At t=0t=0 the system is at an inflamed limit cycle. Six combined doses of only anti-cytokine and MMP inhibition treatment are simulated starting at t=20t=20, with a dose interval of ten time units. The dose magnitude for each of the six doses is 0.4(p) and 0.4(m)

Finally for the bistable case we have considered how an increased rate of Fn-fs clearance could affect treatment options. Research has shown that clearance of macromolecules such as Fn-fs is increased in the cartilage with cyclic loading (Evans and Quinn 2006), so an increase in γfγfalongside reductions in *p* or *m*, may be representative of a course of exercise or physiotherapy in combination with disease-modifying drugs. Increasing the value of γfγf has a similar effect to Fn-fs inhibition and simulations show that if this is raised we no longer need to alter the amount of *f* to bring the system to quiescence (Figure 18). This may mean that combined anti-cytokine and MMP inhibition therapy, alongside physical therapy, could be a viable treatment option.

4.2 Treatment for tristable cases

Where we have tristable behaviour we generally have two inflamed states and one quiescent state. Simulations of treatment options for this type of behaviour show that if the system is at either one of the inflamed states it will act as in the bistable case and can be moved to the quiescent state, with a sufficient number of doses of combined treatments. Additionally if the system is at the higher inflamed state it can be moved to the lower inflamed state with fewer doses of treatment than are required to move the system to quiescence. Figure 19 shows multiple doses of combined treatments of anti-cytokine, MMP inhibition and Fn-fs clearance therapies. Where two doses are given the system returns to the original inflamed state. When four doses are given the system moves to a lower inflammatory state, which in this case is a limit cycle. Six doses are sufficient to move the system to a state of quiescence.

Figure 20, shows a similar pattern of behaviour for anti-inflammatory cytokine therapy. For this parameter set much lower doses of *a* bring about quiescent compared to the bistable case.

4.3 Treatment for persistent inflammation

In cases of persistent inflammation moving to a quiescent state is not possible without parameter changes. However, disease control may still be possible with ongoing doses of disease modifying drugs (Figure 21) which can reduce the cytokine and fibronectin levels to quiescent levels, although this is not self-sustaining. In this case the dose size required is much higher than that needed in the bistable case, to bring the system to low cytokine levels.

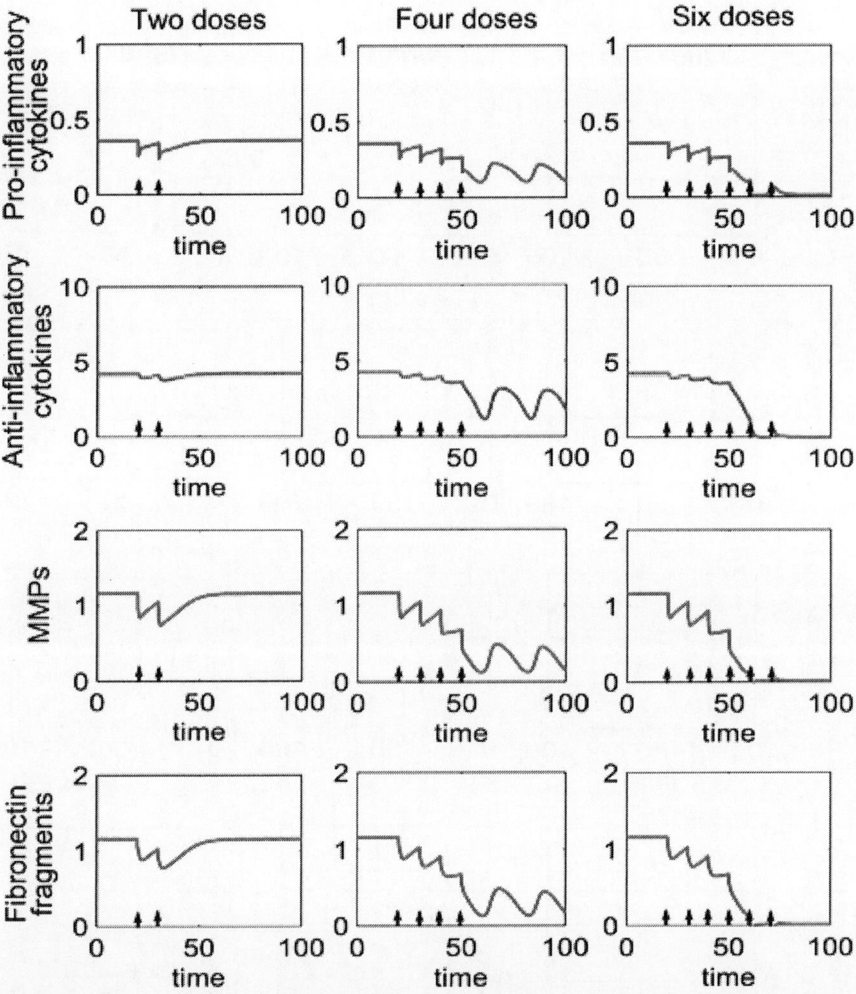

Figure 19. In the tristable system multiple doses may move the system to quiescence and a reduced number of doses may move the system to a less inflamed state. Time course simulations of multiple combined treatments where the system displays tristable behaviour. The *first column* shows two doses of treatment, the *second column* four doses and the *third column* six doses. The doses of anti-cytokine, MMP inhibition and fibronectin fragment inhibition treatment are simulated starting at t=20t=20, with a dose interval of ten time units. The dose magnitude for each dose is 0.1(*p*), 0.2(*m*) and 0.1(*f*)

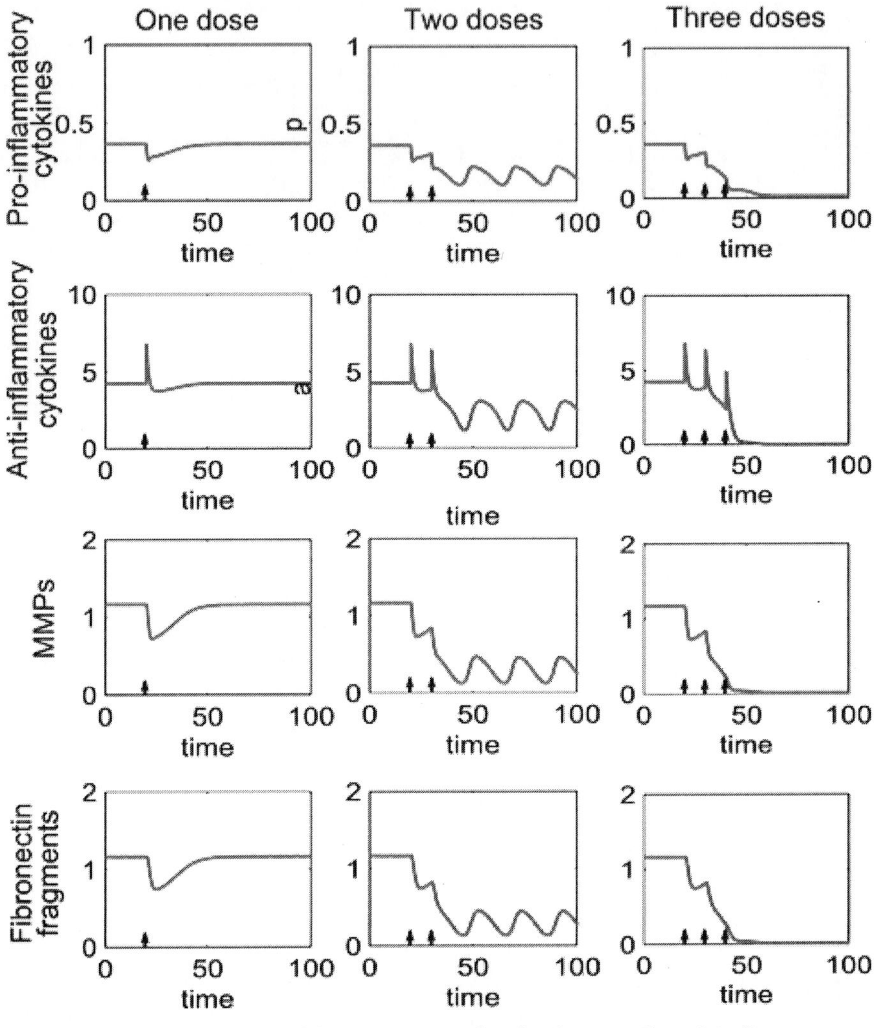

Figure 20. In the tristable system multiple doses of anti-inflammatory cytokine bring the system to either a lower inflamed state or quiescence depending upon the number of doses given. Time course simulations of multiple doses of anti-inflammatory cytokine where we have tristable behaviour in the system. The *first column* shows one dose of treatment, the *second column* two doses and the third column three doses. The doses of anti-inflammatory cytokine have a magnitude of 2.5 and are simulated starting at t=20t=20, with a dose interval of ten time units. The number of doses determines which state the system is moved to

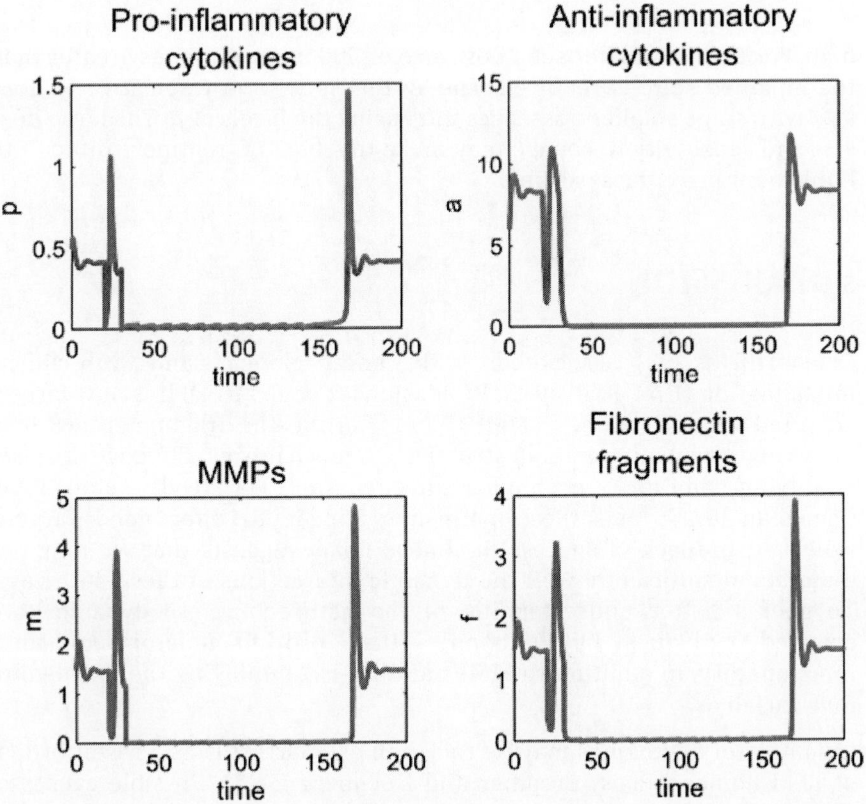

Figure 21. With persistent inflammation combined therapy can induce quiescence but cessation of treatment would allow inflammation to return. Time course for the system showing persistent inflammation. Multiple combined treatments are given ten time units apart starting at time 20. The dose size for the treatments are 0.4(*p*), 1.2(*m*) and 1.1(*f*) representing a 98%(p),82%(m)98%(p),82%(m) and 75%(f)75%(f) reduction from the inflamed state. These lower the system variables to a quiescent level. Parameters used are the reference parameter set as described in Sect. 3, except for $P_{fp} = 40$

There is a trade off between dose size and dose interval, with smaller intervals allowing a decrease in the size of the dose. However, in clinical practice there would be both medical and practical considerations in reducing dose interval.

Repeated doses of anti-inflammatory cytokine can also reduce the cytokine and fibronectin levels. However, the dose size needs to be very high. For this parameter set a dose size of 500 units, 62 times greater than the inflamed state is required. The system is also very sensitive to dose size with some smaller dose sizes increasing the levels of p. The large dose size and sensitivity to change may make this type of treatment difficult to implement in a clinical setting.

5 DISCUSSION

Osteoarthritis is associated with both clinical and sub-clinical inflammation (Hoff et al. 2013; Wojdasiewicz et al. 2014). It is now largely accepted that cytokines, MMPs and fibronectin fragments are key inflammatory mediators in destructive OA mechanisms. Effective disease-modifying therapies are much needed for OA (Qvist et al. 2008; Conaghan 2013), and these pathways appear to offer good targets. However, the lack of success in clinical trials suggests that we may not understand sufficiently well the dynamic interactions of these pathways. To gain a better understanding of the nature of these dynamics we grouped cytokines by function, either pro-inflammatory or anti-inflammatory, in addition to MMPs and Fn-fs, simplifying the problem to four variables.

Inflammatory cytokines may be raised in osteoarthritis as a result of loss of ECM homoeostasis (Creamer and Hochberg 1997). Possible causes of this include increased numbers of cytokine receptors on chondrocytes, making them more sensitive to up-regulation (Wojdasiewicz et al. 2014) and over expression of either cytokines or MMPs as a result of secretory cell senescence where cells over produce cytokines as the result of genomic damage in aging chondrocytes (Berenbaum 2013). Regardless of cause the net result is a loss of the homoeostatic balance of ECM turnover. Bifurcation analysis of the model revealed a range of behaviour types, which in general terms we class as homoeostasis, persistent inflammation, bistable inflammation or tristable inflammation. Individuals in the homoeostasis group would not be susceptible to developing OA, whereas those in the other groups would either be susceptible or have developed OA. In the cytokine-only model for rheumatoid arthritis (Baker et al. 2013), the regions of homoeostasis were relatively large and we suggested that many individuals would, therefore, not be susceptible to RA. However in the OA model presented here, in the parameter ranges we considered, homoeostasis only accounted for very small regions in comparison with the other behaviours (Figs. 6, 7, 9, 10, 13). This suggests that most individuals will be susceptible to developing OA, given an appropriate stimulus. Ageing chondrocytes, and indeed increasing frailty with aging, increase the likelihood of such a stimulus being sufficient to lead to OA changes. This corresponds with the aetiology of OA, given the large percentage of affected individuals including asymptomatic individuals, and the strong association with age.

In the bulk of this paper we have focused on equal Hill coefficients with n=2n=2 for all regulatory functions. There is no experimental evidence to indicate what *n* should be, or indeed whether all the coefficients should be equal. For equal Hill coefficients greater than 2 we have very similar qualitative behaviour to n=2n=2, except for very small regions of parameter space we have a greater number of steady states. This can also be the case with mixed coefficients greater than 2. For example, we have been able to find up to seven steady states with mixed Hill coefficients of 2 and 4. The only major behavioural change in the system comes when all n=1n=1, giving a single steady state which would represent homoeostasis or persistent inflammation, depending on the value of *p* (shown in "Appendix 1"). There would be no bifurcation behaviour in the model and treatment in this case would be as discussed in Sect. 4.3. We have no reason to believe that the special case of all n=1n=1 is likely. Our variables are not single signalling molecules but functional classes and, given the redundancy in cytokine signalling mechanisms many of these are likely to be up-regulated in OA simultaneously. Furthermore, the Hill functions represent pathways involving multiple receptors, and transcriptional regulations which are likely to involve multiple transcription factors per signalling protein (Begitt et al. 2014). Therefore, it seems entirely feasible that the case with n≥2n≥2 is representative of the in vivo situation.

Both oscillatory and steady inflammatory states are present in the model. In the steady case it is likely that such patients would see a gradual increase in inflammation and disability as a result of consistently high cytokine levels. OA progression is generally assessed through patient pain score and decreasing mobility. Some OA patients report intermittent periods of pain in early OA and a pattern of *flare ups* in OA is commonly reported (Creamer and Hochberg 1997; Hawker et al. 2008; Maly and Cott 2009; Allen et al. 2009), which may be a manifestation of the limit cycles we see in the model, or movement between quiescent and inflamed states as levels of cytokines, MMPs or Fn-fs fluctuate. This pattern of behaviour in RA has been linked to cyclic levels of cytokines, and the same may be true for OA, although data is not yet available for OA. The treatments we apply in this model move the system away from a disease state, and these treatments will work both in the case of fixed and oscillatory inflammation. It may be possible, with disease modifying drugs, to move a patient from an oscillatory inflamed state to a fixed inflammatory state, such as in the tristable case illustrated in Figure 20.

In this model, we do not examine which type of inflammatory behaviour is most destructive long term, but it may be possible to explore this with a spatial model of OA cartilage.

Examination of current research and clinical practice gives insight into potential treatment strategies. Currently no disease-modifying treatments are available for OA. Amongst those undergoing clinical review are anti-cytokine and MMP inhibitor therapies. We have considered four different treatment strategies: anti-cytokine therapy, anti-inflammatory cytokines, MMP inhibitors and Fn-fs inhibitors. We found a combined treatment strategy to be the most effective at treating bistable, tristable and persistent inflammation. Our model predictions further suggest dose size and timing are important to treatment outcome and it may be possible to optimise these using control theory. We found that the only effective monotherapy was to use anti-inflammatory cytokines, although this treatment requires very high dose sizes relative to steady state concentrations. Treatment outcome is predicted to be highly sensitive to dose timing and interval since the basin of attraction of the quiescent state is small and local to the state. These issues may make clinical treatment with anti-inflammatory cytokines, such as IL-1Ra, infeasible and may explain the failure of IL-1Ra drugs trials (Chevalier et al. 2009), despite promising experimental results which showed the response to anti-inflammatory therapy in animal models (Caron et al. 1996; Fernandes et al. 1999). For the other three treatment options any one of these alone is predicted to be ineffective, and combined treatments seem to be necessary (Sect. 4). We can see from the phase diagram (Figure 3) why this is the case. Any move from the inflamed state in only one direction would remain in the basin of attraction of the inflamed state. In all the parameter sets that we considered the basin of attraction of the quiescent state was small. Whilst our theoretical model is qualitative in nature, the results suggest that combined treatments offer a much better possibility of success than single treatments, even where the single treatment shows no benefit alone (Figure 14). We saw that in the case of tristability we had an option of treatment to move the system from a higher inflamed state to that of a lower one. If there were additional inflamed states, as considered in Sect. 2, it may be possible, in principle, to move an individual from the higher inflamed state to any of these lower states. However it is unlikely that in the foreseeable future we would be able to identify individuals with such multiple inflamed states or personalise the treatment plan to such a degree.

Some evidence of the feasibility the treatments we have suggested may come from studies into joint lavage therapy. Although controversial due to mixed results, this treatment involves *flushing* an OA joint with saline which is analogous to reducing the value of variables in the model. Some studies have reported relatively long term improvements in patients given this treatment (Arden et al. 2008). The varied results of such treatments with some patients finding it ineffective, corroborates the likely existence of different disease phenotypes.

Clinical and experimental data could provide validation for this theoretical model. Lack of data and the difficulty of obtaining data relating to cytokine and MMP dynamics in OA means that the existence of the behaviour dynamics found in this model has not yet been proven in a clinical or experimental setting. However, studies of cytokine and MMP expression in OA have shown that these are raised both in early and advanced OA (Scanzello et al. 2009). This suggests the presence of an inflamed state where levels of cytokines and MMPs are raised, in comparison to a quiescent state where they remain low as predicted by our model. The study of Homandberg et al. (1997) indicates that bistability exists within normal human cartilage, and could be extended to investigate whether this changes in osteoarthritic cartilage. More experimental work would need to be conducted in this area to show the difference between early and late stage OA, which would help identify which type of treatments would be appropriate. However, since the structure of the cytokine network we model here is well established (Goldring 2000a; Martel-Pelletier et al. 1999; Tetlow et al. 2001; Sandell and Aigner 2001) we are confident that such states could exist, and hence that treatments of the types we have theorised are realistic experimental targets. The lack of biomarkers (easily measurable indicators of disease severity) for OA makes oscillatory behaviour difficult to track, but periods of *flare up* are widely reported anecdotally and considered part of the disease course, particularly in early OA. Associations between these periods of OA flare up, increased MMP expression and increased cartilage degradation have been reported (Manicourt et al. 2005).

Limitations of the model include the lack of a link to clinical disease measures, which include joint space narrowing and radiographic evidence of cartilage deterioration. As better measures of OA disease activity are developed, such as OA biomarkers we may be able to draw more detailed conclusions about OA disease dynamics in future studies. We have not explored spatial and mechanical aspects of the disease, which play a large role in OA progression, and have been explored ematically by others. In this paper we have explored the behaviour of cytokine interactions in the joint and identified potential areas of future research into OA treatment strategies. Although we have been unable to directly validate the model due to a lack of quantitative, comparable experimental or clinical data, we have found that the model mimics many clinical features of the condition. We believe that future work in this area needs to combine all these aspects of OA and joint mechanics, as it is becoming increasingly clear from biological research that the interactions between physical and biochemical factors in OA are significant.

ACKNOWLEDGEMENTS

This work was supported by the Biotechnology and Biological Sciences Research Council [Grant Number BB/I532353/1], by an Industrial CASE Partnership; and Pfizer.

REFERENCES

1. Allen K, Coffman C, Golightly YM, Stechuchak K, Keefe F (2009) Daily pain variations among patients with hand, hip, and knee osteoarthritis. Osteoarthr Cartil 17(10):1275–1282
2. Arden N, Cooper C (2005) Osteoarthr Handb. Taylor & Francis, London
3. Arden N, Reading I, Jordan K, Thomas L, Platten H, Hassan A, Ledingham J (2008) A randomised controlled trial of tidal irrigation vs corticosteroid injection in knee osteoarthritis: the KIVIS study. Osteoarthr Cartil 16(6):733–739
4. Baker M, Denman-Johnson S, Brook BS, Gaywood I, Owen MR (2013) ematical modelling of cytokine-mediated inflammation in rheumatoid arthritis. Med Biol 30(4):311–337
5. Begitt A, Droescher M, Meyer T, Schmid CD, Baker M, Antunes F, Knobeloch KP, Owen MR, Naumann R, Decker T et al (2014) STAT1-cooperative DNA binding distinguishes type 1 from type 2 interferon signaling. Nat Immunol 15(2):168–176
6. Berenbaum F (2013) Osteoarthritis as an inflammatory disease (osteoarthritis is not osteoarthrosis!). Osteoarthr Cartil 21(1):16–21
7. Bogdanov R (1981) Bifurcations of a limit cycle for a family of vector fields on the plane. Sel Sov 1:373–388
8. Caron JP, Fernandes JC, Martel-Pelletier J, Tardif G, Mineau F, Geng C, Pelletier JP (1996) Chondroprotective effect of intraarticular injections of interleukin-1 receptor antagonist in experimental osteoarthritis. suppression of collagenase-1 expression. Arthritis Rheum 39(9):1535–1544
9. Chevalier X, Goupille P, Beaulieu AD, Burch FX, Bensen WG, Conrozier T, Loeuille D, Kivitz AJ, Silver D, Appleton BE (2009) Intraarticular injection of anakinra in osteoarthritis of the knee: a multicenter, randomized, double-blind, placebo-controlled study. Arthritis Care Res 61:344–352
10. Conaghan PG (2013) Osteoarthritis in 2012: parallel evolution of OA phenotypes and therapies. Nat Rev Rheumatol 9(2):68–70
11. Creamer P, Hochberg MC (1997) Osteoarthritis. Lancet 350(9076):503–509
12. Domínguez-Hüttinger E, Ono M, Barahona M, Tanaka RJ (2013) Risk factor-dependent dynamics of atopic dermatitis: modelling multi-scale regulation of epithelium homeostasis. Interface Focus.
13. Evans RC, Quinn TM (2006) Dynamic compression augments interstitial transport of a glucose-like solute in articular cartilage. Biophys J 91(4):1541–1547
14. Felson DT, Lawrence RC, Dieppe PA, Hirsch R, Helmick CG, Jordan JM, Kington RS, Lane NE, Nevitt MC, Zhang Y, Sowers M, McAlindon T, Spector TD, Poole AR, Yanovski SZ, Ateshian G, Sharma L, Buckwalter JA, Brandt KD, Fries JF (2000)

Osteoarthritis: new insights. part 1: the disease and its risk factors. Ann Intern Med 133(8):635–646
15. Fernandes J, Tardif G, Martel-Pelletier J, Lascau-Coman V, Dupuis M, Moldovan F, Sheppard M, Krishnan BR, Pelletier JP (1999) In vivo transfer of interleukin-1 receptor antagonist gene in osteoarthritic rabbit knee joints: prevention of osteoarthritis progression. Am J Pathol 154(4):1159–1169
16. Fernandes JC, Martel-Pelletier J, Pelletier JP (2002) The role of cytokines in osteoarthritis pathophysiology. Biorheology 39(1, 2):237–246
17. Goldring MB (2000a) Osteoarthritis and cartilage: the role of cytokines. Curr Rheumatol Rep 2:459–465
18. Goldring MB (2000b) The role of the chondrocyte in osteoarthritis. Arthritis Rheum 43:1916–1926
19. Graham JM, Ayati BP, Ding L, Ramakrishnan PS, Martin JA (2012) Reaction-diffusion-delay model for EPO/TNF-αα interaction in articular cartilage lesion abatement. Biol Direct 7(1):1–15.
20. Hawker G, Stewart L, French M, Cibere J, Jordan J, March L, Suarez-Almazor M, Gooberman-Hill R (2008) Understanding the pain experience in hip and knee osteoarthritis an OARSI/OMERACT initiative. Osteoarthr Cartil 16(4):415–422
21. Hedbom E, Huselmann HJ (2002) Molecular aspects of pathogenesis in osteoarthritis: the role of inflammation. Cell Mol Life Sci CMLS 59(1):45–53
22. Herald M (2010) General model of inflammation. Bull Biol 72:765–779
23. Hoff P, Buttgereit F, Burmester G, Jakstadt M, Gaber T, Andreas K, Matziolis G, Perka C, Rohner E (2013) Osteoarthritis synovial fluid activates pro-inflammatory cytokines in primary human chondrocytes. Int Orthop 37:145–151
24. Homandberg GA (1999) Potential regulation of cartilage metabolism in osteoarthritis by fibronectin fragments. Front Biosci 4(D7):13–30
25. Homandberg GA, Hui F, Wen C, Purple C, Bewsey K, Koepp H, Huch K, Harris A (1997) Fibronectin-fragment-induced cartilage chondrolysis is associated with release of catabolic cytokines. Biochem J 321(Pt 3):751–757
26. Homandberg GA, Wen C, Hui F (1998) Cartilage damaging activities of fibronectin fragments derived from cartilage and synovial fluid. Osteoarthr Cartil 6(4):231–244
27. Izhikevich E (2006) Dynamical systems in neuroscience: the geometry of excitability and bursting. MIT Press, Cambridge
28. Jit M, Henderson B, Stevens M, Seymour RM (2005) TNF-αα neutralization in cytokine-driven diseases: a ematical model to account for therapeutic success in rheumatoid arthritis but therapeutic failure in systemic inflammatory response syndrome. Rheumatology 44:323–331
29. Kumar R, Clermont G, Vodovotz Y, Chow CC (2004) The dynamics of acute inflammation. J Theor Biol 230:145–155
30. Maly MR, Cott CA (2009) Being careful: a grounded theory of emergent chronic knee problems. Arthritis Care Res 61(7):937–943
31. Manicourt DH, Bevilacqua M, Righini V, Famaey JP, Devogelaer JP (2005) Comparative effect of nimesulide and ibuprofen on the urinary levels of collagen type II c-telopeptide degradation products and on the serum levels of hyaluronan and matrix metalloproteinases-3 and -13 in patients with flare-up of osteoarthritis. Drugs R&D 6(5):261–271
32. Martel-Pelletier J (2004) Pathophysiology of osteoarthritis. Osteoarthr Cartil/OARS Osteoarthr Res Soc 12:31–33
33. Martel-Pelletier J, Alaaeddine N, Pelletier JP (1999) Cytokines and their role in the pathophysiology of osteoarthritis. Front Biosci 4:D694–703
34. Martel-Pelletier J, Boileau C, Pelletier JP, Roughley PJ (2008) Cartilage in normal and osteoarthritis conditions. Best Pract Res Clin Rheumatol 22:351–384
35. O'Hara BP, Urban JP, Maroudas A (1990) Influence of cyclic loading on the nutrition of articular cartilage. Ann Rheum Dis 49(7):536–539
36. Opal SM, DePalo VA (2000) Anti-inflammatory cytokines. Chest 117:1162–1172

37. Pearle AD, Warren RF, Rodeo SA (2005) Basic science of articular cartilage and osteoarthritis. Clin Sports Med 24:1–12
38. Poole AR (1993) Cartilage in health and disease. Arthritis Allied Cond Textb Rheumatol 14:226–284
39. Qvist P, Bay-Jensen AC, Christiansen C, Dam EB, Pastoureau P, Karsdal MA (2008) The disease modifying osteoarthritis drug (DMOAD): is it in the horizon? Pharmacol Res 58(1):1–7
40. Roubille C, Pelletier JP, Martel-Pelletier J (2015) Drug/agent treatments for osteoarthritis: present and future. In: Kapoor M, Mahomed NN (eds) Osteoarthritis: pathogenesis, diagnosis, available treatments, drug safety, regenerative and precision medicine. Springer, Berlin, pp 191–210
41. Sandell LJ, Aigner T (2001) Articular cartilage and changes in arthritis. an introduction: cell biology of osteoarthritis. Arthritis Res 3(2):107–113
42. Scanzello CR, Goldring SR (2012) The role of synovitis in osteoarthritis pathogenesis. Bone 51(2):249–257
43. Scanzello CR, Umoh E, Pessler F, Diaz-Torne C, Miles T, DiCarlo E, Potter HG, Mandl L, Marx R, Rodeo S, Goldring SR, Crow MK (2009) Local cytokine profiles in knee osteoarthritis: elevated synovial fluid interleukin-15 differentiates early from end-stage disease. Osteoarthr Cartil 17(8):1040–1048
44. Seymour RM, Henderson B (2001) Pro-inflammatoryanti-inflammatory cytokine dynamics mediated by cytokinereceptor dynamics in monocytes. Med Biol 18(2):159–192
45. Tetlow LC, Adlam DJ, Woolley DE (2001) Matrix metalloproteinase and proinflammatory cytokine production by chondrocytes of human osteoarthritic cartilage: associations with degenerative changes. Arthritis Rheum 44(3):585–594
46. Vincenti M, Brinckerhoff C (2002) Transcriptional regulation of collagenase (mmp-1, mmp-13) genes in arthritis: integration of complex signaling pathways for the recruitment of gene-specific. Arthritis Res 4(3):157–164
47. Wang X, Brouillette MJ, Ayati BP, Martin JA (2015) A validated model of the pro- and anti-inflammatory cytokine balancing act in articular cartilage lesion formation. Front Bioeng Biotechnol 3:25.
48. Westacott CI, Sharif M (1996) Cytokines in osteoarthritis: mediators or markers of joint destruction? Semin Arthritis Rheum 25(4):254–272
49. Wojdasiewicz P, Poniatowski LA, Szukiewicz D (2014) The role of inflammatory and anti-inflammatory cytokines in the pathogenesis of osteoarthritis. Mediators Inflamm 2014:561459.
50. Zhang L, Szeri A (2005) Transport of neutral solute in articular cartilage: effects of loading and particle size. Proc R Soc A Phys Eng Sci 461(2059):2021–2042

CHAPTER 15

Combination Therapeutics of Nilotinib and Radiation in Acute Lymphoblastic Leukemia as an Effective Method Against Drug-Resistance

Kamran Kaveh[1], Yutaka Takahashi[2,3], Michael A. Farrar[4], Guy Storme[5], Marcucci Guido[6], Jamie Piepenburg[2], Jackson Penning , Jasmine Foo[7], Kevin Z. Leder[8], Susanta K. Hui[2,9]

[1] *Program for Evolutionary Dynamics, Harvard University, Cambridge, Massachusetts, United States of America,*

[2] *Department of Radiation Oncology, University of Minnesota, Minneapolis, Minnesota, United States of America,*

[3] *Department of Radiation Oncology, Osaka University, Osaka, Japan,*

[4] *Department of Laboratory Medicine and Pathology, University of Minnesota, Minneapolis, Minnesota, United States of America,*

[5] *Department of Radiotherapy, Universitair Ziekenhuis Brussel, Brussels, Belgium,*

[6] *Department of Hematology and Beckman Research Institute, City of Hope National Medical Center, Duarte, California, United States of America,*

[7] *Department of ematics, University of Minnesota, Minneapolis, Minnesota, United States of America,*

[8] *Industrial and Systems Engineering, University of Minnesota, Minneapolis, Minnesota, United States of America,*

[9] *Department of Radiation Oncology, City of Hope National Medical Center and Beckman Research Institute, Duarte, California, United States of America*

ABSTRACT

Philadelphia chromosome-positive (Ph+) acute lymphoblastic leukemia (ALL) is characterized by a very poor prognosis and a high likelihood of acquired chemo-resistance. Although tyrosine kinase inhibitor (TKI) therapy has improved clinical outcome, most ALL patients relapse following treatment with TKI due to the development of resistance. We developed an *in vitro*model of Nilotinib-resistant Ph+ leukemia cells to investigate whether low dose radiation (LDR) in combination with TKI therapy overcome chemo-resistance. Additionally, we developed a ematical model, parameterized by cell viability experiments under Nilotinib treatment and LDR, to explain the cellular response to combination therapy. The addition of LDR significantly reduced drug resistance both in vitro and in computational model. Decreased expression level of phosphorylated AKT suggests that the combination treatment plays an important role in overcoming resistance through the AKT pathway. Model-predicted cellular responses to the combined therapy provide good agreement with experimental results. Augmentation of LDR and Nilotinib therapy seems to be beneficial to control Ph+ leukemia resistance and the quantitative model can determine optimal dosing schedule to enhance the effectiveness of the combination therapy.

Author summary

High likelihood of evolution of resistance to therapy is common in most forms of leukemia. This issue persists for tyrosine kinase inhibitor drug treatments as well as other forms of therapies. In the current work, we suggest a combination therapy where Ph+ acute lymphoblastic leukemic cells are treated with low-dose radiation before chemotherapy (Nilotinib). Our in vitro results of the combined therapy accompanied with a ematical model shows successful suppression of resistance to Nilotinib. The ematical model shows a synergistic interaction between Nilotinib and low dose radiation in the chemo dose response function. Beside acute radiation we investigate low dose fractionated therapies with model predicted optimal dosing schedules.

INTRODUCTION

The persistence of chemo-resistant leukemia-initiating cells in Philadelphia-chromosome positive (Ph+) B-cell Acute Lymphoblastic Leukemia (B-ALL) in the bone marrow is a primary mechanism responsible for disease relapse, following treatment, which occurs in the majority of patients. B-ALL is due, in part, to chromosomal translocations (9;22) that result in the generation of a BCR-ABL fusion protein, which fosters the transformation of immature B cells [1]. BCR-ABL+ (i.e., Ph+) leukemia has a poor prognosis; this is particularly true when matched with deletions in Cdkn2a, the gene encoding the tumor suppressor protein ARF, which occurs frequently in B-ALL [2, 3].

A significant breakthrough in the treatment of Ph+ ALL as well as the treatment of chronic myeloid leukemia (CML is associated with p210 isoform, whereas ALL is associated with p190 isoform) was the development of the tyrosine kinase inhibitor (TKI) Imatinib [1]). This drug, and the more potent second generation drugs Dasatinib and Nilotinib, are able to selectively inhibit the BCR-ABL mutant protein and thus significantly reduce Ph+ cell counts [2, 4]. While TKI therapy has long-term efficacy in the treatment of CML, most ALL patients eventually relapse following treatment with TKI due to the development of resistance [5, 6, 7, 8]. Thus a common treatment protocol for ALL patients is TKI therapy until the first remission [9, 10] followed by stem cell transplantation. However, since stem cell transplantation itself carries many risks to patient survival, the ability to extend the efficacy of TKI therapy in Ph+ ALL patients is of great clinical interest. Combination therapy such as Nilotinib with inhibitors of various other pathways (MEK, AKT, and JNK) showed greater reduction in cell viability and lowered risk of resistance [11]. Ionizing radiation has been used for leukemia disease in limited cases, e.g. i) disease involve in the central nervous system (CNS), potential due to ineffective penetration of chemotherapy to CNS [12], (ii) conditioning regimen with high doses of radiation and chemotherapy prior to stem cell transplantation for patients with high risk of relapse [13].

Taking advantage of leukemia radiosensitivity and the benefit of low dose radiation (LDR) in preserving bone marrow functions, we investigated whether the combination of Nilotinib and low dose radiation will be more effective treatment for BCR-ABL+ (i.e., Ph+) leukemia over Nilotinib alone. Furthermore, to optimize the effectiveness of this combination treatment, we developed a ematical model, parameterized via cell viability experiments under Nilotinib treatment and radiation exposure, to predict cellular response to the combination therapy. The optimized ematical model predicts a synergy between LDR and TKI treatment. We propose a combined Nilotinib dose-response function after LDR that accounts for a possible synergistic interaction between LDR and TKI treatment. Model parameters are obtained from in vitro viability measurements in the absence of TKI (Figure 1(a)), with zero LDR (Figure 1(b)) and combination of LDR and TKI for several radiation doses (Figure 1(c)). The model is validated by precise prediction for the drug-dose responses *and* radiation-dose response to combination LDR and TKI treatment.

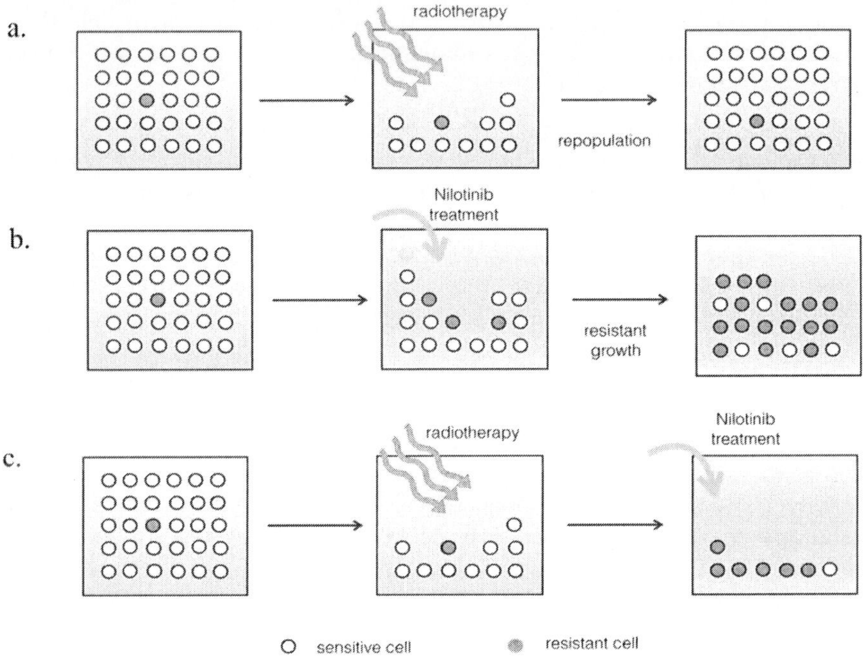

Figure 1. The schematics of sensitive and resistant populations under radiation and/or TKI treatment.a) Soon after acute radiation treatment both cell populations drastically decrease but bounces back through repopulation. (Notice that while a small initial fraction of resistant population is assumed they do not have a growth advantage in the absence of the drug treatment.) b) Upon Nilotinib treatment the sensitive population became disadvantaged and gradually the resistant population outgrows and total population grows back. It does not reach the original population size. c) Combining both therapies both populations are disadvantaged and treatment is much more effective.

It is important to emphasize that our model is focused on the relevance of LDR to prevent small-molecule inhibitor drug resistance. It does not address the efficacy of a successful radiation-drug combination treatment. Answering the question that whether the resistance is pre-existing, selected for or evolves de novo is out the scope of the current work. We do assume a small fraction of pre-existing Nilotinib resistant subtype, as well as possibility to transform into resistant type in the presence of the drug.

MATERIALS AND METHODS

Experiments

Cell line.

Bone marrow was harvested from an Arf-/- (p19-/-) mouse [14]. Red blood cells were lysed and the remaining cells were cultured overnight. Cells were then transduced with a retrovirus containing the BCR-ABL gene, which confers a leukemic phenotype. In that way, B cell acute lymphoblast leukemia cells were obtained and used in this study. Cells were maintained in α-MEM supplemented with 10% FBS, 20 mM L-glutamine, penicillin /streptomycin, and β-mercaptoethanol.

Cell viability assay.

We first performed a set of experiments quantifying cell viability as a function of time, for Nilotinib monotherapy, radiation monotherapy, and combined Nilotinib-radiation therapy. For each culture, 5×10^5 cells were plated to each well of a 6-well plate. For the irradiation-only group, cells were irradiated at 2 Gy or 4 Gy on day 0, by 225 kV X-ray beams using X-RAD 320 orthovoltage biological irradiator with a 0.35 mm cupper filter. In the Nilotinib-only group, 18 nM of Nilotinib was added in fresh media at days 0, 3, 6, and 8. For the combination therapy group, 18nM Nilotinib was added to the cells 4 hours before irradiation of 2 Gy or 4 Gy (Figure 2(a)). A boost of Nilotinib with fresh media was administered at days 3, 6, and 8, maintaining Nilotinib on the cells at all times. Cell viability in the combined therapy of 18 nM Nilotinib with Triciribine, an AKT inhibitor, was also conducted as shown in Figure 2(b). The control cells (without treatment) were plated with dimethyl sulfoxide (DMSO) media because DMSO was used to dissolve nilotinib and triciribine drugs. For all treatments, the cell culture medium was changed at days 0, 3, 6, and 8 to maintain the cells. At days 1, 3, 6, 8 and 10, cell viability was assessed by the Trypan blue exclusion assay with a hemocytometer to count the viable and dead cells. Viability was expressed as the percentage of viable cells of the total cell number. All measurements were performed in triplicate. These experiments were utilized to fit the parameters of the ematical model.

A second set of experiments was performed to determine the dose-response relationship for radiation therapy and Nilotinib monotherapy. For the radiation dose-response assay, under the same plating conditions as described above, cells were administered either 0 or 18 nM Nilotinib for four hours before radiation doses of 0, 1, 2, 4, 6, 8 and 10 Gy on day 0. For the Nilotinib dose-response assay, cells were administered 0, 6, 12.5, or 18

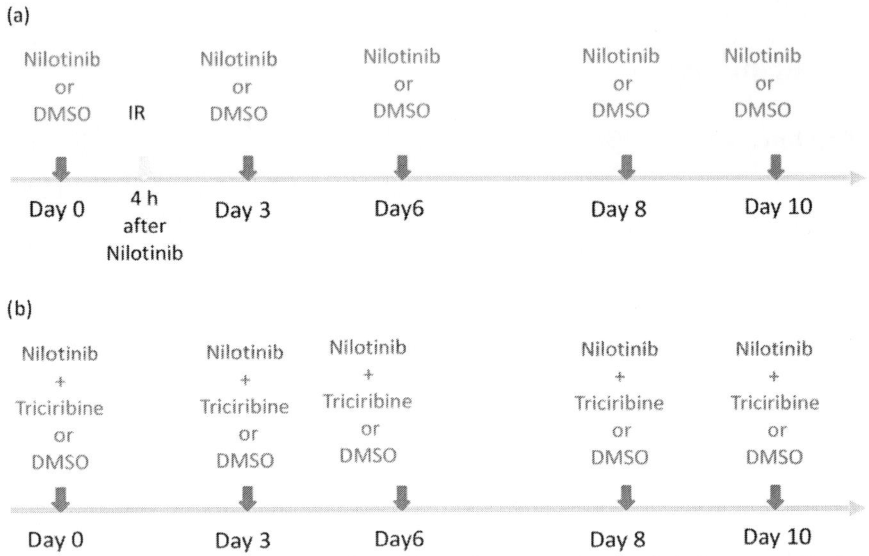

Figure 2. Treatment protocol used in the experiment.(a) Acute radiation dose is administered at day 0 (IR). Doses of Nilotinib is administered every few days to keep the drug concentration in a constant level (day 0, 3, 6, 8, 10). (b) Combined therapy of Nilotinib with Triciribine, an AKT inhibitor, is done for comparison with (a).

nM Nilotinib on day 0. The viability of each cell population was measured 72 hours after administration of Nilotinib. The results of these experiments were compared with model predictions for model validation purposes.

Western blot.

To obtain the whole-cell lysates from the cells, the cells were washed in PBS and then lysed in 10 mM Tris-HCl (pH 7.5), 150 mM NaCl, 1 mM EDTA, 1% Triton X-100, 1 mM phenylmethylsulfonyl chloride/isopropanol, and 1 g/ml pepstatin/methanol for 60 min on ice. During incubation, cells were homogenized every 30 minutes using disposable pestles. After centrifugation for 10 min at 14000rpm, the protein was obtained and quantified using BCA Protein Assay kit (ThermoFischer Scientific, Waltham, MA). For each sample, 20ug of total protein were then separated by 10% SDS-PAGE and electroblotted on polyvinylidene difluoride membranes (Bio-Rad, Hercules, CA). After blocking for 1 h in PBS supplement with 5% skim milk, immunodetection of AKT and p-AKT were performed with anti-rabbit monoclonal antibody (1:1000) or antirabbit polyclonal antibody (1:2000;). Anti-rabbit IgG (1:3000) and enhanced chemiluminescence

(Amersham Pharmacia Biotech, Aylesbury, United Kingdom) were used for detection.

Mathematical model

Our computational model is a coupled system of ordinary differential equations representing two populations. A Nilotinib sensitive and a Nilotinib resistant population. The use of ordinary differential equations is very common to describe the population dynamics and emergence of a new trait [15, 16] Logistic terms impose limitation on growth of each population, and an additional term allows for conversion from Nilotinib sensitive to Nilotinib resistant phenotype in the presence of non-zero concentration of the drug. More specifically, we assumed the following for the dynamics of the two subpopulations of the sensitive and resistant phenotypes:

- At the beginning of the experiment, the majority of cells are Nilotinib sensitive. In the absence of the drug, the division rates (and apoptosis rates) of both phenotypes are considered to be the same. Both sensitive and resistant cells are assumed to have similar growth patterns and carrying capacities.
- While we assume a small initial fraction of resistant cells, only in the presence of Nilotinib, resistant cells have a proliferative advantage over Nilotinib sensitive cells. This can be due to depletion of the division rate or increase in the apoptosis rate of the sensitive cells relative to that of the resistant cells.
- In the presence of the drug, sensitive cells can transform at a constant rate into resistant cells. This is due to the fact that random mutations among the sensitive population can give rise to a much more adaptive phenotype for high concentrations of Nilotinib.
- The radiosensitivity of both phenotypes was set to be the same. Furthermore, the effect of TKI treatment on radiosensitivity of both cell types is ignored. Furthermore, Malignant mutations due to low dose ionizing radiation are ignored.

The dynamical model that describes the above mechanisms is detailed in the Supplementary Information (SI). We refer to this model as the proliferation-mutation model.

To identify the response of Ph+ ALL cells to Nilotinib treatment, radiation, and the combination of both therapies, we proposed a simple functional form of this dependence and evaluated the fit against a large set of dose response data from experiment. This simple linear dose-response model can be written as follows:

$$\begin{aligned}\text{(proliferation rate)} &= r_0 - (\text{Nilotinib dose}) \times (r_1 + \text{radiation dose} \times r_2),\\ \text{(apoptosis rate)} &= d_0 + (\text{Nilotinib dose}) \times (d_1 + \text{radiation dose} \times d_2).\end{aligned} \tag{1}$$

where the growth rate coefficients r_i and d_i are constants that need to be identified based on the experimental results. It is common to model dose response functions using a Hill function structure (3- or 4-parameter logistic function). The Hill function imposes a low level of drug efficacy at low doses and a saturation of drug efficacy at high doses. A linear response is most accurate to approximate a Hill function dose-response near IC_{50} does—which is the case here. In the SI section, we show that how the above linear dose-response functions can be derived from a more common Hill -function.

Eq (1) can be rewritten as

$$\begin{aligned} r_S &= r_{S,0} - (r_{S,1} + r_{S,2}D)c,\\ r_R &= r_{R,0} - (r_{R,1} + r_{R,2}D)c,\\ d_S &= d_{S,0} + (d_{S,1} + d_{S,2}D)c,\\ d_R &= d_{R,0} + (d_{R,1} + d_{R,2}D)c, \end{aligned} \tag{2}$$

where c and D represent the Nilotinib and radiation doses respectively. The constants $r_{S,0}$ and $r_{R,0}$ denote the proliferation rates of sensitive and resistant populations in the absence of therapy, and $d_{S,0}$ and $d_{R,0}$ denote the death rates of sensitive and resistant populations in the absence of therapy. The coefficients $r_{S,1}$, $r_{R,1}$ represent the dose-response relationship between Nilotinib and proliferation rate of sensitive and resistant cells, respectively. Similarly, the coefficients $d_{S,1}$, $d_{R,1}$ describe how Nilotinib impacts the death rate of sensitive and resistant cells, respectively. Lastly, the coefficients $r_{S,2}$, $r_{R,2}$, $d_{S,2}$, $d_{R,2}$ determine the strength of the radiation-drug interaction on proliferation and death rates sensitive and resistant cells, respectively. These coefficients were fit to experimental data sets studying proliferation and death rates at a variety of Nilotinib and radiation doses. For example, non-negligible positive fitted values of $r_{S/R,2}$ and $d_{S/R,2}$ reveal synergistic interaction between the therapies.

We also incorporated an immediate cell-kill term after each radiation dose in accordance with the standard linear- quadratic model [17, 18, 19, 20]. According to the LQ model the effects of radiation cell kill is given by the survival fraction after the radiation exposure

$$\text{Surviving fraction} = \exp[-\alpha \times (\text{radiation dose}) - \beta \times (\text{radiation dose})^2)] \tag{3}$$

where α and β are the radio-sensitivity parameters to be determined from the experimental data. In short, α represents the rate of cell kill due to single tracks of radiation and β represents cell kill due to two independent radiation tracks. The linear quadratic model is widely used due to its excellent agreement with empirical data for a wide range of radiation doses. In SI section we explain how to incorporate the linear quadratic framework for surviving fraction into our mutation-proliferation model.

The above ematical framework can predict the population fraction (cell viability) at each day given the initial values of Nilotinib and irradiation doses. We fit the model parameters in an iterative fashion. All parameter estimation are obtained by finding optimal parameter set that minimizes the square root distance of solution of Eq. S3 (in SI) with respect to the time series viabilities at days 0, 3, 6, 8, 10. The steps are as follows:

1. Using the cell viability time point data series, in absence of Nilotinib and three radiation doses (0, 2, and 4 Gy), we set the ?control? values of the growth parameters, that is, $r_{S,0} = r_{R,0}$, $d_{S,0} = d_{R,0}$, K (carrying capacity) as well as radio-sensitivity coefficients $\alpha$and β.
2. Setting these values, we use viability time series data in the presence of Nilotinib (dose = 18nM) to fit for coefficients $r_{S,1}$, $r_{R,1}$, $d_{R,2}$ and $d_{S,2}$ as well as v. For this case we use the values $r_{S,0} = r_{R,0}$, $d_{S,0} = d_{R,0}$, K obtained in previous step.
3. The values for $r_{S,2}$, $r_{R,2}$, $d_{R,2}$ and $d_{S,2}$ are obtained from fitting the viability time series for the combination therapy case with doses 2Gy and 4Gy. For this case we use the values $r_{S,0} = r_{R,0}$, $d_{S,0} = d_{R,0}$, K as well as $r_{S,1}$, $r_{R,1}$, $d_{R,2}$ and $d_{S,2}$ obtained in previous steps.

RESULTS

In vitro experiments

We first measured the time-dependent cell viability of Ph+ ALL cells *in vitro* in response to Nilotinib, radiation, and combination therapy with both Nilotinib and radiation (Figure 3(a)). For the radiation-only arm, we observed an initial large reduction inviability by day 6 for 2 Gy and day 8 for 4 Gy. Experiments using Nilotinib monotherapy showed an incremental reduction in cell viability over the first 6 days to about 57.2%. Subsequently the cell populations began to develop a resistance to the drug and viabilities began to increase. When used in combination, Nilotinib + radiation induced a more effective initial cell killing and the cancer cell population was controlled at very low numbers (under 10% viability) for the duration of the experiment. In other words, there was nearly no cell population recovery, and resistance to Nilotinib was not developed. Under the Nilotinib + 4 Gy treatment arm, the cell population was entirely eliminated. Thus, the combination therapies were able to not only elicit a larger reduction in cell population, but also to maintain control of low cell viabilities without the development of resistance over a longer time period. The use of Triciribine was able to keep the cell viability as low as Nilotinib+ radiation group (Figure 3(b)).

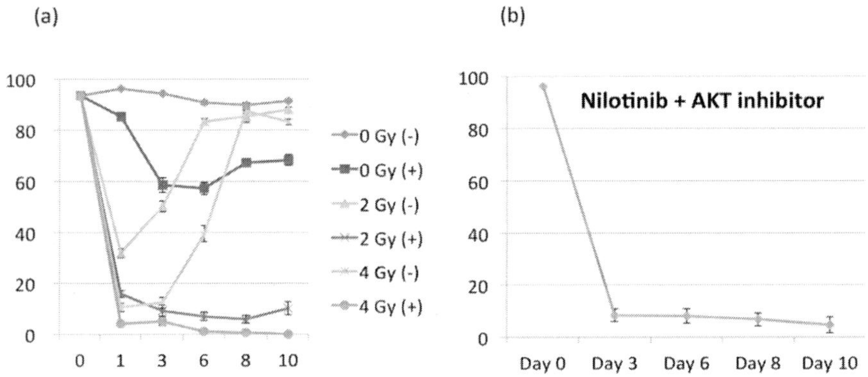

Figure 3. In vitro viability measurements for BCR-ABL+ ALL cell lines in different treatment regimes. (The y-axis is the cell viability in percent and x-axis is time (days).) (a) Cell viability measurements for cell lines treated with low dose radiation and Nilotinib. Plus sign denotes application of Nilotinib (18 nM). b) Similar set up but with AKT inhibitors applied instead of LDR. The results of both low dose radiation and AKT inhibition seems to be the same in preventing the evolution of Nilotinib resistance.

The radiation dose-response assay revealed an LD50 of 2 Gy in the absence of Nilotinib, and 0.5 Gy in the presence of 18 nM Nilotinib. The Nilotinib dose-response assay revealed an IC_{50}oof 18nM in the absence of radiation. To investigate how the combined therapy provides the synergistic effect, we evaluated one of the key pathways associated with cell proliferation, and chemoresistance. At day 3 after treatment, p-AKT was slightly increased by adding either Nilotinib or radiation (Figure 4(a)). However, the expression level was gradually decreased at large dose of radiation (6 Gy) with or without Nilotinib (Figure 4(b)). Notably, the combined therapy eventually (day 7 and10) decreased the p-AKT even at 2 Gy (Figure 4(c)).

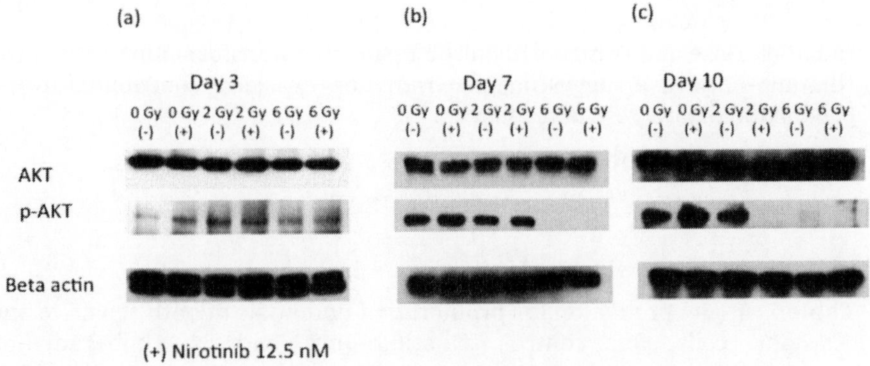

Figure 4. At day 3, AKT was activated by both Nilotinib and radiation. However, after 7 days, p-AKT was down regulated compared to day 3. At day 10, however, Nilotinib without radiation again upregulated p-AKT while combination with radiation kept downregulation. Phospholiation of the AKT may be a critical role in Nilotinib resistant and combination with radiation may killed leukemia cells more efficiently because of p-AKT downreguration.

Fitting the model to experimental data

Using the numerical solutions of the proliferation-mutation model optimized for the best fit with experimental data points, we determined the coefficients in the Nilotinib dose-response function, r_i and d_i, as well as the transformation rate (Table 1). The optimized coefficients indicate an increase in fitness of resistant cells relative to that of sensitive cells in the presence of Nilotinib. (Fitness is defined as the difference between division rate and apoptosis rate in the cell population.) Using a numerical sensitivity analysis, we confirmed that these parameter estimates are robust to perturbations in the initial frequency of resistant cells in the population. Since the experiments demonstrated an initially strong sensitivity of the populations to Nilotinib treatment, we set the frequency of resistant cells to be small. We set this to be 0.1% of the total population for the remainder of investigations. The optimized parameters are reported in Table 2 with the carrying capacity set to $K = 4.2$ in all cases. These results indicate the presence of synergistic effects between Nilotinib and radiation. In particular, the radiation tyrosine-kinase inhibitor interaction was strongest in the resistant population. This is not surprising as in the presence of Nilotinib the sensitive population is already highly disadvantaged in terms of growth. The value of transformation rate ν in this case is independent of radiation dose and is higher than the respective transformation rate in the Nilotinib-only case suggesting that radiation exposure contributes to the production of new resistant cells.

Table 1. Table summary of death rate dose-response coefficients.

$r_{S,0}$	$r_{R,0}$	$r_{S,1}$	$r_{R,1}$	$r_{S,2}$	$r_{R,2}$
2.5369	2.5369	0.0155	0.0	0.0140	0.0
$d_{S,0}$	$d_{R,0}$	$d_{S,1}$	$d_{R,1}$	$d_{S,2}$	$d_{R,2}$
2.0550	2.0550	0.0	0.0	0.0025	0.0114

Table 2. Model prediction for proliferation potentials of both sensitive and resistant cells for control, Nilotinib-only and Nilotinib+radiation treatments.

	r_s	r_R	d_s	d_R	v
control	2.5369	2.5369	2.0550	2.0550	0.0
nilo-only	2.2579	2.5369	2.0550	2.0550	0.0409
nilo+ 2Gy	1.7539	2.5369	2.1450	2.4654	0.1768
nilo+ 4Gy	1.2499	2.5369	2.2350	2.8758	0.1768

The results for population fractions (cell viabilities) as a function of time (days) were plotted in Figure 5. The agreement between ematical model predictions and the in vitro measurements suggests that the proposed radiation-drug interaction term in the linear dose-response is a plausible choice. All the above results of the parameter estimation process are summarized in the Tables 1 and 2.

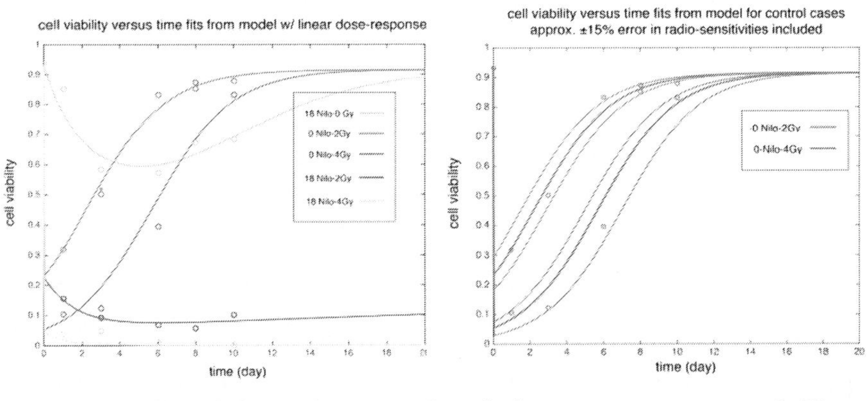

Figure 5. Cell viabilities for control and the treatment cases. a) Time-dependent cell viability *in vitro* under Nilotinib treatment with and without radiation (0-10 days). Solid lines are model fit and circles are experimental results. Red and magenta correspond to 4Gy and 2Gy radiation without Nilotinib (control). b) Time dependent cell viability after 2Gy (magenta) and 4Gy (red) irradiation (day 0). Solid lines show model fit and circles are the in vitro experimental data. Thin lines show model results for varying radiosensitivity parameters by approximately ±15 per cent. The values for parameters are reported in Table 2, $\alpha = 0.6647$, $\beta = 0.079$ and $\Delta\alpha = \pm 0.10$, $\Delta\beta = \pm 0.012$.

Model validation

We next tested predictions of the fitted model with an independent set of experimental results. In particular, we compared model predictions to the experimental results of the dose-response assays for Nilotinib (in absence of radiation) and radiation (at 0 and 18 nM Nilotinib). See Figure 6for these comparisons. As can be seen, the results are in very good agreement. We then used the model to predict the Nilotinib dose-response curve with 1, 2, 3 and 4 Gy radiation. The results are plotted for 0-22nM of Nilotinib. Note that at higher doses we see a discrepancy with the experiment (25nM 0Gy). This is due to the fact that the linear approximation of the Hill dose-response curve only applies to the initial section of the curve and is expected to break down at higher doses.

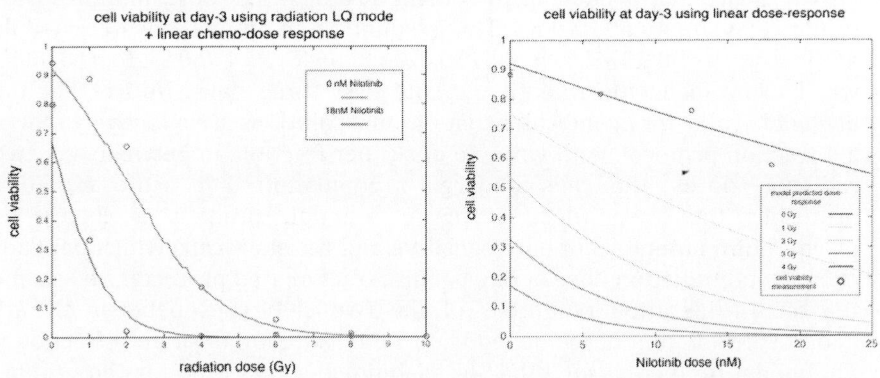

Figure 6. Radiation and Nilotinib dose responses. a) Radiation dose-response curve under no Nilotinib treatment (magenta) and 18 nM Nilotinib (red). Solid lines are model predictions with independently fitted parameters (from Tables 1 and 2). Experimental data shown in circles. Cell viabilities are measured after three days. b) Nilotinib dose-response curves in absence of radiation. Solid lines are model predictions with independently fitted parameters, experimental data shown in circles. Cell viabilities are measured after three days. Other lines represent model predictions for efficacy of Nilotinib treatment combined with various radiation doses of 1, 2, 3 and 4 Gy.

Optimization of combination therapies

The combination therapy above was shown to reduce tumor cell viability, and through computational techniques an optimal schedule of dosing may be approached. Given the potential toxicity of Nilotinib to normal tissues, we first assumed a standard, constant dose of 18 nM Nilotinib. Next, we considered a five-day radiation treatment protocol, where the summed dose of the protocol is constant. As an example, we use a total dose of 2Gy; which, as can be seen in Figure 5, has room for improvement with the combination therapies and does not irradiate the cells to a viability that is not interesting ematically (4Gy in Figure 5, for example). The aforementioned result will serve as a comparison for our proposed protocol, which will attempt to minimize the total tumor cell viability at day 10. The control parameters are the radiation doses given at days 0, 1, 2, 3 and 4 denoted by $D_i(i = 1, 2, 3, 4, 5)$. Placed under the constraint of the total allowable dose, the control parameters were varied by a nonlinear constrained minimization protocol that sought to minimize the tumor cell viability at day 10. Under this minimization protocol, the minimum tumor cell viability was determined by searching the potential dose regimen in the answer space and producing the dosing protocol that best minimized the tumor cell viability at 10 days. This resulted in an optimal dosing schedule of (in Gy) $D_1 = 0.9371, D_2 = 0.5139, D_3 = 0.6445, D_4 = 0.045, D_4 = 0.0064$, which front-loads the radiation in the first three days. Notice that the negligible value for D_4 indicates that optimal solutions are effectively that of a 4-fraction protocol, with variable doses per fraction. In between any two radiation doses, the cells undergo repopulation. For acute radiation protocols, the linear dose response function predicts a change in proliferation potentials of both sensitive and resistant cells which depends on the total radiation dose at day 0. For a fractionated protocol, we assume the the proliferation potentials of the two cell types between the kth and k+1th radiation doses depends on total radiation dose until fraction k. The model prediction for total cell viabilities versus time for the optimal fractionation protocol is plotted in Figure 7 and compared with a constant 5-fraction radiation protocol ($D_i = 0.4$) and 2Gy acute radiation treatment ($D_1 = 2$Gy, $D_2 = D_3 = D_4 = D_4 = 0$ Gy). At larger time, that is, greater than 10 days, the optimal protocol has the lower cell viability more importantly in a downward trend beyond the 10 day point, whereas the acute treatment begins an upward trend and the constant fractionation protocol does not suppress the cell viability to the levels of the optimal or acute protocols. This suggests the synergistic interactions between the therapies is heightened with larger initial doses of fractionated radiation, and the optimal protocol for long term suppression falls between acute dosing and constant fraction protocols, which results in the front-loading protocol. This dose dependent fractionation may be generalized to various scenarios in the clinical setting as determined by clinical status. Further, the computational model may be extended to consider alternative Nilotinib dosing strategies in combination of the radiation protocol to minimize the cell viability at day 10. Concurrent use of the 5-day optimal fractionation protocol with a varied Nilotinib dosing strategy shows an alternative, though equally efficacious (similar cell viability at 10 day) strategy. This alternative Nilotinib dosing strategy maintains the average daily dose of 18 nM, however, it begins at lower concentration and increases throughout the course of treatment. Specifically, the Nilotinib dose over the first 3 days is 10 nM, followed by 18 nM for 4 days, and finally 26 nM for 3 days, essentially back-loading the Nilotinib onto the irradiated cells. After 10 days, the dose returns to 18 nM daily. This protocol is visualized in Figure 7 as well, where at larger times the new strategy of Nilotinib dosing reduces the cell viability as well as the previous protocol. This is clinically relevant as dose titration is common, if needed, and there is no trade off with treatment efficacy should this dosing schedule be indicated.

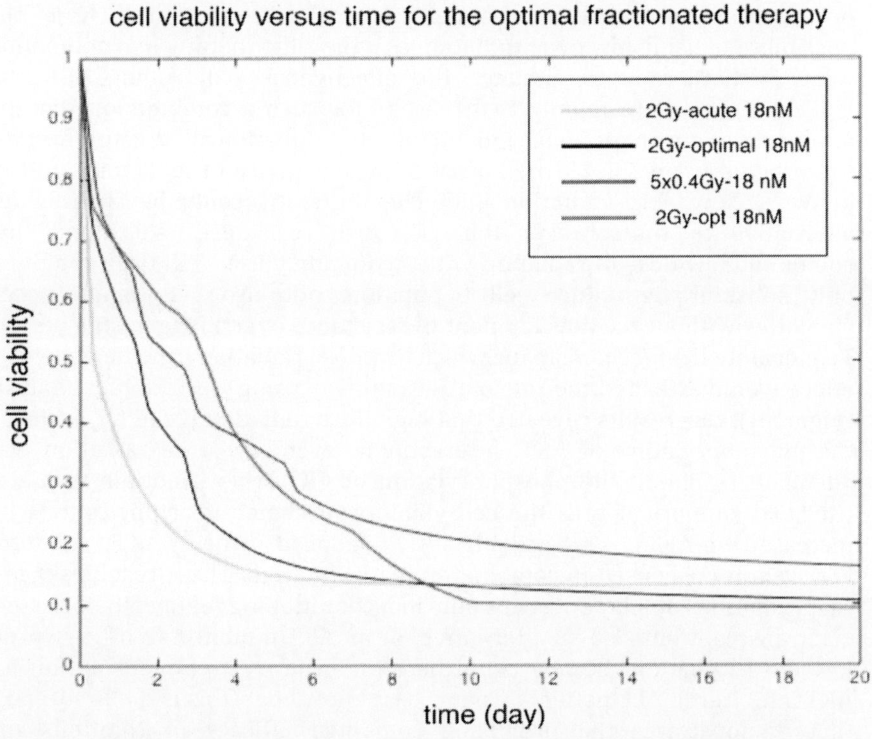

Figure 7. Comparison between three fractionation protocols: 5-fraction with 0.4 Gy per fraction (blue), optimal radiation dose fractions (black) and acute (green) and optimal fractionated radiation + Nilotinib protocols (red).

The total cell viabilities is clearly lowest for day 9-10 in the optimal protocol. (See text for details.)

DISCUSSION

Although the clinical outcome of BCR-ABL leukemia has been improved with the advent of TKI therapy, overcoming chemo-resistance has been a major challenge. This study demonstrated that low dose irradiation combined with Nilotinib provided enhanced and prolonged efficacy for leukemic cells in vitro. A companion theoretical model provided good agreement with experimental results with an opportunity for further optimization to enhance treatment efficacy. We explored here the possibility of using low dose radiation as a first line therapy in combination with chemotherapy to enhance the effectiveness controlling BCR-ABL leukemia. There are mainly two reasons for such a combination therapy approach: a) advent of image guided targeted radiation allow more focused dose delivery [21, 22, 23] b) known radiosensitivity of leukemia but not known if low dose radiation with chemotherapy could be an effective alternative to control over the TKI drug resistance. We found that combining low dose of radiation with a commonly used TKI drug could not only substantially reduce cell population, but also maintain low cell viabilities without the development of resistance over a longer time period. To elucidate the mechanism for which low dose radiation provide beneficial effect, we investigated the role of AKT pathway using western blot analysis (Figure 4). Our results revealed that high dose radiation (i.e. 6 Gy) reduced the phosphorylation of AKT. Interestingly, even low dose radiation was adequate to inhibit the phosphorylation of AKT when Nilotinib was also used. To support of this, the cell viability of chemo-resistant cells (with increased p-AKT) was significantly reduced when AKT inhibitor (Triciribine) was used in combination with Nilotinib. These results suggest that radiation can play roles not only in a conditioning regimen before stem cell transplant but also an alternative for an AKT inhibitor. In other words, low dose radiation therapy could be an alternative treatment option for AKT inhibitor in ALL patients. To our best knowledge, this is the first report that demonstrates the promising combined efficacy of Nilotinib and radiation for ALL, with minimum damage to vital organs. Because of low dose with limited toxicity, it can be delivered to the whole body. Since our study was performed in normal cell culture, however, in vivo study is required to take into consideration of microenvironmental factors.

We also constructed a dynamical model to explain the observations and to predict response to additional combination therapy schedules. To tease apart the responses of Ph+ ALL cells to Nilotinib, radiation, and the combination of both therapies, we proposed a simple functional form of the combination dose-response relationship and evaluated the fit against a large set of dose response data. In particular, we proposed a simple Nilotinib dose-response function in which radiation dose may alter the strength of the Nilotinib response in a dose-dependent fashion. Parameter-fitting revealed an optimal parameter set that showed very good agreement with experimental results. Analysis of this optimal parameter set revealed dose-dependent synergistic effects between Nilotinib and radiation response. In particular the radiation tyrosine-kinase inhibitor interaction is strongest in the resistant subpopulation of cells. Our analysis also demonstrated that the model predictions are robust to variation in the initial frequency of resistant cells. We compared fitted model predictions with an independently generated second set of experimental data and found good agreement.

We next utilized the validated model to investigate optimal combination strategies for Nilotinib and radiation in Ph+ ALL. As a simple test, we assumed a standard 18 nM Nilotinib dose and investigated strategies allowing up to 2 Gy over the course of a 5-day treatment. We determined that the optimal therapeutic schedule, given these constraints, spread most of the radiation dose over the first three days. Thus, a 'sweetspot' exists between acute radiation protocols and constant treatment protocols. These results suggest a promising direction for investigation of new treatment strategies in Ph+ ALL and for providing an optimal treatment regimen. However, we note that all experiments (and thus parametrization of the model) was done in vitro. Further in vivo studies are needed to determine treatment schedules for the clinic with more accuracy. To summarize, augmentation of LDR-Nilotinib therapy may be beneficial to control Ph+ve leukemia resistance and the model can determine optimal dosing schedule to enhance the effectiveness of the combination therapy.

REFERENCES

1. Deininger M, Buchdunger E, Druker BJ. The development of imatinib as a therapeutic agent for chronic myeloid leukemia. Blood. 2005;105(7):2640–2653. pmid:15618470
2. Talpaz M, Shah NP, Kantarjian H, Donato N, Nicoll J, Paquette R, et al. Dasatinib in imatinib-resistant Philadelphia chromosome–positive leukemias. New England Journal of Medicine. 2006;354(24):2531–2541. pmid:16775234
3. Shimoni A, Volchek Y, Koren-Michowitz M, Varda-Bloom N, Somech R, Shem-Tov N, et al. Phase 1/2 study of nilotinib prophylaxis after allogeneic stem cell transplantation in patients with advanced chronic myeloid leukemia or Philadelphia chromosome–positive acute lymphoblastic leukemia. Cancer. 2015;121(6):863–871. pmid:25387866
4. Kantarjian H, Giles F, Wunderle L, Bhalla K, O'Brien S, Wassmann B, et al. Nilotinib in imatinib-resistant CML and Philadelphia chromosome–positive ALL. New England Journal of Medicine. 2006;354(24):2542–2551. pmid:16775235
5. Gaynon PS. Childhood acute lymphoblastic leukaemia and relapse. British journal of haematology. 2005;131(5):579–587. pmid:16351633
6. Pui CH, Evans WE. Treatment of acute lymphoblastic leukemia. New England Journal of Medicine. 2006;354(2):166–178. pmid:16407512

7. Den Boer ML, van Slegtenhorst M, De Menezes RX, Cheok MH, Buijs-Gladdines JG, Peters ST, et al. A subtype of childhood acute lymphoblastic leukaemia with poor treatment outcome: a genome-wide classification study. The lancet oncology. 2009;10(2):125–134. pmid:19138562
8. Yang K, Fu Lw. Mechanisms of resistance to BCR–ABL TKIs and the therapeutic strategies: A review. Critical reviews in oncology/hematology. 2015;93(3):277–292. pmid:25500000
9. Ottmann O. Management of Philadelphia chromosome-positive acute lymphoblastic leukemia. Leukemia supplements. 2012;1:S7–S9. pmid:27175253
10. Ottmann OG, Pfeifer H. Management of Philadelphia chromosome–positive acute lymphoblastic leukemia (Ph+ ALL). ASH Education Program Book. 2009;2009 (1):371–381.
11. Feldhahn N, Arutyunyan A, Stoddart S, Zhang B, Schmidhuber S, Yi SJ, et al. Environment-mediated drug resistance in Bcr/Abl-positive acute lymphoblastic leukemia. Oncoimmunology. 2012;1(5):618–629. pmid:22934254
12. .Kwon JH, Koh Yi, Yoon Ss, Park S, Kim I. Clinical outcome and efficacy of current anti-leukemic therapy for leptomeningeal involvement in acute myeloid leukemia. International journal of hematology. 2016;104(5):574–581. pmid:27431487
13. Wong JY, Forman S, Somlo G, Rosenthal J, Liu A, Schultheiss T, et al. Dose escalation of total marrow irradiation with concurrent chemotherapy in patients with advanced acute leukemia undergoing allogeneic hematopoietic cell transplantation. International Journal of Radiation Oncology* Biology* Physics. 2013;85(1):148–156.
14. Kamijo T, Zindy F, Roussel MF, Quelle DE, Downing JR, Ashmun RA, et al. Tumor suppression at the mouse INK4a locus mediated by the alternative reading frame product p19 ARF. Cell. 1997;91(5):649–659. pmid:9393858
15. Komarova NL, Wodarz D. Targeted cancer treatment in silico. Springer; 2014.
16. Panetta JC. A ematical model of drug resistance: heterogeneous tumors. ematical biosciences. 1998;147(1):41–61. pmid:9401351
17. Chadwick K, Leenhouts H. A molecular theory of cell survival. Physics in medicine and biology. 1973;18(1):78. pmid:4803965
18. Dale RG, Jones B, et al. Radiobiological modelling in radiation oncology. British Institute of Radiology; 2007.
19. Hall EJ, Giaccia AJ. Radiobiology for the Radiologist. Lippincott Williams and Wilkins; 2006.
20. Brenner DJ, Hlatky LR, Hahnfeldt PJ, Hall EJ, Sachs RK. A convenient extension of the linear-quadratic model to include redistribution and reoxygenation. International Journal of Radiation Oncology* Biology* Physics. 1995;32(2):379–390.
21. Hui SK, Kapatoes J, Fowler J, Henderson D, Olivera G, Manon RR, et al. Feasibility study of helical tomotherapy for total body or total marrow irradiationan. Medical physics. 2005;32(10):3214–3224. pmid:16279075
22. Magome T, Froelich J, Takahashi Yea. Evaluation of functional marrow irradiation based on skeletal marrow composition obtained using dual-energy CT. international Journal of Radiation Oncology. 2016;.
23. Rosenthal J, Wong J, Stein A, Qian D, Hitt D, Naeem H, et al. Phase 1/2 trial of total marrow and lymph node irradiation to augment reduced-intensity transplantation for advanced hematologic malignancies. Blood. 2011;117(1):309–315. pmid:20876852

CHAPTER 16

Evolutionary Dynamics in the Two-Locus Two-Allele Model with Weak Selection

Martin Pontz[1,2] · Josef Hofbauer[1] · Reinhard Bürger

[1] Institut für ematik, Universität Wien, Oskar-Morgenstern-Platz 1, 1090 Wien, Austria

[2] Vienna Graduate School of Population Genetics, Vienna, Austria

ABSTRACT

Two-locus two-allele models are among the most studied models in population genetics. The reason is that they are the simplest models to explore the role of epistasis for a variety of important evolutionary problems, including the maintenance of polymorphism and the evolution of genetic incompatibilities. Many specific types of models have been explored. However, due to the ematical complexity arising from the fact that epistasis generates linkage disequilibrium, few general insights have emerged. Here, we study a simpler problem by assuming that linkage disequilibrium can be ignored. This is a valid approximation if selection is sufficiently weak relative to recombination. The goal of our paper is to characterize all possible equilibrium structures, or more precisely and general, all robust phase portraits or evolutionary flows arising from this weak-selection dynamics. For general fitness matrices, we have not fully accomplished this goal, because some cases remain undecided. However, for many specific classes of fitness schemes, including additive fitnesses, purely additive-by-additive epistasis, haploid selection, multilinear epistasis, marginal overdominance or underdominance, and the symmetric viability model, we obtain complete characterizations of the possible equilibrium structures and, in several cases, even of all possible phase portraits. A central point in our analysis is the inference of the number and stability of fully polymorphic equilibria from the boundary flow, i.e., from the dynamics at the four marginal single-locus subsystems. The key ematical ingredient for this is index theory. The specific form of epistasis has both a big influence on the possible boundary flows as well as on the internal equilibrium structure admitted by a given boundary flow.

Keywords: election Recombination Epistasis Linkage disequilibrium Equilibrium structure Phase portrait

1 INTRODUCTION

One of the central goals of the pioneers of population genetics was to demonstrate that the inheritance and evolution of continuously varying traits could be explained on the basis of Mendelian genetics (Fisher 1918, 1930). Haldane (1931) and Wright (1935) were apparently the first who formulated explicit dynamical models for the evolution of gene frequencies if selection acts on more than one locus. Under various assumptions about dominance, Haldane considered two loci at each of which a wild type and a deleterious variant segregate. Motivated by empirical examples, he assumed that if both variants occur in the same genotype, they have a selective advantage. Wright investigated a model in which finitely many loci contribute to a quantitative trait that is under quadratic selection toward an intermediate optimum. Both Haldane and Wright assumed that gene frequencies at the loci are probabilistically independent, i.e., they are in linkage equilibrium, and derived the stable equilibrium states as well as other properties of their models.

In an investigation designed to show that selection can lead to tighter linkage, Kimura (1956) derived and studied a full two-locus two-allele model, i.e., one that takes into account linkage disequilibrium. The general (deterministic) two-locus two-allele model for the evolutionary dynamics under selection and recombination was derived and investigated by Lewontin and Kojima (1960). They deduced both general properties as well as properties of special cases, such as additive gene action. Among others, they showed that strong epistasis together with linkage disequilibrium can lead to significantly different outcomes than would occur for independent loci.

Extensive analyses of a special class of fitness patterns, the so-called symmetric viability model (which originated from Wright's and from Kimura's work), were performed by Bodmer and Felsenstein (1967) and Feldman and Karlin (1970). The latter authors derived all fifteen possible equilibria and determined their stability for several special cases. Later, Feldman and Liberman (1979) showed that as many as four boundary equilibria and two polymorphic equilibria can be simultaneously stable, and Hastings (1985) demonstrated that up to four stable internal equilibria may coexist. The complexity of this model is also underlined by the finding of Ewens (1968) that there is a gap in the range of recombination rates for which a pair of internal equilibria is stable. A comprehensive review of this model and its extension to multiple loci can be found in Christiansen (1999).

Other important special classes of fitness patterns are the additive and the multiplicative model, in which fitnesses of multilocus genotypes are obtained by adding or multiplying the fitnesses of the constituent single-locus genotypes. In the former, additive epistasis is absent, in the latter, multiplicative epistasis is absent. For additive fitnesses, mean fitness is a Lyapunov function (Ewens 1969), all equilibria are in linkage equilibrium, and, generically, every trajectory converges to an equilibrium point (Karlin and Liberman 1978, 1990; Nagylaki et al. 1999). The multiplicative model is much more complicated, although the linkage-equilibrium manifold is invariant. In this model and away from the linkage-equilibrium manifold, mean fitness may decrease and, for intermediate recombination rates, asymptotically stable equilibria may exist that are in linkage disequilibrium (e.g., Moran 1964, 1968; Bodmer and Felsenstein 1967; Nagylaki 1977; Karlin and Feldman 1978; Hastings 1981a). A detailed review of the theory of two-locus and multilocus models is given in Bürger (2000, Chap. 2).

With general fitnesses, the dynamics in the two-locus two-allele model can be complex. The existence of stable limit cycles has been demonstrated both for the continuous-time model (Akin 1979, 1982) and the discrete-time model (Hastings 1981b; Hofbauer and Iooss 1984). Such complex behavior cannot occur if loci are assumed to be independent, i.e., if linkage equilibrium is imposed. Then the dynamics is gradient-like and mean fitness is a global Lyapunov function (Nagylaki 1989).

However, even the case of two independent, diallelic loci has never been analyzed systematically. Although a Lyapunov function exists, the equilibrium structure can still be quite complicated. For instance, Moran (1963) showed that, apart from degenerate cases, the maximum number of internal (polymorphic) equilibria is five, and up to three can be asymptotically stable.

In this study we perform a systematic analysis of the two-locus two-allele model with constant fitnesses under the assumption of linkage equilibrium. The goal is to determine and classify all possible equilibrium structures and phase portraits (Sects. 3, 4). We assume continuous time for reasons outlined below. We have not fully accomplished our goal, however, we identified all 42 possible (equivalence classes of) boundary flows and 190 potentially possible extended boundary flows, i.e., flows at or close to the boundary. Of these 190 extended boundary flows, the existence of 185 could be proved; the other cases remain undecided. In Section S1 of the Supplementary Information (SI), we present corresponding phase portraits. A large number of extended boundary flows admits not only several non-equivalent phase portraits, but also more than one equilibrium structure, as characterized by the number and stability of boundary and internal equilibria.

We use this general analysis to obtain a detailed classification of equilibrium structures and phase portraits for a number of important special cases that have received considerable attention in the literature. These include the case of marginal overdominance or underdominance (Sect. 5); linear isoclines, which turn out to exist if and only if fitnesses among loci are additive or the only epistatic interactions are additive-by-additive (Sect. 6); multilinear epistasis (Sect. 7); equivalent loci (Sect. 8); and the symmetric viability model (Sect. 9). These results provide considerable insight into the interplay of dominance and epistasis in maintaining genetic polymorphism.

The analysis of this simplified model has immediate implications for the full two-locus two-allele model. This is a consequence of a general theorem by Nagylaki et al. (1999), which applies to multilocus systems. These authors proved under weak technical assumptions that if selection is much weaker than recombination, then after an evolutionarily short period, in which linkage disequilibrium decays to close to zero, the dynamics of the full model (either in discrete or in continuous time) is governed by this weak-selection limit. The model investigated in this paper is the weak-selection limit of the two-locus two-allele model with selection and recombination.

2 MODEL

We start with the standard two-locus two-allele model with viability selection and discrete time. Thus, we assume a randomly mating, diploid population with discrete and non-overlapping generations in which viability selection acts on two diallelic, recombining loci. Therefore, gametes are in Hardy–Weinberg proportions. Mutation, random drift, and other evolutionary forces are absent.

Let A_1 and A_2 be the alleles at locus B_1 and B_2 those at locus B. Let the frequencies of the four gametes A_1B_1, A_1B_2, A_2B_1, and A_2B_2 be denoted by x_1, x_2, x_3, and x_4, respectively, where $\sum_{i=1}^{4} x_i = 1$, and let $w_{ij} > 0$

denote the (constant) viability of an individual with genotype *ij*. In addition to positing absence of sex effects, i.e., $w_{ij} = w_{ji}$, we posit absence of position effects, i.e., $w_{14} = w_{23}$. The recombination probability is denoted by *r*. The frequencies of alleles A1A1 and B1B1 are denoted by $p = x_1 + x_2$ and $q = x_1 + x_3$, respectively. Letting $D = x_1x_4 - x_2x_3$ be the classical measure of linkage disequilibrium, we obtain

$$\begin{aligned} x_1 &= pq + D, \\ x_2 &= p(1-q) - D, \\ x_3 &= (1-p)q - D, \\ x_4 &= (1-p)(1-q) + D. \end{aligned} \tag{2.1}$$

Under the above symmetry assumptions, the fitnesses of genotypes are completely specified by the following matrix:

	B_1B_1	B_1B_2	B_2B_2
A_1A_1	w_{11}	w_{12}	w_{22}
A_1A_2	w_{13}	w_{14}	w_{24}
A_2A_2	w_{33}	w_{34}	w_{44}

(2.2)

Then evolution of gamete frequencies is given by (Lewontin and Kojima 1960)

$$x_i' = \frac{x_i w_i - \eta_i w_{14} r D}{\bar{\omega}}, \quad i = 1, 2, 3, 4, \tag{2.3}$$

where $\eta_1 = \eta_4 = -\eta_2 = -\eta_3 = 1,\ w_i = \sum_{j=1}^{4} w_{ij}x_j$ is the marginal fitness of gamete *i*, and $\bar{\omega} = \sum_{j=1}^{4} w_j x_j$ is the mean fitness of the population. This is a dynamical system on the simplex S4S4 which has received much attention in the literature but is well understood only in special cases (see Sect. 1). For a review consult Chapter 2 in Bürger (2000).

If the assumption of linkage equilibrium, i.e., D=0D=0, is imposed, the dynamics (2.3) simplifies to the following system of difference equations defined on the unit square $[0, 1] \times [0, 1]$,

$$\Delta p = p(1-p)\frac{1}{2\bar{w}}\frac{\partial \bar{w}}{\partial p}, \qquad (2.4a)$$

$$\Delta q = q(1-q)\frac{1}{2\bar{w}}\frac{\partial \bar{w}}{\partial q}, \qquad (2.4b)$$

where by (2.1) mean fitness $\bar{w} = \bar{\omega}(D = 0)$ is only a function of p and q; cf. Haldane (1931) and Wright (1935, 1942). As shown more generally for multiple multiallelic loci by Nagylaki (1989), $\bar{w}\bar{w}$ is monotone increasing along trajectories of (2.4) and constant only at equilibria. Therefore, if all equilibria are isolated points, every trajectory converges to an equilibrium.

In general, the manifold $D=0D=0$ is not invariant under (2.3). However, assuming weak selection, i.e., setting $w_{ij}=1+sm_{ij}w_{ij}=1+sm_{ij}$, rescaling time according to $t = \lfloor \tau/s \rfloor$, and letting $s \downarrow 0$, the so called *weak-selection limit* of (2.3) is obtained:

$$\dot{p} = p(1-p)\frac{1}{2}\frac{\partial \bar{m}}{\partial p}, \qquad (2.5a)$$

$$\dot{q} = q(1-q)\frac{1}{2}\frac{\partial \bar{m}}{\partial q}. \qquad (2.5b)$$

Here,

$$\bar{m} = m_1 pq + m_2 p(1-q) + m_3(1-p)q + m_4(1-p)(1-q) \qquad (2.6)$$

is the mean (Malthusian) fitness of the population and

$$m_i = m_{i1} pq + m_{i2} p(1-q) + m_{i3}(1-p)q + m_{i4}(1-p)(1-q) \qquad (2.7)$$

the marginal (Malthusian) fitness of gamete *i*. We note that the dynamics (2.5) remains unchanged if the same constant is added to every $m_{ij}m_{ij}$. If every $m_{ij}m_{ij}$ is multiplied by the same positive constant, only a change in time scale results. Therefore, in (2.5) we could substitute $w_{ij}w_{ij}$ for $m_{ij}m_{ij}$ without changing the phase portrait. In particular, two of the nine parameters in the fitness scheme (2.2) could be set to fixed, different values.

Nagylaki et al. (1999) proved (for multiple multiallelic loci) that if $r > 0$ is given, selection is sufficiently weak, i.e., $s>0s>0$ is sufficiently small, and if all equilibria of (2.5) are hyperbolic, then every trajectory of

(2.3) converges to an equilibrium point on an invariant manifold, ΛsΛs, which is contained in an $O(s)$ neighborhood of the linkage-equilibrium manifold D=0. The dynamics on Λs is a small perturbation of the time-s map of the weak-selection limit (2.5), which is gradient-like. In particular, it is easy to show directly for (2.5), but also follows from a result by Nagylaki (1989) for (2.4), that $\dot{\bar{m}} \geq 0$ for every $(p, q) \in [0, 1]^2$ and $\dot{\bar{m}} = 0$ if and only if (p, q) is an equilibrium. Therefore, m¯m is a strict Lyapunov function for (2.5).

Because $\bar{w} = 1 + s\bar{m}$, the equilibria of (2.4) and (2.5) are the same, and so are their stability properties (since w¯≥mini,jwij>0w¯≥mini,jwij>0). Therefore, if selection is sufficiently weak and after some (usually short) time has passed (Nagylaki 1993), the dynamics of the full two-locus system (2.3) is closely approximated by the dynamics of the weak-selection limit (2.5). Therefore, each solution of (2.3) converges to an equilibrium point, and this equilibrium point is in an $O(s)$ neighborhood of an equilibrium of (2.5) (Theorem 3.1 in Nagylaki et al. 1999).

3 EQUILIBRIA AND THEIR STABILITY

The four monomorphic equilibria CiCi of (2.5), corresponding to fixation of gamete i, exist always. They represent the corners of the state space [0,1]2[0,1]2 (Figure 1). The eigenvalues of the Jacobian at CiCi are easily calculated and are as follows:

	p direction	q direction	
$C_1 = (1, 1)$	$m_{13} - m_{11}$	$m_{12} - m_{11}$	
$C_2 = (1, 0)$	$m_{24} - m_{22}$	$m_{12} - m_{22}$	(3.1)
$C_3 = (0, 1)$	$m_{13} - m_{33}$	$m_{34} - m_{33}$	
$C_4 = (0, 0)$	$m_{24} - m_{44}$	$m_{34} - m_{44}$	

An equilibrium is called linearly stable, or a sink, if all eigenvalues (of its Jacobian) have negative real part. If at least one eigenvalue has positive real part, it is linearly unstable. It is called a source if all eigenvalues have positive real part, and it is a saddle if eigenvalues with positive and negative real part occur. Obviously, the corner equilibrium CiCi is linearly stable if and only if the fitness of the homozygous genotype ii is higher than each of the two 'neighboring' single-locus heterozygous genotypes ij, where j differs from i by a single allele.

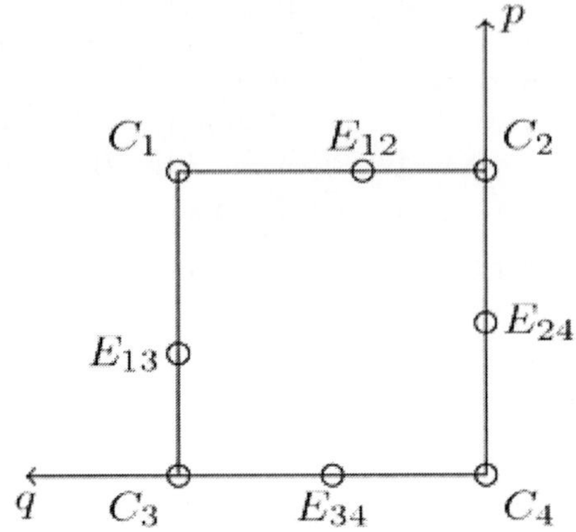

Figure 1. The state space with all possible boundary equilibria. The orientation is such that it corresponds to the fitness matrix in (2.2)

Next there may exist up to four equilibria at which one locus is polymorphic, so-called single-locus polymorphisms (SLPs). They are located on the edges of the state space. We denote the equilibrium on the edge connecting CiCi with CjCj by EijEij (Figure 1). The coordinates of these edge equilibria are easily calculated and are given by

$$E_{ij}: q = \frac{m_{ij} - m_{jj}}{2m_{ij} - m_{ii} - m_{jj}} \text{ and } p = 0 \text{ if } ij = 34, \text{ and } p = 1 \text{ if } ij = 12; \tag{3.2a}$$

$$E_{ij}: p = \frac{m_{ij} - m_{jj}}{2m_{ij} - m_{ii} - m_{jj}} \text{ and } q = 0 \text{ if } ij = 24, \text{ and } q = 1 \text{ if } ij = 13. \tag{3.2b}$$

Therefore, the edge equilibrium EijEij exists, i.e., is in the interior of the edge, if and only if $(m_{ij} - m_{ii})(m_{ij} - m_{jj}) > 0.$ Obviously, this is just the well-known condition of either overdominance or underdominance applied to a one-locus boundary system. To distinguish this notion of overdominance (underdominance) from that of marginal overdominance (underdominance) introduced below, we call this overdominance (underdominance) on an edge.

The eigenvalues of the edge equilibria, hence their stability, can also be determined quite straightforwardly, because the Jacobian is in triangular form, so that the eigenvalues appear on the diagonal. One eigenvalue, called internal, determines stability of this equilibrium within its edge. For Eij, it is given by

$$\mu_{ij} = \frac{(m_{ij} - m_{ii})(m_{ij} - m_{jj})}{m_{ii} + m_{jj} - 2m_{ij}}. \tag{3.3}$$

Therefore, as is well known, the SLP EijEij is linearly stable (unstable) within its edge if and only if there is overdominance (underdominance) on this edge.

The other eigenvalue, called external, determines stability of the edge equilibrium transversal to the boundary. If the external eigenvalue is negative, the equilibrium is called *strictlysaturated* or *externally stable* (Hofbauer and Sigmund 1998; Karlin 1980). This has the interpretation that the allele missing on this edge cannot invade the population near this boundary equilibrium. The external eigenvalues are of more complicated form than the internal eigenvalues and are presented in "Appendix A.1". The distinction between external and internal eigenvalues will be essential for our analysis.

Thus, there are at least four boundary equilibria, but there may be up to eight. We ignore the degenerate cases, in which every point on an edge is an equilibrium. This occurs if and only if $m_{ii} = m_{ij} = m_{jj}$, where *i* and *j* are gametes differing by one allele.

Finally, we turn to the internal, or fully polymorphic, equilibria. By (2.5), they are the solutions of the two equations $\partial\bar{m}/\partial p = 0$ and $\partial\bar{m}/\partial q = 0$ that satisfy $0 < p < 1$ and $0 < q < 1$. Equivalently, these equations can be written as

$$p = f(q) = \frac{m_{A_1A_2} - m_{A_2A_2}}{2m_{A_1A_2} - m_{A_1A_1} - m_{A_2A_2}}, \tag{3.4a}$$

$$q = g(p) = \frac{m_{B_1B_2} - m_{B_2B_2}}{2m_{B_1B_2} - m_{B_1B_1} - m_{B_2B_2}}, \tag{3.4b}$$

where

$$m_{A_1A_1} = m_{11}q^2 + 2m_{12}q(1-q) + m_{22}(1-q)^2, \tag{3.5a}$$

$$m_{A_1A_2} = m_{13}q^2 + 2m_{14}q(1-q) + m_{24}(1-q)^2, \quad (3.5b)$$

$$m_{A_2A_2} = m_{33}q^2 + 2m_{34}q(1-q) + m_{44}(1-q)^2, \quad (3.5c)$$

and

$$m_{B_1B_1} = m_{11}p^2 + 2m_{13}p(1-p) + m_{33}(1-p)^2, \quad (3.6a)$$

$$m_{B_1B_2} = m_{12}p^2 + 2m_{14}p(1-p) + m_{34}(1-p)^2, \quad (3.6b)$$

$$m_{B_2B_2} = m_{22}p^2 + 2m_{24}p(1-p) + m_{44}(1-p)^2, \quad (3.6c)$$

are the marginal fitnesses of the one-locus genotypes at *A* and *B*, respectively. (In the presence of linkage disequilibrium, the expressions in (3.5) and (3.6) need to be normalized; see Ewens and Thomson 1977.) Thus, the internal equilibria are the intersection points of the isoclines $p = f(q)$ and $q = g(p)$.

The following results were proved by Moran (1963). We will give a slightly different proof of the first statement.

Theorem 3.1

(a)If all equilibria are isolated, then (2.5) has at most five internal equilibria.

(b)Five internal equilibria can be realized, and up to three can be sinks.

(c)Sinks correspond to local maxima of m¯m¯.

Proof

(a)To determine the intersection points of the isoclines (3.4a) and (3.4b), we have to solve the fixed point equation

$$p = f(g(p)). \quad (3.7)$$

Because the numerators and denominators of $f(q)$ and $g(p)$ are polynomials of degree two or less, numerator and denominator of the rational function $f(g(p))$ are polynomials of degree four or less. Therefore the intersection points, hence the internal equilibria, are the zeros of a polynomial of degree five or less.

(b)See Moran (1963) and panel 6 in Figure S1d of the SI, which shows the phase portrait of Moran's example. (c) is also shown by Moran (1963). It follows immediately from the fact that m¯m¯ is a strict Lyapunov function.

To determine the stability of an internal equilibrium, we need the Jacobian.

Lemma 3.2

The Jacobian at an internal equilibrium (p, q) is given by

$$J = J(p, q) = \begin{pmatrix} p(1-p)m_A & 2p(1-p)\tilde{m} \\ 2q(1-q)\tilde{m} & q(1-q)m_B \end{pmatrix}, \quad (3.8)$$

where $m_A = m_{A_1A_1} + m_{A_2A_2} - 2m_{A_1A_2}$, $m_B = m_{B_1B_1} + m_{B_2B_2} - 2m_{B_1B_2}$, *and* $\tilde{m} = m_1 - m_2 - m_3 + m_4$. *The eigenvalues of* J *are real.*

Proof

A simple calculation shows that

$$\frac{1}{2}\frac{\partial \bar{m}}{\partial p} = pm_A + m_{A_1A_2} - m_{A_2A_2}. \quad (3.9)$$

Therefore, (2.5a) yields

$$\frac{\partial \dot{p}}{\partial p} = (1-2p)\frac{1}{2}\frac{\partial \bar{m}}{\partial p} + p(1-p)\frac{1}{2}\frac{\partial^2 \bar{m}}{\partial p^2} = p(1-p)m_A, \quad (3.10)$$

because $\frac{\partial \bar{m}}{\partial p} = 0$ at an internal equilibrium and $m_A = \frac{1}{2}\frac{\partial^2 \bar{m}}{\partial p^2}$. The other derivatives in J are calculated in a similar way.

Finally, it is straightforward to compute the discriminant of J, which is the square of the trace minus four times the determinant:

$$(\mathrm{tr}\, J)^2 - 4\det J = [p(1-p)m_A - q(1-q)m_B]^2 + 16pq(1-p)(1-q)\tilde{m}^2 > 0.$$

The inequality holds because $(p, q) \in (0, 1)^2$. Therefore, all eigenvalues are real. □?

For the rest of this paper, we impose the assumption

$$\text{All equilibria of (2.5) are hyperbolic.} \qquad (\mathcal{H})$$

Therefore, eigenvalues at equilibria are negative or positive, but not zero. This assumptions excludes not only curves of equilibria, but also complete dominance or recessivity of an allele.

The stability of an internal equilibrium is most easily determined by employing the planar Routh–Hurwitz criterion. It states that an equilibrium is (i) a saddle point if detJ<0detJ<0, (ii) a sink if $\det J < 0$, (ii) a sink if $\det J > 0$ and $\mathrm{tr}\, J < 0$,

Motivated by Lewontin and Kojima (1960), we say that a locus (e.g., *A*) exhibits marginal, or induced, overdominance at (p, q) if the inequalities

$$m_{A_1A_2} > m_{A_1A_1} \text{ and } m_{A_1A_2} > m_{A_2A_2} \qquad (3.11)$$

hold. If both inequality signs are reversed, one obtains marginal, or induced, underdominance. The following result was proved by Kojima (1959). We will give a similar, but more direct, proof.

Corollary 3.3

An internal equilibrium (p, q) is linearly stable, i.e., a sink, if and only if both loci exhibit marginal overdominance at (p, q) and mAmB>4m~2mAmB>4m~2.

Proof

By the Routh–Hurwitz criterion, the equilibrium (p, q) is a sink if and only if $p(1-p)m_A + q(1-q)m_B < 0$ and

$$\det J(p, q) = pq(1-p)(1-q)(m_A m_B - 4\tilde{m}^2) > 0. \qquad (3.12)$$

Therefore, both mAmA and mBmB must be negative. Because $0 < p < 1$ must hold, (3.4) implies that m_A, $m_{A_2A_2} - m_{A_1A_2}$, and $m_{A_1A_1} - m_{A_1A_2}$ all have the same sign. This together with an analogous argument for locus *B* proves that marginal overdominance is necessary. The sufficiency condition follows immediately from (3.12). □

Remark 3.4

(i)As pointed out by Kojima (1959), (3.12) is equivalent to the condition that the geometric mean of the dominance variances of each locus exceeds the additive-by-additive variance of the two-locus system.

(ii)The proof of Corollary 3.3 shows that, at equilibrium, locus *A* displays marginal overdominance if and only if (cf. Hastings 1982)

$$m_{A_1A_2}^2 > m_{A_1A_1} m_{A_2A_2}. \tag{3.13}$$

4 EQUILIBRIUM STRUCTURE AND FLOWS

Since the internal equilibria are obtained from the solutions of the quintic polynomial equation (3.7), it is generally difficult to determine their number, position, or stability. Therefore, we first study the possible flows on the boundary of the state space. Subsequently, we extend the flows to a neighborhood of the boundary. Using index theory, we will be able to shed light on the equilibrium structure, i.e., the number and stability properties of equilibria (for a precise definition, see Sect. 4.3), and on the possible phase portraits, i.e., the topological structures of the flow (2.5). Finally, we exclude several potentially possible equilibrium structures and generate phase portraits for most of the remaining cases.

To facilitate the characterization of equilibrium structures and phase portraits, we identify flows that are topologically equivalent or obtained by symmetry operations corresponding to a relabeling of alleles at a locus $(A_1 \text{ or } A_2, B_1 \text{ or } B_2)$ or of the loci (*A* or *B*). More precisely, we call two systems of the form (2.5), or the corresponding fitness schemes M and $\tilde{M}$ in $\mathsf{R}^{3\times 3}$ (2.2), or the flows, *equivalent* if there exists an edge-preserving homeomorphism h of [0,1]2[0,1]2 onto itself that maps orbits of (2.5) generated by M onto orbits generated by $\tilde{M}$ (preserving the arrow of time), i.e., the phase portraits are topologically equivalent. Here, edge-preserving means that each of the four edges is mapped onto an edge, not necessarily onto itself.

An edge-preserving homeomorphism is a composition of one of the eight symmetry operations of the square with a 'proper' or 'pinned' homeomorphism of the square, i.e., a homeomorphism that leaves the four corners fixed and maps each edge onto itself. The symmetry group of the square, the dihedral group D4D4, consists of four reflections and four rotations (including the identity). Therefore, each equivalence class is invariant under rotations by multiples of 90∘90∘, reflections about the diagonal or antidiagonal, and reflections about the middle vertical or horizontal axis of the matrix (2.2), or of the unit square (Figure 1).

Finally, we call a fitness scheme M [or its induced flow (2.5), or its phase portrait] *robust* if it has a neighborhood of equivalents in $\mathbb{R}^{3\times3}$. This is essentially the concept of structural stability, adapted to the selection Eq. (2.5). For the single-locus two-allele model, i.e., on every edge, there are three robust equivalence classes: they correspond to the classical selection patterns of overdominance, underdominance, or intermediate dominance (directional selection).

The goal of our paper is to find all robust (equivalence classes of) phase portraits arising from (2.5). A necessary condition for robustness is that all equilibria are hyperbolic, i.e., condition (HH) holds. A classical characterization of structural stability in two-dimensional systems due to Andronov and Pontryagin (1937) implies:

*If (2.5) satisfies (*HH*) and there is no saddle connection in the interior of* [0,1]2[0,1]2, *then this system is robust.*

A saddle connection is an orbit which has one saddle as its αα-limit and an other as its ωω-limit. Most phase portraits shown in Section S1 have no saddle connection. Exceptions occur for the symmetry classes ss, bb, and ee (defined below), where the phase portraits were generated by matrices satisfying the corresponding symmetry condition. In all these cases, however, breaking the symmetry yields phase portraits that are members of the same class. Thus, they are robust in this sense. We now explain how to obtain all phase portraits. This requires three steps.

4.1 Flows on the boundary

As shown in Sect. 3, on each edge there is either no or one equilibrium, and each edge equilibrium can be internally stable or unstable. This gives 44=25644=256 different types of (nondegenerate) flows on the boundary. This type can be easily determined from the selection scheme (2.2), by observing the order relations in each boundary row and boundary column, i.e., overdominance, underdominance, or intermediate dominance. Applying the symmetry operations, their number drops to 42 different boundary flows or, more precisely, boundary-flow equivalence classes. They are displayed in Figure 2.

We recall that a matrix is called centrosymmetric if it is invariant under reflections about the central entry, and bisymmetric if it is symmetric and centrosymmetric. A centrosymmetric fitness matrix (2.2) gives rise to the well-studied symmetric viability model (Sect. 9). A symmetric fitness matrix (2.2) represents a model in which the loci *A* and *B* are equivalent (Sect. 8).

The following code was used to label boundary flows: $E^n_m C_k x$ Ckx represents a flow of symmetry class $x \in \{b, c, s, e, a\}$ with *n* edge equilibria, *m* of them internally stable, and *k* linearly stable corner equilibria. Here, bb, cc, or ss means that at least one flow in this class is generated by a bisymmetric, centrosymmetric, or symmetric matrix, respectively. This does not imply that all matrices generating a flow in such a class have the respective symmetry property. The letter eerefers to a flow for which the flows at (at least) one pair of opposite edges belong to the same single-locus class and have the same direction if there is intermediate dominance. This is the case if the entries in the two boundary columns (or rows) of the fitness scheme have the same order relation. The letter aa indicates asymmetry, i.e., the boundary flow cannot be generated by a matrix with one of the properties bb, cc, ss, or ee. In one case, $E^2_1 C_1 a$, the above code is not sufficient to identify the equivalence class; therefore, we used $E^2_1 C_1 a', E^2_1 C_1 a'', E^2_1 C_1 a'''$.

By reversing all arrows in a flow, the reversed flow is obtained. Equivalently, the signs of all entries in the fitness matrix are reversed. It turns out that 10 of the 42 cases are invariant under flow reversal. In Figure 2, the other 32 cases are placed such that pairs obtained by reversal are arranged vertically.

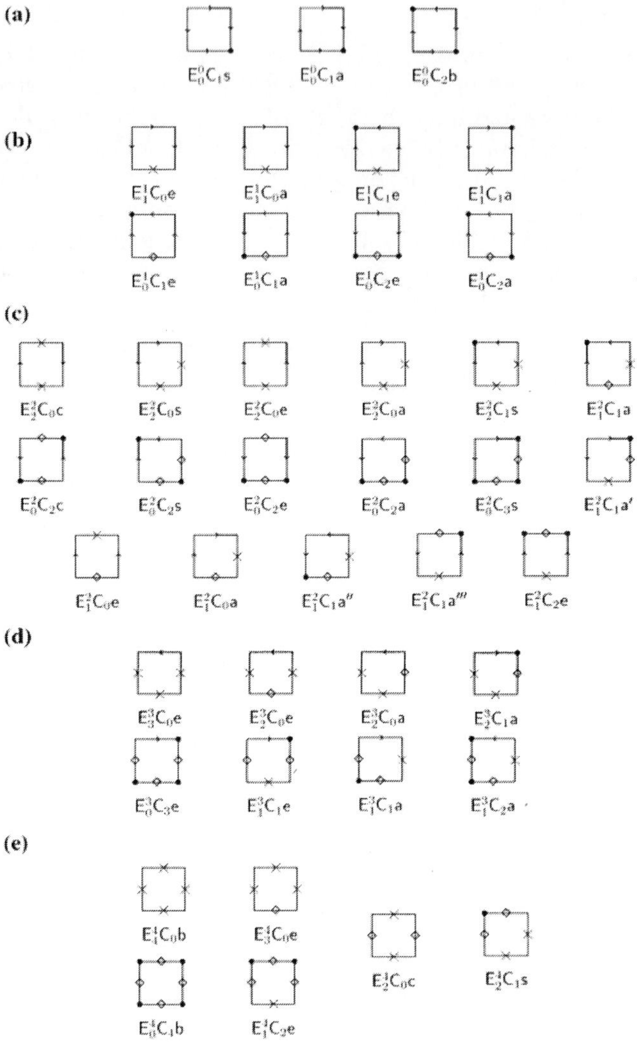

Figure 2. The 42 possible boundary flows, or boundary-flow equivalence classes, for (2.5). The 16 flow-reversal pairs are arranged *vertically*. For instance, $E_0^2\hat{C}_2c$ is the reversed boundary flow of $E_0^2\check{C}_2c$, whereas $E_1^2C_0e$ is self inverse. Internal stability is indicated by a *solid dot*(at a *corner*) or a cross (at an edge equilibrium). A *rhombus* indicates an internally unstable edge equilibrium

4.2 Extended boundary flows

Now we consider flows not only on the boundary, but in a sufficiently small neighborhood of the boundary. Here, external stability of the edge equilibria plays a central role (Figure 3). We need some preparation and recall from below Eq. (3.3) when an equilibrium in our two-dimensional model is strictly saturated. Because we impose the hyperbolicity assumption (HH) throughout, we simply use the term saturated equilibrium.

Therefore, a corner equilibrium is *saturated* if and only if both eigenvalues are negative. An edge equilibrium is saturated if and only if the external eigenvalue is negative. An internal equilibrium is, by definition, always saturated. For a hyperbolic equilibrium x^x^, the index is defined by

$$\mathrm{ind}(\hat{x}) = \mathrm{sgn}(\det(-J_{\hat{x}})) = (-1)^k, \qquad (4.1)$$

where *k* denotes the number of positive eigenvalues of the Jacobian Jx^ ^ at x^. It follows, that in a planar system an equilibrium with index −1−1 is a saddle point; sources and sinks have index +1+1. The sum of the indices of all saturated boundary equilibria is called the *boundary index sum*and is denoted by δδ. If there are no saturated boundary equilibria, then δ=0δ=0.

In general, δδ is not uniquely determined by the boundary flow because external eigenvalues of edge equilibria may be positive or negative for a given boundary flow (cf. Figure 3). However, δδ is uniquely determined by the *extended boundary flow*, which we use as a short hand for the equivalence class (in the above sense) of flows with a given boundary flow (class) together with the signs of the external eigenvalues of all edge equilibria. Different extended boundary flows of the same boundary-flow class may have the same δδ.

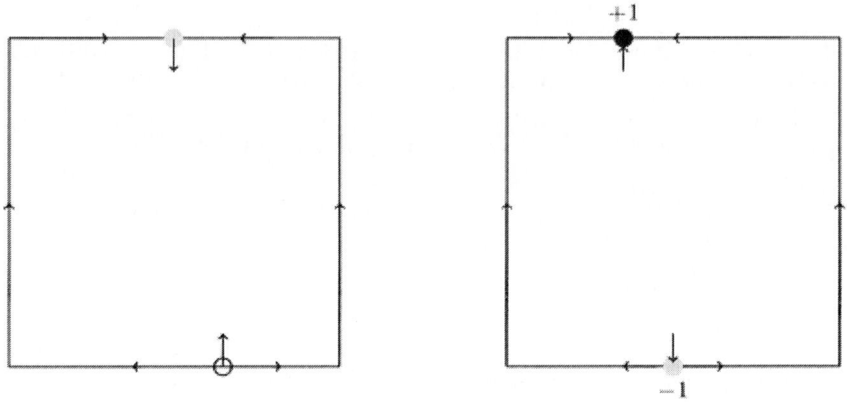

Figure 3. The boundary-flow class $E_1^2C_0e$ with positive (*left*) and negative (*right*) external eigenvalues at the edge equilibria. In the *left panel*, the edge equilibria are not saturated. In the *right panel*, E12E12 is asymptotically stable and saturated with index +1+1, and E34E34 is a saturated saddle with index −1−1. Therefore, δ=0δ=0 in both panels. If the flow in the *left panel* is reversed, i.e., all *arrows* are reversed, the flow in the *right panel* is obtained after a rotation by 180∘180∘. Extended boundary flows, where the flow reversal has the same boundary flow and δδ but a different external stability, can occur only for boundary-flow classes that are invariant under flow reversal

Lemma 4.1

For a boundary flow of type $E_m^n C_k$, the boundary index sum δδ can assume at most the (integer) values satisfying

$$k + m - n \leq \delta \leq k + m. \tag{4.2}$$

Every boundary flow in Figure 2 satisfies $k + m - n \geq -2 \; and \; k + m \leq 4.$

Proof We have to sum the indices of the saturated equilibria. Obviously, there are *k* stable, hence saturated, corner equilibria and at most *m* saturated, internally stable edge equilibria. This yields the upper bound k+mk+m. **Because the number of saturated, internally unstable edge equilibria is at most** $n - m$, the lower bound $k - (n - m)$ **results.**

The second statement follows easily by checking Figure 2. □

By systematic construction of the extended boundary flows for each of the 42 boundary flows, we obtain 200 potential extended boundary flows (see Table S1). It is important to recall that for given δδ more than one extended boundary flow may exist. In the following we show that not all of these 200 extended boundary flows can be realized for (2.5).

Theorem 4.2

Assume (2.5) and (HH).

(a)The boundary index sum δ=−2δ=−2 cannot occur.

(b)The boundary index sum δ=−1δ=−1 cannot occur for any of the boundary flows $E_2^4C_0c$, $E_2^3C_0e$, $E_1^3C_1e$, $E_1^2C_0e$, *or* $E_1^2C_0a$.

(c)The boundary index sum δ=0δ=0 can neither occur for E_{34} *and* E_{24} if the edge equilibria E_{34} *and* E_{24} are saturated and the others are not.

(d)The boundary index sum δ=0δ=0 cannot occur for $E_1^3C_1e$ 1e if the edge equilibrium E13E13is saturated and the others are not.

The proof of (a) is given in "Appendix A.2". The other statements can be proved in a similar manner.

Remark 4.3

(a)The extended boundary flow described in Theorem 4.2(d) is the reversed flow of the first extended boundary flow in Theorem 4.2(c).

(b)Table S1 informs us that there is exactly one potential extended boundary flow for each of the cases excluded by Theorem 4.2, except for $E_2^4C_0c$ and $\delta = -1$. Then the two potentially possible extended boundary flows are the reversed flows of each other and thus both are excluded by Theorem 4.2.

(c)We could not exclude the boundary index sum δ=−1δ=−1 for the cases $E_2^3C_0a$, $E_1^3C_1a$, $E_2^4C_1s$, $E_3^4C_0e$, and $E_1^4C_2e$. However, we conjecture that $\delta = -1$ does not occur (see Sect. 4.3).

(d)As a consequence of Theorem 4.2 and statement (b) of this remark, the number of potentially possible extended boundary flows reduces to 190.

(e)It is important to note that statements (c) and (d) of Theorem 4.2 exclude the existence of specific extended boundary flows and do not imply that the boundary index sum $\delta = 0$ does not occur in these cases.

(f)The above theorem constrains the range (4.2) of possible values of δδ to

$$\max\{-1, k + m - n\} \le \delta \le k + m. \tag{4.3}$$

An important tool for drawing conclusions about internal equilibria is the following index theorem:

Theorem 4.4

(Hofbauer 1990) Assume (2.5) and (HH). Then

$$\sum_{\hat{x} \text{ saturated}} \operatorname{ind}(\hat{x}) = +1, \tag{4.4}$$

where the sum runs over all saturated equilibria.

This theorem has a number of important consequences. The simple proofs are left to the reader.

Corollary 4.5

Assume (2.5) and (HH).

(a)The number of saturated equilibria is odd;

(b)If δ is odd (even), the number of internal equilibria is even (odd);

(c)The number of internal equilibria is at least $|1-\delta|$and at most five;

(d)If $\delta>1$, there are at least $\delta-1$ internal saddle points;

(e)If $\delta<1$, the total number of sinks and sources is at least $1-\delta$;

(f)The boundary index sum δδ is invariant under flow reversal.

4.3 Established equilibrium structures and phase portraits

We show that 185 of the 190 theoretically possible extended boundary flows exist. This task is simplified by taking flow reversals into account.

Importantly, flow reversal neither changes the number of internal equilibria nor their index. Therefore, the boundary index sum $\delta\delta$ is invariant under flow reversal; cf. Corollary 4.5(f). If we identify flow-reversal pairs, it is sufficient to consider 16+101boundary flow classes. Here, 16 is the number of representatives of boundary flows in Figure 2 that are not invariant under flow reversal, and 10 is the number of boundary flows that are invariant. These 16 flows give rise to 79 theoretically possible extended boundary flows (SI, Table S1). The 10 self-inverse boundary flows give rise to 42 extended boundary flows, 22 of which form flow-reversal pairs (Table S1). Thus, up to flow reversal, there are 79+31 different extended boundary flows. By Theorem 4.2, 2+5 of these boundary flows cannot exist. Overall, we present phase portraits for 75+25extended boundary flows (Section S1 of the SI). Applying flow reversal, this proves existence of $2\times75+35=185$ different extended boundary flows. All these flows have boundary index sum $\delta\geq0\delta\geq0$.

We could not realize 2+1 potential extended boundary flows (five, including flow reversal), all of them having boundary index sum $\delta=-1$. Therefore, their existence remains undecided. For the following reasons, we conjecture that they do not exist. (i) Theorem 4.2(b) excludes the other five extended boundary flows with $\delta=-1$. (ii) A boundary index sum of $\delta=-1$ requires that the number of saddles in the interior is less than the total number of sinks and sources in the interior minus one. Despite extensive search no such equilibrium pattern could be found (see Section S1 of the SI).

Two flows yielding the same extended boundary flow (up to equivalence in the sense defined above) may still differ in their *equilibrium structure* because they differ in the number of internal equilibria or, if they have the same number of internal equilibria, in the number of sinks, sources, or saddles. In addition, some equilibrium structures can be realized by non-equivalent phase portraits; see below.

We show existence of an extended boundary flow or equilibrium structure by presenting a numerical fitness matrix (mijmij) yielding a flow on the entire state space, [0,1]2, that has the given extended boundary flow (Section S1 of the SI). We represent each equivalence class of flows on [0,1]2by a phase portrait. In general, the phase portrait is not uniquely determined by the extended boundary flow, because already number and stability of internal equilibria are often not uniquely determined.

We present more than one phase portrait for several extended boundary flows. We do this for cases for which we found more than one equilibrium structure. In addition, we give examples of fitness matrices that yield the same equilibrium structure but non-equivalent phase portraits. One particularly interesting case is $E_2^3C_1a$ with $\delta = 1$, which admits three extended boundary flows and seven different phase portraits. One of these extended boundary flows has two equilibrium structures, each with two phase portraits (Figure S2i, panels 1, 2, 3, 4). Another remarkable case is $E_3^4C_0e$ with $\delta = 1$, which admits four different extended boundary flows (Figure S2j, panels 2, 3, 4, 5). However, there is also a considerable number of boundary flows for which the value of δδ uniquely determines the extended boundary and also the equilibrium structure and the phase portrait (up to equivalence). See Table S1 and the phase portraits in Section S1 for detailed information.

4.4 Permanence

In ematical modeling of biological systems, the notion of *permanence* is very important. The dynamical system (2.5) is permanent if there exists a compact subset of (0,1)2 such that every solution starting at (p0,q0)∈(0,1)2enters this subset and remains there (consult Hofbauer and Sigmund 1998 for an account on permanence theory). Permanence is equivalent to the boundary being a repeller. It can be shown that it is sufficient that all stationary points on the boundary are repelling.

We want to identify which extended boundary flows can be permanent. First, δ≠0 implies that at least one boundary equilibrium is saturated and, therefore, attracts at least one orbit from the interior. Second, not all extended boundary flows with δ=0 can be permanent, because there are extended boundary flows for which one equilibrium on the boundary is saturated with index −1 and an other has index +1+1. Therefore, an extended boundary flow is permanent if and only if no boundary equilibrium is saturated. Thus, we obtain:

Theorem 4.6

A full flow consistent with a boundary flow of type $E_i^j C_0 x$ is permanent if and only if no edge equilibrium is saturated.

Corollary 4.7

Assume (2.5) is permanent. Then

(a)the extended boundary flow is of one of the 14 classes $E_1^1C_0e$, $E_1^1C_0a$, $E_2^2C_0c$, $E_2^2C_0s$, $E_2^2C_0e$, $E_2^2C_0a$, $E_1^2C_0e$, $E_1^2C_0a$, $E_3^3C_0e$, $E_3^3C_0e$, $E_2^3C_0e$, $E_2^3C_0a$, $E_4^4C_0b$, $E_3^4C_0e$, $E_2^4C_0c$, and has no saturated edge equilibrium;

(b)there exists an internal equilibrium that is a sink, for example the global maximum of the mean fitness.

Proof

(a)follows from Theorem 4.6 upon comparison with Figure 2.

(b)follows because $\bar{m}$ is a (strict) Lyapunov function.□

We note that each of the extended boundary flows in Corollary 4.7 could give rise to one, three, or five internal equilibria. However, the boundary flow E44C0bE44C0b is the only one for which we found a fitness matrix such that (2.5) has five internal equilibria (see Figure S1d, panel 6).

5 CONTINUOUS ISOCLINES: MARGINAL OVERDOMINANCE OR UNDERDOMINANCE

In this and the next section we study which equilibrium structures and phase portraits are obtained by imposing specific properties on the isoclines. The structure of the two isoclines is important because the internal equilibria are their intersection points. As special cases, we will encounter well known models.

We start by investigating the equilibrium structure if the isoclines are continuous and map [0, 1] into (0, 1). From (3.11), we recall the definitions of marginal overdominance and underdominance.

Lemma 5.1

The isocline $f(q)$ $(g(p))$, defined in (3.4), is continuous and maps [0, 1] into (0, 1) if and only if locus A (B) exhibits either marginal overdominance or underdominance for every $q \in [0, 1]$ $(p \in [0, 1])$.

Proof

From the definition of the isocline $p = f(q)$, we obtain that $f(q)$ is continuous if and only if its denominator $(2m_{A_1A_2} - m_{A_1A_1} - m_{A_2A_2})$ does not change sign on [0, 1]. The requirement $0 < f(q) <$ 1is then satisfied if and only if both $m_{A_1A_2} - m_{A_1A_1}$ $m_{A_1A_2} - m_{A_2A_2}$ –mA1A1 and mA1A2−mA2 A2mA1A2−mA2A2 have the same sign as the denominator. The argument for $g(p)$ is analogous. □

This lemma implies that the graph of $f(q)$ intersects the boundary of the state space (precisely) at the equilibria E24E24 and E13E13, and the graph of $g(p)$ does so at E34E34 and E12E12. If both isoclines are continuous and map [0, 1] into (0, 1), we conclude from Lemma 5.1 and Theorem 3.1 that the number of interior equilibria is between one and five, and both bounds can be assumed.

Next we want to derive the possible equilibrium structures if both loci exhibit marginal overdominance everywhere, i.e., at every (p,q)∈[0,1]2. Obviously, this is a much stronger assumption than marginal overdominance at an equilibrium, as used in Corollary 3.3. We will need the following lemma:

Lemma 5.2

Locus *A* exhibits marginal overdominance everywhere if and only if

$$m_{13} > \max\{m_{11}, m_{33}\} \; and \; m_{24} > \max\{m_{22}, m_{44}\}, \quad (5.1)$$

i.e., overdominance holds on the edges $q = 0 \; and \; q = 1$, and

$$m_{14} > \max\left\{m_{12} - \sqrt{(m_{13} - m_{11})(m_{24} - m_{22})},\; m_{34} - \sqrt{(m_{13} - m_{33})(m_{24} - m_{44})}\right\}. \quad (5.2)$$

Proof

We observe that

$$ax^2 + 2bx(1-x) + cx^2 > 0 \text{ for every } x \in [0, 1] \quad (5.3)$$

if and only if $a > 0, c > 0, \text{ and } b > -\sqrt{ac}$. Therefore, the conditions in the lemma follow immediately because $m_{A_1A_2} - m_{A_1A_1}$ and $m_{A_1A_2} - m_{A_2A_2}$ can be written in the form of (5.3) and need to be positive. □

Theorem 5.3

Assume that both loci exhibit marginal overdominance everywhere. Then the following holds:

(a)The boundary flow is of $E_4^4C_0b$.

(b)No boundary equilibrium is saturated, whence $\delta = 0$ and the system is permanent.

(c)There may exist one, three, or five internal equilibria.

(d)If there is a unique internal equilibrium, it is globally asymptotically stable.

(e)If there are three internal equilibria, two are sinks and one is a saddle.

(f)If there are five internal equilibria, three are sinks and two are saddles.

Proof

Statement (a) is clear from Lemma 5.2. Combining (3.11) with (A.1) and Theorem 4.6 yields (b). Statement (b) together with Corollary 4.5(b) gives claim (c). Theorem 4.4 yields the desired numbers of saddles for (d), (e), and (f). Because the trace of *J* (3.2) is negative due to (3.11), the internal equilibria with index +1+1 are sinks. □

Phase portraits with one or three internal equilibria are shown in Figure S1d, panels 4 and 5. We could neither find an example for case (f) in Theorem 5.3 nor could we exclude it.

Theorem 5.4

Assume that both loci exhibit marginal underdominance everywhere. Then the following holds:

(a)The boundary flow is of type $E_0^4C_4b.$.

(b)All boundary equilibria are saturated, whence δ=0.

(c)There may exist one, three, or five internal equilibria.

(d)If there is a unique internal equilibrium, it is a source.

(e)If there are three internal equilibria, two are sources and one is a saddle.

(f)If there are five internal equilibria, three are sources and two are saddles.

Remark 5.5

The possible flows on [0,1]2[0,1]2 for Theorem 5.4 are obtained by flow reversal from flows occurring by Theorem of 5.3.

Finally, the following can be shown. We leave the simple proof to the reader.

Theorem 5.6

Assume one locus exhibits marginal overdominance everywhere and the other marginal underdominance everywhere.

(a) The boundary flow is of type $E_2^4C_0c.$ ....

(b)The two internally stable edge equilibria are saturated, whence δ=2.

(c)There exists a unique internal equilibrium, and it is a saddle.

Remark 5.7

It seems interesting that there exist fitness matrices that do not satisfy the assumption of the above theorem and generate a phase portrait with three internal equilibria and the boundary flow $E_2^4C_0c.$ .... Thus, the assumptions of marginal overdominance and underdominance everywhere do not only constrain the potential boundary flows substantially, but also the phase portraits that can occur for a given boundary flow.

6 LINEAR ISOCLINES

Here, we study the dynamics of systems obtained by fitness matrices yielding linear isoclines. This leads to a system of differential equations that has been investigated by Schuster et al. (1981) to study the evolution of two strategies in asymmetric animal contests, for example between two species. It also turns out to be equivalent to a model used by Zhivotovsky and Gavrilets (1992) to study quantitative genetic variation under epistatic selection. A number of well known models emerge as special cases, for instance the additive model or the haploid model.

It will be convenient to reparameterize the fitnesses mijmij as follows:

	B_1B_1	B_1B_2	B_2B_2
A_1A_1	$2a_1 + 2a_2 + e_{11}$	$2a_1 + a_2 + d_2 + e_{12}$	$2a_1$
A_1A_2	$a_1 + 2a_2 + d_1 + e_{21}$	$a_1 + a_2 + d_1 + d_2 + e_{22}$	$a_1 + d_1$
A_2A_2	$2a_2$	$a_2 + d_2$	0

(6.1)

Here, aiai are allelic effects, didi dominance effects, and eijeij epistatic effects. According to the remark below (2.7), one additional parameter could be specified (e.g., by setting a1=1a1=1), but we refrain from doing so. The general assumption (HH) implies that we exclude complete dominance or recessivity of an allele, i.e., $|d_i| \neq a_i$ for $i = 1, 2$, and analogous restrictions apply to e12e12 and e21e21.

A straightforward calculation shows that the dynamics (2.5) becomes

$$\dot{p} = p(1-p)[a_1 + d_1(1-2p) + 2e_{22}q + 2(e_{12} - 2e_{22})pq + (e_{21} - 2e_{22})q^2 + (e_{11} - 2e_{12} - 2e_{21} + 4e_{22})pq^2], \quad (6.2a)$$

$$\dot{q} = q(1-q)[a_2 + d_2(1-2q) + 2e_{22}p + 2(e_{21} - 2e_{22})pq + (e_{12} - 2e_{22})p^2 + (e_{11} - 2e_{12} - 2e_{21} + 4e_{22})p^2q]. \quad (6.2b)$$

We note that the right-hand side of (6.2a) is simply the additive effect of allele A1A1 multiplied by *p*, and analogously the right-hand side of (6.2b) yields the additive effect of allele B1B1.

Remark 6.1

A standard decomposition of the total genetic variance (e.g., Kempthorne 1955) shows that dominance-by-dominance interactions are absent if and only if $e_{11} - 2e_{12} - 2e_{21} + 4e_{22} = 0$, additive-by-dominance interactions are absent if and only if

$$e_{11} = 2e_{12} = 2e_{21} = 4e_{22}, \tag{6.3}$$

additive-by-additive interactions are absent if and only if $e_{11} = e_{12} = e_{21} = e_{22} = 0$, and dominance is absent if and only if $d_1 = d_2 = 0$ and (6.3) holds.

We infer from (6.2) that both isoclines are linear if and only if (6.3) holds. With this assumption, (6.2) simplifies to

$$\dot{p} = p(1 - p)(a_1 + d_1 - 2d_1 p + 2e_{22}q), \tag{6.4a}$$

$$\dot{q} = q(1 - q)(a_2 + d_2 - 2d_2 q + 2e_{22}p). \tag{6.4b}$$

Interestingly, this is a model that was studied independently and in different contexts by Schuster et al. (1981) and Zhivotovsky and Gavrilets (1992). The latter authors introduced an *n*-locus version of the fitness scheme (6.1) with a generalized version of constraint (6.3), as 'the simplest generalization of the additive model to include dominance and pairwise additive-by-additive epistasis'. Schuster et al. (1981) identified all possible (non-degenerate) phase portraits for (6.4). Because they were interested in a game-theoretical context, they studied a more general model in which one of the coefficients e22e22 in (6.4) was substituted by a sixth, independent coefficient. Then limit cycles can occur because the corresponding 4×44×4 fitness matrix (mij)(mij) is not symmetric. Here, we present the results of Schuster et al. (1981) that apply to our model (6.4) in our terminology and complement them. They also follow directly from our results in Sect. 4.

Since the isoclines are linear, there is either no or one internal equilibrium. Therefore, by Corollary 4.5(c), the boundary index sum can assume only the values δ=0,1δ=0,1, or 2. Furthermore, if two edge equilibria exist on opposite edges, they must have the same internal stability. This rules out the following 10 of the 42 boundary flows in Figure 2: $E_1^2C_0e$, $E_1^2C_1a'''$, $E_1^2C_2e$, $E_2^3C_0a$, $E_2^3C_1a$, $E_1^3C_1a$, $E_1^3C_2a$, $E_3^4C_0e$,

$E_1^4C_2e$, and $E_2^4C_1s$. C1s. In addition, the following boundary flows are easily ruled out: $E_1^1C_0a$, $E_0^1C_1a$, $E_2^2C_0a$, $E_0^2C_2a$, $E_1^2C_0a$ and $E_1^2C_1a''$.

Finally, the planar Routh–Hurwitz criterion implies that the internal equilibrium (p^,q^)(p^,q^)resulting from (6.4) is

- a saddle if $d_1d_2 < e_{22}^2$;
- a sink if $d_1d_2 > e_{22}^2$ and $d_1\hat{p}(1-\hat{p}) + d_2\hat{q}(1-\hat{q}) > 0$; and
- a source if $d_1d_2 > e_{22}^2$ and $d_1\hat{p}(1-\hat{p}) + d_2\hat{q}(1-\hat{q}) < 0$.

We summarize the results:

Corollary 6.2

Assume (6.4).

(a)The following eight boundary flows can have $\delta = 0$: $E_2^2C_0c$, $E_2^2C_0s$, $E_3^3C_0e$, $E_4^4C_0b$ (in each case the internal equilibrium is globally attracting); $E_0^2C_2c$, $E_0^2C_2s$, $E_0^3C_3e$, $E_0^4C_4b$ (in each case the internal equilibrium is a source). The last four cases are the flow reversals of the first four cases.

(b)The following ten boundary flows and their flow reversals can have $\delta = 1$: $E_1^1C_0e$, $E_1^1C_1e$, $E_1^1C_1a$, $E_2^2C_0c$, $E_2^2C_0s$, $E_2^2C_0e$ (if only E12E12 is saturated), $E_2^2C_1s$, $E_1^2C_1a$ (if both edge equilibria are not saturated), $E_3^3C_0e$ (if only E13E13 is saturated), $E_2^3C_0e$ (if only E13E13 is saturated). In addition, $E_0^0C_1s$ *and* $E_0^0C_1a$ *have* $\delta = 1$. In all these 22 cases, an internal equilibrium does not exist.

(c)The following five boundary flows and their flow reversals can have $\delta = 2$: $E_1^1C_1a$, $E_2^2C_0c$, $E_2^2C_0s$, $E_1^2C_1a$, $E_2^3C_0e$. In addition, $E_0^0C_2b$ *and* $E_2^4C_0c$ *have* $\delta = 2$. In each of these twelve cases, the internal equilibrium is a saddle.

Therefore, a phase portrait is uniquely determined by its boundary flow and boundary index sum δδ. However, for 12 boundary flows, δδ is not uniquely determined by the boundary flow. For $E_2^2C_0c$, $E_0^2C_2c$, $E_2^2C_0s$, *and* $E_0^2C_2s$, δ can assume all three possible values, 0, 1, and 2.

Zhivotovsky and Gavrilets (1992) derived results about the maintenance of a multilocus polymorphism, i.e., an internal equilibrium, and showed that several models investigated previously, mainly in a quantitative-genetic context, can be obtained as special cases of their model. Among others, they deduced the conditions for the existence and for the stability of a fully polymorphic equilibrium if all loci have equal effects on fitness and selection is sufficiently weak that linkage disequilibrium can be ignored. The assumption of equal effects is equivalent to a symmetric fitness matrix (2.2), i.e., ai=aai=a and di=ddi=d for every *i*.

A more detailed study of this symmetric model was performed by Gavrilets (1993). If the fitness matrix is symmetric (ai=aai=a and di=ddi=d), the conditions for the existence and stability of an internal equilibrium become very simple. An internal (nondegenerate) equilibrium exists if and only if $0 < \frac{a+d}{2(d-e_{22})} < 1$. It exists and is linearly stable if and only if $-d < e_{22} < (d-a)/2$. This condition can be satisfied only if $d > a/3$. It also implies that there are internally stable equilibria on the edges $p = 1$ and $q = 1$. The corresponding boundary flow is $E_2^2C_0s$ with $\delta = 0$. We will briefly return to this model in Sect. 8, where we treat general symmetric fitness matrices.

Now we turn to two important special cases of the model (6.4).

6.1 The additive fitness model

This classical model (Bodmer and Felsenstein 1967; Ewens 1969) assumes that loci contribute additively to fitness. Thus there is no epistasis, but dominance is admitted. It is obtained from (6.1) by assuming

$$e_{11} = e_{12} = e_{21} = e_{22} = 0. \tag{6.5}$$

In fact, this model has been analyzed without ignoring linkage disequilibrium, i.e., the full four-gamete system (2.3), and indeed for any number of loci. Ewens (1969) proved for an arbitrary number of multiallelic loci that mean fitness is a Lyapunov function. Moreover, every equilibrium is in linkage equilibrium (for non-zero recombination, as we assume here). For diallelic loci, an internal equilibrium exists if and only if at every locus there is either overdominance or underdominance. An internal equilibrium is unique if it is isolated; it is globally asymptotically stable if there is overdominance on both loci (Karlin and Liberman 1978, 1990). All trajectories converge exponentially to an equilibrium on the linkage-equilibrium manifold $D = 0$ (Lyubich 1992). For a more detailed review, see Bürger 2000 (pp. 48–50, 76–78). Therefore, the flows derived by assuming $D = 0$ are representative for the full dynamics after a sufficiently long time has passed and also for every trajectory that starts close to $D = 0$

The isocline $\dot{p} = 0$ simplifies to a horizontal straight line (in a representation as in Figure 1), and the isocline $\dot{q} = 0$ to a vertical straight line. It follows that locus *A* (*B*) exhibits marginal overdominance if and only if it exhibits overdominance on one of the respective edges, i.e., if and only if $d_1 > a_1$ ($d_2 > a_2$);); analogously for underdominance. Therefore, in accordance with the above mentioned results, we obtain:

Corollary 6.3

For the additive fitness model the following boundary flows with boundary index sum δδ occur:

(a) $\mathrm{E}_0^0\mathrm{C}_1\mathrm{s}$ *with* $\delta = 1$ 1 if dominance is intermediate at both loci, i.e., $-a_i < d_i < a_i$;

(b) $\mathrm{E}_2^2\mathrm{C}_0\mathrm{e}$ ($\mathrm{E}_0^2\mathrm{C}_2\mathrm{e}$) *with* $\delta = 1$ if dominance is intermediate at one locus and the other locus is overdominant (underdominant);

(c) $\mathrm{E}_4^4\mathrm{C}_0\mathrm{b}$ ($\mathrm{E}_0^4\mathrm{C}_4\mathrm{b}$) *with* $\delta = 0$ if both loci are overdominant (underdominant);

(d) $E_2^4C_0c$ *with* $\delta = 2$ if one locus is overdominant, the other underdominant.

An internal equilibrium exists and is unique for the boundary flows $E_4^4C_0b$, $E_0^4C_4b$, and $E_2^4C_0c$. It is globally attracting, a source, and a saddle, respectively. Each of the above boundary flows admits only one phase portrait.

6.2 The haploid model

If selection acts on haploids instead of diploids, fitnesses can be assigned directly to gametes. Denoting the fitness of gamete *i* by vivi, straightforward calculations show that under the assumption $D = 0$ the following dynamics is obtained (cf. Haldane 1931):

$$\dot{p} = p(1-p)[v_2 - v_4 + (v_1 - v_2 - v_3 + v_4)q], \quad (6.6a)$$

$$\dot{q} = q(1-q)[v_3 - v_4 + (v_1 - v_2 - v_3 + v_4)p]. \quad (6.6b)$$

Comparison with (6.4) reveals that this is obtained from the fitness scheme (6.1) if one sets $d_1 = d_2 = 0,\ v_4 = 0$ (without loss of generality), $v_3 = a_2,\ v_2 = a_1,\ v_1 = a_1 + a_2 + 2e_{22}$, and assumes (6.3). Therefore, there is additive-by-additive epistasis but no dominance.

The isocline $\dot{p} = 0$ is given by the vertical line $q = a_1/(2e_{22})$, and the isocline q˙=0q˙=0 is given by the horizontal line $p = a_2/(2e_{22})$. Hence, there are no edge equilibria, but there may exist one internal equilibrium given by these values. The only possible boundary flows are $E_0^0C_1s$, $E_0^0C_1a$, and $E_0^0C_2b$. Their boundary index sum is uniquely determined and $\delta = 1, 1, \text{and } 2$, respectively. It is easy to show that all three cases can be realized. An internal equilibrium exists only in the third case, and it is a saddle. These results are in accordance with results obtained previously for the following more general models.

In game theory, (6.6) is known as the replicator dynamics for 2×2partnership games, which is generalized by the replicator dynamics for 2×2bimatrix games. There, the coefficients of *p*and *q* in (6.6) may differ. Schuster and Sigmund (1981) proved that periodic orbits occur if these coefficients have opposite signs (see Hofbauer and Sigmund 1998, Sections 10 and 11, for a treatment of bimatrix games and the replicator dynamics).

The complete haploid selection model, i.e., without the assumption of linkage equilibrium, was investigated by Felsenstein (1965), Feldman (1971), Rutschman (1994), and Bank et al. (2012). The latter authors proved that there exists at most one internal equilibrium and, if it exists, it is unstable. Then two vertex equilibria are asymptotically stable. This corresponds to the boundary flow $\mathrm{E}_0^0\mathrm{C}_2\mathrm{b}$. Otherwise, one vertex is globally asymptotically stable and the boundary flow is either $\mathrm{E}_0^0\mathrm{C}_1\mathrm{s}$ or $\mathrm{E}_0^0\mathrm{C}_1\mathrm{a}$. Bank et al. (2012) also identified the parameter combinations that lead to the respective equilibrium structures.

7 THE MULTILINEAR EPISTASIS MODEL

Hansen and Wagner (2001) introduced a model of gene interaction that assumes that the effect of gene substitutions due to changes in the genetic background (the other loci) can be described by a linear transformation. Although their model is formulated in terms of genetic effects on quantitative traits, it can be applied to our context if we consider fitness as the trait. A two-locus version of this model, and its applications to the maintenance of genetic variation, was studied by Hermisson et al. (2003). Following their formulation, the fitness of the two-locus genotype AiAjBkBℓAiAjBkBℓ can be written as

$$w(ij, k\ell) = \mu + \alpha(ij) + \beta(k\ell) + \gamma\alpha(ij)\beta(k\ell). \quad (7.1)$$

Comparison with the fitness scheme (6.1) shows, after some calculation, that their model is the special case of (6.1) obtained by assuming

$$e_{11} = \frac{4a_1a_2e_{22}}{(a_1+d_1)(a_2+d_2)}, \text{ and } e_{12} = \frac{2a_1e_{22}}{a_1+d_1}, \text{ and } e_{21} = \frac{2a_2e_{22}}{a_2+d_2}. \quad (7.2)$$

It follows that the isoclines $\partial \bar{m}/\partial p = 0$ and $\partial \bar{m}/\partial q = 0$ take the form

$$\frac{(a_2 + d_2 - 2d_2 p)\varphi_2(q)}{(a_1 + d_1)(a_2 + d_2)} = 0, \tag{7.3a}$$

$$\frac{(a_1 + d_1 - 2d_1 q)\varphi_1(p)}{(a_1 + d_1)(a_2 + d_2)} = 0, \tag{7.3b}$$

where

$$\varphi_i(x) = (a_1 + d_1)(a_2 + d_2) + 2e_{22}x(a_i + d_i - d_i x). \tag{7.3c}$$

Therefore, they are in product form. It is not difficult to show that the isoclines of (6.2) are in product form if and only if (7.2) holds. In this case, which we assume below, the dynamics has separated variables.

We define

$$\tilde{p} = \frac{a_2 + d_2}{2d_2} \quad \text{and} \quad \tilde{q} = \frac{a_1 + d_1}{2d_1}, \tag{7.4}$$

and observe that $0 < \tilde{p} < 1$ $(0 < \tilde{q} < 1)$ if and only if there is overdominance or underdominance at locus *A* (*B*). If $0 < \tilde{p} < 1$ and $0 < \tilde{q} < 1$, we call the equilibrium $(\tilde{p}, \tilde{q})$ the central equilibrium. We note that $p = \tilde{p}$ and $q = \tilde{q}$ are invariant lines of the dynamics (6.2) with (7.2). As a consequence, saddle connections occur robustly in this class, e.g., for the boundary flow $\mathrm{E}_4^4\mathrm{C}_0\mathrm{b}$ with $\delta = 2$ (Figure S3h, panel 6).

The isoclines may be composed of up to three straight lines. For $\dot{p} = 0$ (and in a representation as in Figure 1), these are the horizontal line $p = \tilde{p}$ and the two vertical lines that are given by the solutions of $\varphi_2(q) = 0$. If $0 < \tilde{p} < 1$, then the two edge equilibria E13E13 and E24E24 exist and have the same value, $p = \tilde{p}$ Thus, edge equilibria occur always in pairs on opposite edges. If in addition $0 < \tilde{q} < 1$, the other two edge equilibria and the central equilibrium $(\tilde{p}, \tilde{q})$ exist. Furthermore, the intersection points of $\varphi_2(q) = 0$ and $\varphi_1(p) = 0$ 0 may yield up to four additional internal equilibria. Thus, in total there may be up to five internal equilibria.

The case of five internal equilibria can be realized for overdominance at both loci as well as for underdominance at both loci, but not if one locus exhibits overdominance and the other underdominance (Theorem 7.1 and Table S3). The corresponding boundary flows are $E_4^4C_0b$ with $\delta = 4$ and its flow reversal $E_0^4C_4b$.

Because opposite edge equilibria occur pairwise, flows of type E_m^1 or E_m^3 cannot occur. The following theorem lists all possible equilibrium structures. Only 16 of the 42 boundary flows in Figure 2 can occur. In particular, it states the extent to which the equilibrium structure can be inferred from the boundary flow. The proof is given in "Appendix A.3". Parameter combinations that yield all possible equilibrium structures are given in Table S3.

Theorem 7.1

Assume (6.2) and (7.2).

(a)All possible equilibrium structures in the interior are given in Table 1.

(b)For the top twelve boundary-flow classes EnmCkxEmnCkx in Table 1, the number of asymptotically stable edge equilibria is min{m,δ}min{m,δ}. For the other four classes, this is in general wrong.

Remark 7.2

Because mean fitness is a strict Lyapunov function (Sect. 2), there is a globally asymptotically stable equilibrium if and only if there is precisely one sink. This occurs in the following cases: $E_0^0C_1s$ and $E_0^0C_1a$ (a corner equilibrium is globally attracting); $E_2^2C_0c$, $E_2^2C_0e$, and $E_1^2C_0e$ (an edge equilibrium is globally attracting $E_4^4C_0b$ and $E_3^4C_0e$, each with $\delta = 0$ (the central equilibrium is globally attracting); $E_2^4C_0c$ with $\delta = 0$ (in this case the central equilibrium is a sink, hence globally attracting).

Table 1. Internal equilibrium structures for the multilinear epistasis model

Boundary flow	$\delta = 0$	$\delta = 1$	$\delta = 2$	$\delta = 3$	$\delta = 4$
$E_0^0C_1s$	–	0	–	–	–
$E_0^0C_1a$	–	0	–	–	–
$E_0^0C_2b$	–	–	1 saddle	–	–
$E_2^2C_0c$	–	0	–	–	–
$E_0^2C_2c$	–	0	–	–	–
$E_2^2C_0e$	–	0*	–	–	–
$E_0^2C_2e$	–	0*	–	–	–
$E_1^2C_0e$	–	0	–	–	–
$E_1^2C_1a'''$	–	–	1 saddle	–	–
$E_1^2C_2e$	–	0	–	2 saddles	–
$E_4^4C_0b$	1 sink	–	1 saddle	–	1 source, 4 saddles
$E_0^4C_4b$	1 source	–	1 saddle	–	1 sink, 4 saddles
$E_3^4C_0e$	1 sink	–	1 saddle or 1 source and 2 saddles	–	–
$E_1^4C_2e$	1 source	–	1 saddle or 1 sink and 2 saddles	–	–
$E_2^4C_0c$	1 sink or 1 source	–	1 saddle	–	–
$E_2^4C_1s$	–	1 sink or source, 1 saddle	–	–	–

A '–' indicates that this value of δ does not occur, and '0' indicates that the number of internal equilibria is zero. A comma means 'and'. * Indicates that this case can be realized by matrices with different extended boundary flows; thus there are two different equilibrium structures. For the four boundary flows at the bottom, which admit two different internal equilibrium structures for one value of δ, each of the internal equilibrium structures is generated by a different, but unique, extended boundary flow

We observe from (7.2) that the multilinear model reduces to that with linear isoclines in Sect. 6if and only if $d_1 = d_2 = 0$ or $e_{22} = 0$. Thus, the multilinear model of epistasis coincides with the epistatic model of Zhivotovsky and Gavrilets (1992) if and only if dominance is absent, when it simplifies to the haploid model (Sect. 6.2), or if epistasis is absent, when it reduces to the additive model (Sect. 6.1). It is easy to show that if dominance and epistasis are present, the multilinear model and that of Zhivotovsky and Gavrilets are different and none is a special case of the other. The reason is that in the model of Zhivotovsky and Gavrilets there are no additive-by-dominance or dominance-by-dominance interactions, whereas they occur in the multilinear model if dominance is present (as follows from Eq. 7.2 and Remark 6.1).

Remark 7.3

(a)The following (extended) boundary flows occur in the multilinear epistasis model but not in the model of Zhivotovsky and Gavrilets (1992): $E_1^2C_0e$, $E_1^2C_1a'''$, $E_1^2C_2e$, $E_3^4C_0e$, $E_1^4C_2e$, $E_2^4C_1s$; $E_4^4C_0b$ and $E_0^4C_4b$ (each with $\delta = 2$ and $\delta = 4$) and $E_2^4C_0c$ with $\delta = 0$ (see Table S2).

(b)The extended boundary flows that occur both in the multilinear model and in the model of Zhivotovsky and Gavrilets are those occuring in the haploid model (no edge equilibria), those occuring in the absence of dominance and epistasis (listed in Corollary 6.3), and $E_2^2C_0c$ and $E_0^2C_2c$, each with $\delta = 1$ (see Table S2). However, in the latter case the fitness matrices are different.

Finally, we note that the above model was investigated by Bomze et al. (1983) in the context of Mendelian game dynamics of asymmetric contests. Because then the fitness matrix (2.2) is in general not symmetric, periodic solutions may occur. Bomze (1982) provided a classification of all possible phase portraits.

8 EQUAL LOCUS EFFECTS

If both loci contribute to fitness equally, the 3×33×3 matrix in (6.1) is symmetric, i.e.,

$$a_1 = a_2 = a,\ d_1 = d_2 = d,\ \text{and } e_{21} = e_{12}. \tag{8.1}$$

Under this assumption, which we impose throughout this section, only the nine boundary flows in Figure 2 occur that have a code ending by an ss or bb. However, additional restrictions on the possible phase portraits and equilibrium structures occur as we will demonstrate now.

The symmetry implies that edge equilibria occur pairwise, i.e., (E_{24}, E_{34}) and (E_{12}, E_{13}). Equilibria of a pair have the same internal and the same external stability. Therefore, boundary flows with an odd number of stable corner equilibria give rise to odd values δδ; with an even number of stable corner equilibria δδ is even.

Using the symmetry (8.1), we infer from (6.2) that the isoclines take the form

$$h(x) = \frac{a + d + 2e_{22}x + (e_{12} - 2e_{22})x^2}{2d - 2(e_{12} - 2e_{22})x - (e_{11} - 4e_{12} + 4e_{22})x^2}, \qquad (8.2)$$

where for the p˙=0p˙=0 isocline we have h(p)=qh(p)=q, and for the q˙=0q˙=0 isocline we have h(q)=ph(q)=p. Therefore, we obtain up to three symmetric internal equilibria. They satisfy p=qp=q and are the solutions of the cubic equation h(x)=xh(x)=x. In addition, there may exist a pair of internal equilibria that are mirror images of each other with respect to the diagonal p=qp=q. Since the internal equilibria are the zeros of the quintic polynomial (3.7), this pair of equilibria is given by the zeros of a quadratic polynomial.

The following theorem determines the number of internal equilibria for all possible boundary flows. We recall from Corollary 4.5 that the number of internal equilibria is odd if and only if δδis even.

Theorem 8.1

Assume (6.2) and (8.1). Possible boundary flows are of class $E_{2m}^{2n}C_k s$, where $n, m \in \{0, 1, 2\}$, $m \leq n$, and $k \in \{0, 1, 2, 3, 4\}$. The boundary index sum δδ is odd for boundary-flow classes with an odd number of stable corner equilibria, and δδ is even otherwise. In addition, the following holds:

(a)For the boundary-flow classes $E_2^2C_1s$ and $E_0^2C_3s$, each with $\delta = 1$ or $\delta = 3$, and for $E_2^4C_1s$ with $\delta = 3$, the number of internal equilibria can be zero (only for $\delta = 1$), two, or four.

(b)For the boundary-flow classes $E_4^4C_0b$ and $E_0^4C_4b$, each with $\delta = 0$ or $\delta = 4$, the number of internal equilibria can be one (only for $\delta = 1$), three, or five.

(c)For the boundary-flow classes $E_0^0C_1s$ (which has $\delta = 1$), $E_0^0C_2b$ $E_0^0C_2b$ (which has $\delta = 2$), $E_2^2C_0s$ and $E_0^2C_2s$ (which both have $\delta = 0$ *or* $\delta = 2$), $E_4^4C_0b$ and $E_0^4C_4b$ *with* $\delta = 2$, *and* $E_2^4C_1s$ *with* $\delta = 1$, the respective maximum number of internal equilibria (four or five) cannot be assumed. Any smaller number of internal equilibria admitted by Corollary 4.5 (1 or 3 *if* $\delta = 0$ *or* $\delta = 2$, *0 or 2 if* $\delta = 1$) can be assumed.

The proof is given in "Appendix A.4". Without proof (which is simple), we note that in this symmetric case, δ=−1δ=−1 can be excluded for E42C1sE24C1s. Therefore, the only possible values are $\delta = 1$ and $\delta = 3$. .

Finally, we briefly treat two special cases. For the model of Gavrilets (1993) mentioned in Sect. 6, i.e., Eq. (6.4) with (8.1), the following boundary flows and values δδ can occur: All eight boundary flows from Corollary 6.2 ending with an ss or bb occur. For $E_0^0C_1s$, $E_0^0C_2b$, $E_2^2C_1s$, $E_0^2C_3s$, $E_4^4C_0b$, and $E_0^4C_4b$, the boundary index sum is already uniquely determined ($\delta = 1, 2, 1, 1, 0$, and 0, respectively). For $E_2^2C_0s$ and $E_0^2C_2s$ the case $\delta = 1$ is easily excluded, whence only $\delta = 0$ or $\delta = 2$ are possible.

If the symmetry assumption (8.1) is imposed on the multilinear epistasis model treated in Sect. 7, the following boundary flows and equilibrium structures occur.

$E_0^0C_1s$ ($\delta = 1$) and $E_0^0C_2b$ ($\delta = 2$); in both cases, the equilibrium structure is unique.

$E_2^4C_1s$ ($\delta = 1$): both cases (sink and saddle, source and saddle) can be realized.

$E_4^4C_0b$ and $E_0^4C_4b$: each with $\delta = 0$ and $\delta = 4$. $\delta = 2$ cannot occur because this would require neighboring edge equilibria to differ in their external stability.

Interestingly, there are no equilibrium structures with two edge equilibria in the multilinear model if loci have equal effects.

9 THE SYMMETRIC VIABILITY MODEL

As already outlined in the Introduction, the so-called symmetric viability model has received much attention in the literature. One reason is that special cases of it arise naturally when two diallelic loci determine a quantitative character that is under stabilizing selection toward an intermediate optimum (e.g., Wright 1935, 1952; Hastings 1987; Nagylaki 1989; Gavrilets and Hastings 1993; Bürger and Gimelfarb 1999; Willensdorfer and Bürger 2003). For a detailed review, consult Bürger (2000, Chap. 6.2).

It has the property that fitnesses of genotypes are invariant under the simultaneous exchange of A1A1 with A_1 with A_2 and B_1 with B_2. Therefore the resulting fitness matrix (2.2) is centro-symmetric and depends only on four parameters. We shall use the following parametrization (cf. Nagylaki 1989):

$$\begin{array}{c|ccc} & B_1B_1 & B_1B_2 & B_2B_2 \\ \hline A_1A_1 & r_1+r_2+m & r_1-l & r_1+r_2-m \\ A_1A_2 & r_2-l & 0 & r_2-l \\ A_2A_2 & r_1+r_2-m & r_1-l & r_1+r_2+m \end{array} . \qquad (9.1)$$

This is equivalent to the most general form, as first introduced by Bodmer and Felsenstein (1967). The model studied by Lewontin and Kojima (1960) corresponds to the special case m=0m=0. Interestingly, this is also a special case of the multilinear epistasis model treated in Sect. 7 [by setting $a_1 = a_2 = e_{11} = e_{12} = e_{21} = 0$ in (6.1)]. Bodmer and Parson (1962) assumed $r_1 = r_2$, which makes the matrix symmetric, in addition to being centro-symmetric. Thus, only boundary flows ending with bb can be realized. The most general models of stabilizing selection (among those referred to above) require all four parameters in (9.1). However, because the double heterozygote has the highest fitness and fitness of trait values decays symmetrically with distance from the optimum, the four parameters have to satisfy certain inequalities (see Nagylaki 1989).

We analyze the general model (9.1). Instead of $(p, q) \in [0, 1]^2$, we use the coordinates $(x, y) \in [-1, 1]^2$ defined by

$$p = \frac{1+x}{2}, \; q = \frac{1+y}{2}. \tag{9.2}$$

This transforms the system (2.5) into

$$\dot{x} = (1 - x^2)(r_1 x + my + lxy^2), \tag{9.3a}$$

$$\dot{y} = (1 - y^2)(r_2 y + mx + lyx^2), \tag{9.3b}$$

which will form the basis for the subsequent analysis.

It is immediate that the origin $O = (0,0)$ is an equilibrium and the dynamics is point-symmetric with respect to it. In particular, if $(-\hat{x}, -\hat{y}),$ is an equilibrium of (9.3), so is $(-\hat{x}, -\hat{y}),$ and both have the same eigenvalues because the respective Jacobian matrices are equal. We observe that the isoclines of (9.3) are in product form if m=0m=0, whence a special case of the multilinear epistasis model emerges. The isoclines are linear if l=0l=0.

If the isoclines are not linear, they are given by

$$x = \psi_1(y) = \frac{-my}{r_1 + ly^2}, \tag{9.4a}$$

$$y = \psi_2(x) = \frac{-mx}{r_2 + lx^2}. \tag{9.4b}$$

Theorem 9.1

(a)The central, or symmetric, equilibrium *O* exists always.

(b) $If\, l \neq 0 \; and\, r_1 r_2 \neq 0,$, the following pairs of unsymmetric internal equilibria may exist:

$$y_{1,2} = \pm\sqrt{\frac{-r_1 - m\sqrt{\frac{r_1}{r_2}}}{l}}, \quad x_{1,2} = \sqrt{\frac{r_2}{r_1}}\, y_{1,2}, \tag{9.5a}$$

$$y_{3,4} = \pm\sqrt{\frac{-r_1 + m\sqrt{\frac{r_1}{r_2}}}{l}}, \quad x_{3,4} = -\sqrt{\frac{r_2}{r_1}}\, y_{3,4}. \qquad (9.5b)$$

In particular, the following holds:

(i) If $r_1 = \operatorname{sgn} r_2 = \operatorname{sgn} l,$, then at most one of these pairs is admissible.

(ii) If $r_1 = \operatorname{sgn} r_2 = -\operatorname{sgn} l,$, then both pairs may be admissible.

(iii) If $r_1 = -\operatorname{sgn} r_2$, *then* O is the only internal equilibrium.

The *proof* is given in “Appendix A.5.1”.

It is straightforward to compute the Jacobian of (9.3). At an internal equilibrium $(\hat{x}, \hat{y})$ it simplifies to

$$J(\hat{x}, \hat{y}) = \begin{pmatrix} (1-\hat{x}^2)(r_1 + l\hat{y}^2) & (1-\hat{x}^2)(m + 2l\hat{x}\hat{y}) \\ (1-\hat{y}^2)(m + 2l\hat{x}\hat{y}) & (1-\hat{y}^2)(r_2 + l\hat{x}^2) \end{pmatrix}; \qquad (9.6)$$

cf. Lemma 3.2. Evaluation at the central equilibrium O yields

$$J_O = \begin{pmatrix} r_1 & m \\ m & r_2 \end{pmatrix}. \qquad (9.7)$$

From the Routh–Hurwitz criterion we infer immediately:

Lemma 9.2

Let

$$\rho = \det J_O = r_1 r_2 - m^2, \qquad (9.8a)$$

$$\operatorname{tr} J_O = r_1 + r_2. \qquad (9.8b)$$

Then O is

(a)a saddle with index $-1 \ if \ \rho < 0,$

(b)a sink with index $+1 \ if \ \rho > 0 \ and \ \mathrm{tr} \ J_O < 0,$

(c)a source with index $+1 \ if \ \rho > 0 \ and \ \mathrm{tr} \ J_O > 0.$.

Note that the sign of trJOtrJO can be determined immediately from the fitness scheme (9.1). If the fitness of the double heterozygous genotpye $A_1A_2B_1B_2$ exceeds the arithmetic mean fitness of, for instance, the homozygous genotypes $A_iA_iB_2B_2 \ (i = 1 \text{ or } 2)$, then $\mathrm{tr} \ J_O < 0.$.

Remark 9.3

(a)Lemma 9.2 settles the internal equilibrium structure for the case l=0l=0, in which *O*is the only internal equilibrium.

(b)Let $r_1r_2 = 0$. If $r_1 = r_2 = 0$, the equilibrium OO is a saddle, and there exists either no other internal equilibrium $(\text{if } |m/l| > 1)$ or the curve of equilibria $y = -m/(lx)$, yielding a degenerate flow. If $r_1 = m = 0$, then $y = 0$ is a line of equilibria, hence again degenerate. If $r_1 = 0$ and $m \neq 0$ and $l \neq 0$, then O is the unique internal equilibrium, and it is a saddle.

For the remainder of this section, we assume l $l \neq 0$ and $r_1r_2 \neq 0$. As a corollary to the above theorem, we obtain

Corollary 9.4

If $\mathrm{sgn} \ r_2 = -\mathrm{sgn} \ r_1$, then *O* is the unique internal equilibrium. It is a saddle point and δ=2δ=2.

Proof

We already know from Theorem 9.1 that O is unique. If $\operatorname{sgn} r_2 = -\operatorname{sgn} r_1$, then $\rho < 0$ and O is a saddle. Since the index of a saddle point is −1−1, Theorem 4.4 shows that the boundary index sum is $\delta = 2.$

Here is the main result of this section.

Theorem 9.5

(a)The internal equilibrium structures that can occur in the symmetric viability model are given in Table 2.

(b)The stability of the boundary equilibria can be inferred from the boundary flow type EnmCkEmnCk as follows:

(b1)The number of asymptotically stable corner equilibria is k.

(b2)The number of asymptotically stable edge equilibria is min{m,δ}min{m,δ}.

Table 2. Internal equilibrium configurations for the symmetric viability model

Boundary flow	$\delta = 0$		$\delta = 2$		$\delta = 4$	
	$\rho < 0$	$\rho > 0$	$\rho < 0$	$\rho > 0$	$\rho < 0$	$\rho > 0$
$E_0^0C_2b$	–	–	1 saddle	1 sink or source[b], 2 saddles	–	–
$E_2^2C_0c$	1 saddle, 2 sinks	1 sink	1 saddle	1 sink or source[a], 2 saddles	–	–
$E_0^2C_2c$	1 saddle, 2 sources	1 source	1 saddle	1 sink or source[a], 2 saddles	–	–
$E_4^4C_0b$	1 saddle, 2 sinks	1 sink	1 saddle	1 source, 2 saddles	3 saddles	1 source, 4 saddles
$E_0^4C_4b$	1 saddle, 2 sources	1 source	1 saddle	1 sink, 2 saddles	3 saddles	1 sink, 4 saddles
$E_2^4C_0c$	×	1 sink or 1 source[b]	1 saddle	1 sink or source[b], 2 saddles	–	–

A '–' indicates that this value of δ does not occur (see Table S1) and '×' indicates that this combination of ρ and δ cannot occur. A comma means 'and'. [a] Whether O is a sink or a source needs to be determined from the sign of $r_1 + r_2$. [b] The stability of O switches under flow reversal, which (in these classes) does not alter the extended boundary-flow class

The *proof* is given in "Appendix A.5.2". Examples of all possible flows can be found in Section S1 of the SI.

From Table 2, we infer immediately that O is the unique internal equilibrium if either $\delta = 0$ and $\rho > 0$ or if $\delta = 2$ and $\rho < 0$. In fact, the table shows that the number of internal equilibria can always be determined from the extended boundary flow and ρρ.

Nagylaki (1989) identified all equilibrium structures and phase portraits for the special case of the symmetric viability model that arises from stabilizing selection toward an optimum situated on the genotypic value of the double heterozygote. He admitted arbitrary functions decaying from the optimum monotonically and symmetrically, and he assumed absence of linkage equilibrium. Thus, his model is a special cases of (9.3) (after transformation between (p, q) and (x, y) coordinates). He proved that only the two phase portraits for $E_0^0C_2b$ and $\rho > 0$, then O can occur and four of the five listed for $E_2^2C_0c$ (if $\delta = 2$, then O is a sink in his model).

10 DISCUSSION

The analysis of the classical two-locus two-allele selection-recombination model is notoriously difficult. As outlined in the Introduction, despite considerable efforts only special cases are well understood, for instance, the models with additive or multiplicative fitnesses. We investigated a simplification of the full two-locus two-allele model (2.3) by assuming that the two loci are independent, i.e., in linkage equilibrium. This simplified model is given by (2.5) and, essentially, goes back to Wright (1942), although special cases had been studied earlier (Haldane 1931; Wright 1935).

The model (2.5) is not only more accessible to ematical analysis than (2.3), it is also of biological relevance because it has been derived as the weak-selection limit of the full model. Therefore, it provides a good approximation if selection is not too strong and the two loci are unlinked or only weakly linked. Indeed, under the non-degeneracy assumption (HH), a theorem by Nagylaki et al. (1999) demonstrates that in the full model (2.3) and for sufficiently weak selection relative to recombination, every trajectory converges to an equilibrium point, and every such equilibrium point is a perturbation of an equilibrium point of the weak-selection limit (2.5). Throughout this paper, we imposed condition (HH).

Importantly, mean fitness is a strict Lyapunov function for (2.5). Hence, every solution converges to an equilibrium point (Sect. 2). This is not always the case in the full model, neither for (2.3) nor its continuous-time analog. For each the existence of stable limit cycles was demonstrated (Akin 1979, 1982; Hastings 1981b; Hofbauer and Iooss 1984) if selection and recombination are of similar strength. For the weak-selection limit (2.5), Moran (1963) showed that, in addition to the eight possible boundary equilibria, there may exist up to five internal equilibria, and three can be simultaneously stable (Theorem 3.1). For the full model (2.3), the maximum number of equilibria is 15, seven of them being internal equilibria. This is an immediate consequence of a result of Altenberg (2010), who proved the conjecture of Feldman and Karlin (1970) that the maximum number of equilibria in a selection-recombination model with *n* gametes is 2n−12n−1.

10.1 Boundary flows, extended boundary flows, and phase portraits

Although we could not fully accomplish our goal of deriving and classifying all possible equilibrium structures and (equivalence classes of) phase portraits of the weak-selection limit (2.5), we identified the extended boundary flows and determined all equilibrium structures for several important types of fitness patterns. These results yield interesting insights into the role of epistasis and dominance in generating equilibrium structures.

For general fitnesses, we identified all possible boundary flows, i.e., flows on the boundary of the state space [0,1]2[0,1]2 (Figure 1). The four corners correspond to the monomorphic equilibria. The dynamics on the edges correspond to the single-locus dynamics when the other locus is fixed for one or the other allele. There are 42 (topologically) different boundary flows or, more precisely, boundary-flow classes because boundary flows that are obtained by relabeling loci or alleles are identified (Sect. 4, Figure 2). Of these 42 boundary flows, there are 16 pairs for which a member of a pair is obtained by reversing the flow of the other member. The other 10 boundary-flow classes are self inverse under flow reversal.

These boundary flows are by far not sufficient to describe all possible phase portraits, i.e., the topological structure of the flow on [0,1]2[0,1]2. As an intermediate step, we studied the extended boundary flows (Sect. 4.2). They describe the dynamics in a small neighborhood of the boundary, in particular, the external stability of the edge equilibria. A key ingredient for deriving the possible equilibrium structures and phase portaits on the full state space [0,1]2[0,1]2from a given extended boundary flow is the boundary index sum δδ (defined below Eq. 4.1). The possible values of δδ are constrained by Lemma 4.1, and Theorem 4.2 shows that for specific boundary flows additional values δδ can be excluded. Thus, we are still left with 190 potentially possible extended boundary flows. The most important tools for drawing conclusions about the internal equilibrium structure are Theorem 4.4, which is a special case of a more general index theorem (Hofbauer 1990), and Corollary 4.5. Overall, we showed existence of 185 extended boundary flows by providing a fitness matrix and a phase portrait generating such an extended boundary flow (SI, Figs. S1–S5, Table S1).

For a given boundary-flow class and a value of δδ, there may still exist more than one extended boundary flow yielding this boundary-flow class and this δδ. Moreover an extended boundary flow may be compatible with more than one equilibrium structure, and an equilibrium structure may be generated by non-equivalent phase portraits (see Sect. 4.3 and Table S1, as well as the phase portraits in Section S1). Apart from characterizing all possible equilibrium structures or even all possible phase portraits, also more specific problems remain unresolved. For instance, can a sink and a source in the interior coexist? There are also five extended boundary flows, all with δ=−1δ=−1, whose existence we could not exclude. We conjecture that they do not exist (see Sect. 4.3).

10.2 Permanence

An important notion in modeling biological systems is permanence. This is a generalization of the notion of a protected polymorphism, which is mainly used for one-locus two-allele models in (spatially) structured populations. Loosely speaking, permanence means that no type or species will be lost because its frequency will remain above a certain threshold. For our model (2.5), this implies that a permanent system can exhibit only 14 types of extended boundary flows (Corollary 4.7), all having δ=0δ=0. At least one of these extended boundary flows $(E_4^4C_0b)$ can be generated by permanent flows with one, three, or five internal equilibria (Figure S1d, panels 4, 5, 6). Thus, the phase portrait is not uniquely determined by the extended boundary flow.

For every given fitness matrix generating one of the 14 boundary-flow classes of Corollary 4.7and parametrized as in (6.1), a sufficiently strong increase of the epistasis parameter e22 yields a permanent flow. This is a consequence of Corollaries 4.7 and A.1, since in the parametrization

(6.1), e22 is the only parameter in m14m14 that is independent of the given boundary flow.

10.3 Marginal overdominance or underdominance

In Sect. 5, we assumed that each locus exhibits either marginal overdominance at every point (p, q) or marginal underdominance; see (3.11) and Lewontin and Kojima (1960). This is equivalent to having continuous isoclines (3.4) that map [0, 1] into (0, 1). If there is marginal overdominance at both loci, then $E_4^4C_0b$ is the only possible boundary-flow class and no boundary equilibrium is saturated. Therefore, the extended boundary flow is uniquely determined, δ=0δ=0, and the system is permanent. This is compatible with having one, three, or five internal equilibria, of which one, two, or three, respectively, are sinks (Theorem 5.3). We could find phase portraits with one or three internal equilibria, but not with five. However, we could not prove that five internal equilibria cannot be realized. Although Moran's (1963) example (Figure S1d, panel 6) has boundary-flow class E44C0bE44C0b and δ=0δ=0, it does not satisfy the assumptions of Theorem 5.3.

A result analogous to Theorem 5.3 holds if both loci exhibit marginal underdominance because then every flow can be obtained by flow reversal. If one locus exhibits marginal overdominance and the other marginal underdominance, then the boundary-flow class is E42C0cE24C0c, there exists a unique internal equilibrium, which is a saddle, and the two internally stable edge equilibria are linearly stable (Theorem 5.6). Hence, the phase portrait is uniquely determined by the boundary-flow class.

Table S2 lists the extended boundary flows and indicates by which of the special fitness patterns they can be generated.

10.4 Linear isoclines, or additive-by-additive epistasis

A particularly interesting special class arises if linear isoclines are posited (Sect. 6). With the fitness parameterization (6.1) and the assumption (6.3), the dynamics (2.5) simplifies to (6.4). This type of model has been studied independently, in different generality and in different contexts, by Schuster et al. (1981) and by Zhivotovsky and Gavrilets (1992) and Gavrilets (1993); see Sect. 6 for a more detailed appraisal. In population genetics terms, (2.5) has linear isoclines if and only if epistatic interactions are absent or there are only additive-by-additive interactions. The assumption of linear isoclines rules out 16 of the 42 possible boundary flows and greatly reduces the number of possible phase portraits because there can be at most one internal equilibrium. Also the stability of the internal equilibrium, if it exists, is easily determined. Corollary 6.2 lists all possible flows. In particular, the phase portraits of (6.4) are uniquely determined by the boundary flow and by $\delta\delta$. For 12 boundary flows, however, $\delta\delta$ may assume more than one value.

Corollary 6.2 shows also that a stable internal equilibrium cannot exist unless there is at least one internally stable edge equilibrium, i.e., there is overdominance in at least one single-locus boundary system. Analogously, an internal source can occur only if there is underdominance in at least one single-locus boundary system.

As discussed in Sect. 6.1 on the additive model, in the absence of epistasis all trajectories of the full model (2.3) converge to the linkage-equilibrium manifold if the recombination rate is positive. Therefore, the phase portraits derived for (2.5) are representative for the full model. In addition, every possible boundary flow determines the phase portrait uniquely (Corollary 6.3).

Another important special case with linear isoclines is the haploid selection model (6.6). Since there is no dominance, the only possible boundary flows are E00C1sE00C1s, E00C1aE00C1a, and E00C2bE00C2b. As in the additive case, the phase portraits of (6.6) are uniquely determined by the boundary flows: in the first two cases there is no internal equilibrium, in the third case there is a saddle. The weak-selection limit (6.6) of the haploid model captures all possible equilibrium structures and phase portraits of the full haploid model with linkage disequilibrium, at least for continuous time (Bank et al. 2012 and Sect. 6.2).

10.5 Multilinear epistasis

The so-called multilinear model of epistasis was introduced by Hansen and Wagner (2001). It assumes that the effects of gene substitutions due to changes in the genetic background can be described by a linear transformation (Sect. 7). In this model, all types of epistatic interactions can occur. In the two-locus case, these are additive-by-additive, additive-by-dominance, and dominance-by-dominance interactions (cf. Remark 6.1). Interestingly, the multilinearity assumption (7.1) turns out to be equivalent to assuming that the isoclines are in product form; see (7.3). Therefore, they are composed of vertical or horizontal straight lines. In the absence of dominance, the multilinear model and that of Zhivotovsky and Gavrilets (1992) coincide. In this case, both models reduce formally to the haploid model (6.6). Otherwise, they differ. In fact, up to five internal equilibria may occur in the multilinear model. It is also possible to have overdominance on every edge, i.e., in each marginal one-locus system, but no stable internal equilibrium (Table 1, boundary flow $E_4^4C_0b$ with $\delta = 4$). All possible equilibrium structures could be identified (Theorem 7.1). Seven of the 16 possible boundary flows determine the phase portrait uniquely.

10.6 Equivalent loci

In Sect. 8, we briefly treat the equilibrium structures generated by symmetric fitness matrices. This is equivalent to assuming that both loci are equivalent, an assumption made in many investigations. Only the nine boundary flows of Figure 2 ending with an ss or bb can occur. The symmetry properties of this model greatly simplify its analysis, so that for every possible boundary flow the possible values of δδ and the possible number of internal equilibria can be determined (Theorem 8.1). Moran's (1963) example of a flow with three stable internal equilibria is symmetric. Thus, the assumption of symmetry reduces the complexity of the two-locus model only in certain aspects.

10.7 The symmetric viability model

As indicated in the Introduction and in Sect. 9, the symmetric viability model may be one of the best studied dynamical systems in population genetics. Although it depends on only four parameters, its complexity seems to preclude a comprehensive ematical analysis. Even the subclass arising from models of stabilizing selection on a quantitative trait toward an intermediate optimum is well understood only for special cases, such as quadratic or Gaussian stabilizing selection (Bürger 2000, Chap. 6.2; Willensdorfer and Bürger 2003).

The weak-selection limit, Eq. (9.3) in our parameterization (9.1) and in the transformed coordinates, is still sufficiently complex to admit a wide variety of equilibrium structures and phase portraits. Nevertheless, it is simple enough to admit the identification of all possible equilibrium structures (Theorem 9.5 and Table 2). This requires knowledge of the boundary-flow class (by definition, the types ending with bb or cc occur), the boundary index sum, and the sign of two simple compound parameters (ρ=r1r2−m2ρ=r1r2−m2 and r1+r2r1+r2).

Theorem 9.5 shows that the maximum number of internal equilibria of (9.3) is five and the maximum number of stable internal equilibria is two.

This is different in the full two-locus symmetric viability model with linkage disequilibrium. Then the maximum number of internal equilibria is seven (Feldman and Karlin 1970), and four can be simultaneously stable (Hastings 1985). Interestingly, the same maximum numbers of internal and of stable internal equilibria occur in the special case arising from Gaussian stabilizing selection on a quantitative trait (Willensdorfer and Bürger 2003). In the full two-locus model, three of the seven internal equilibria are so called symmetric equilibria whose gamete frequencies satisfy $x_1 = x_4$ and $x_2 = x_3$. In the weak-selection limit, these symmetry conditions collapse to $p = \frac{1}{2}$ and $q = \frac{1}{2}$ or, in the (x, y) coordinates of Sect. 9, to $x = y = 0$. Thus, in the weak-selection limit, the manifold defined by $x_1 = x_4$ and $x_2 = x_3$ collapses to the single point $(x, y) = (0, 0)$. The four unsymmetric equilibria determined by Feldman and Karlin (1970) for the full model correspond to the four unsymmetric internal equilibria determined by Theorem 9.1.

10.8 Multiplicative fitnesses

The reader may have noticed that we did not consider multiplicative fitnesses, although the multiplicative viability model has received great attention in the literature. This model has always been treated in discrete time because then multiplicate fitnesses have an immediate biological meaning. In contrast to the additive viability model, it has the property that the linkage-equilibrium manifold $(D = 0)$ is invariant under the map (2.3). Also in contrast to the additive model, solutions do not necessarily approach linkage equilibrium, and stable equilibria with $D \neq 0$ may exist. By performing the weak-selection limit, multiplicative fitnesses become additive because $(1 + \epsilon a)(1 + \epsilon b) \approx 1 + \epsilon(a + b)$. Thus, the weak-selection limits of the additive and the multiplicative model coincide.

10.9 Inferring stable two-locus polymorphisms

An interesting, old question is whether the maintenance of a stable two-locus polymorphism requires some form of overdominance at the individual loci. Kojima (1959) showed for independent loci, i.e., our model (2.5), that marginal overdominance (3.11) on both loci at the equilibrium point is necessary for the maintenance of a stable two-locus polymorphism. However, this condition is not sufficient (Corollary 3.3).

For the full two-locus model, Hastings (1982) showed that for a small range of recombination rates stable internal equilibria (sinks) may display marginal underdominance. Theorem 5.3 and Figure S1d (panel 5) show that unstable internal equilibria may exist even if both loci display marginal overdominance on the whole state space.

A simpler and more general question is whether inferences about existence and stability of internal equilibria can be drawn from knowledge of the boundary flow. In general, the answer is negative. Without further restrictions on the fitness scheme, (2.2) or (6.1), there is no boundary flow that determines the phase portrait uniquely. The boundary-flow class $E_0^0C_2b$ is the only one for which an internal equilibrium exists always. All other boundary-flow classes can have δ=1δ=1 and all can be realized without an internal equilibrium (Figs. S2a – S2k). It is even possible to maintain an internal sink if no edge equilibrium exists, i.e., if all four marginal one-locus systems display intermediate dominance. Indeed, all three boundary-flow classes E_0^0 admit an internal sink (Figure S2a, panels 2 and 4; Figure S3a, panel 2), but none of our special fitness patterns does so.

In the model with linear isoclines, a stable internal equilibrium can occur only if there is overdominance at least at two edges (boundary flows $E_2^2C_0c$, $E_2^2C_0s$, $E_3^3C_0e$). Overdominance at all four edges ($E_4^4C_0b$) is sufficient for the existence of an internal sink (Corollary 6.2). For multilinear epistasis, which admits not only additive-by-additive but also additive-by-dominance and dominance-by-dominance interactions (though not in their most general form), an internal sink can exist only if there are four edge equilibria. However, for each of the six resulting boundary flows, there may exist either an internal sink or an internal source (Theorem 7.1). For the symmetric viability model, at least two edge equilibria are necessary for the existence of an internal sink. However, for each of the five resulting boundary flows, there may exist either an internal sink or an internal source (Theorem 9.5). Nevertheless, for the models with linear isoclines (i.e., only additive-by-additive epistasis) and with multilinear epistasis, many boundary-flow classes determine the equilibrium structure uniquely (Corollary 6.2 and Theorem 7.1).

Our treatment of special fitness patterns is not exhaustive. For instance, Hastings and Hom (1990) determined the equilibrium structure of a model with stabilizing selection toward an optimum with arbitrary position by assuming absence of linkage disequilibrium. More recently, Feldman and Puniyani (2006) examined an extension of the Lewontin–

Kojima version of the full two-locus symmetric viability model, in which substitutions at locus *A* do not depend on the background alleles at locus *B*, but substitutions at locus *B* depend on the background locus *A*. The boundary flow classes that arise in this framework are $E_2^2C_0e$, $E_1^2C_0e$, $E_1^2C_2e$, $E_4^4C_0b$, $E_3^4C_0e$, and $E_2^4C_0c$.
and E42C0c.E24C0c.

ACKNOWLEDGEMENTS

Open access funding provided by Austrian Science Fund (FWF). Financial support by the Austrian Science Fund (FWF) through the Vienna Graduate School of Population Genetics (Grant DK W1225-B20) to MP and Grant P25188-N25 to RB is gratefully acknowledged. Many thanks to Professor Thomas Nagylaki and an anonymous reviewer for helpful comments.

REFERENCES

1. Akin E (1979) The geometry of population genetics. Lecture notes in bioematics 31. Springer, Berlin
2. Akin E (1982) Cycling in simple genetic systems. J Biol 13:305–324
3. Altenberg L (2010) Proof of the Feldman–Karlin conjecture on the maximum number of equilibria in an evolutionary system. Theor Popul Biol 77:263–269
4. Andronov A, Pontryagin L (1937) Systèmes grossiers. Dokl Akad Nauk SSSR 14:247–251
5. Bank C, Bürger R, Hermisson J (2012) Limits to parapatric speciation: Dobzhansky-Muller incompatibilities in a continent-island model. Genetics 191:845–863
6. Bodmer WF, Felsenstein J (1967) Linkage and selection: theoretical analysis of deterministic two locus random mating model. Genetics 57:237–265
7. Bodmer WF, Parson PA (1962) Linkage and recombination in evolution. Adv Genet 11:1–100
8. Bomze IM (1982) Über einige spieldynamische Modelle der Evolutionstheorie. Dissertation, Universität Wien
9. Bomze IM, Schuster P, Sigmund K (1983) The role of Mendelian genetics in strategic models on animal behaviour. J Theor Biol 101:19–38
10. Bürger R (2000) The ematical theory of selection, recombination, and mutation. Wiley, Chichester
11. Bürger R, Gimelfarb A (1999) Genetic variation maintained in multilocus models of additive quantitative traits under stabilizing selection. Genetics 152:807–820
12. Christiansen FB (1999) Population genetics of multiple loci. Wiley, Chichester
13. Ewens WJ (1968) A genetic model having complex linkage behavior. Theor Appl Genet 38:140–143
14. Ewens WJ (1969) Mean fitness increases when fitnesses are additive. Nature 221:1076

15. Ewens WJ, Thomson G (1977) Properties of equilibria in multilocus genetic systems. Genetics 87:807–819
16. Feldman MW (1971) Equilibrium studies of two locus haploid populations with recombination. Theor Popul Biol 2:299–318
17. Feldman MW, Karlin S (1970) Linkage and selection: two locus symmetric viability model. Theor Popul Biol 1:39–71
18. Feldman MW, Liberman U (1979) On the number of stable equilibria and the simultaneous stability of fixation and polymorphism in two-locus models. Genetics 92:1355–1360
19. Feldman MW, Puniyani A (2006) A semi-symmetric two-locus model. Theor Popul Biol 69:211–215
20. Felsenstein J (1965) The effect of linkage on directional selection. Genetics 52:349–363
21. Fisher RA (1918) The correlation between relatives on the supposition of Mendelian inheritance. Trans R Soc Edinb 52:399–433
22. Fisher RA (1930) The genetical theory of natural selection. Clarendon Press, Oxford
23. Gavrilets S (1993) Equilibria in an epistatic viability model under arbitrary strength of selection. J Biol 31:397–410
24. Gavrilets S, Hastings A (1993) Maintenance of genetic variability under strong stabilizing selection: a two-locus model. Genetics 134:377–386
25. Haldane JBS (1931) A ematical theory of natural selection. Part VIII. Metastable populations. Proc Camb Philos Soc 27:137–142
26. Hansen T, Wagner G (2001) Modeling genetic architecture: a multilinear theory of gene interactions. Theor Popul Biol 59:61–86
27. Hastings A (1981a) Simultaneous stability of D=0D=0 and D≠0D≠0 for multiplicative viabilities at two loci: an analytical study. J Theor Biol 89:69–81
28. Hastings A (1981b) Stable cycling in discrete-time genetic models. Proc Natl Acad Sci USA 78:7224–7225
29. Hastings A (1982) Unexpected behavior in two locus genetic systems: an analysis of marginal underdominance at a stable equilibrium. Genetics 102:129–138
30. Hastings A (1985) Four simultaneously stable polymorphic equilibria in two-locus two-allele models. Genetics 109:255–261
31. Hastings A (1987) Monotonic change of the mean phenotype in two-locus models. Genetics 117:583–585
32. Hastings A, Hom C (1990) Multiple equilibria and maintenance of additive genetic variance in a model of pleiotropy. Evolution 44:1153–1163
33. Hermisson J, Hansen T, Wagner G (2003) Epistasis in polygenic traits and the evolution of genetic architecture under stabilizing selection. Am Nat 161:708–734
34. Hofbauer J (1990) An index theorem for dissipative semiflows. Rocky Mt J 20:1017–1031
35. Hofbauer J, Iooss G (1984) A Hopf bifurcation theorem for difference equations approximating a differential equation. Monatsh 98:99–113
36. Hofbauer J, Sigmund K (1998) Evolutionary games and population dynamics. Cambridge University Press, Cambridge
37. Karlin S (1980) The number of stable equilibria for the classical one-locus multiallele selection model. J Biol 9:189–192
38. Karlin S, Feldman MW (1978) Simultaneous stability of D=0D=0 and D ≠≠ 0 for multiplicative viabilities at two loci. Genetics 90:813–825
39. Karlin S, Liberman U (1978) The two-locus multi-allele additive viability model. J Biol 5:201–211
40. Karlin S, Liberman U (1990) Global convergence properties in multilocus viability selection models: the additive model and the Hardy-Weinberg law. J Biol 29:161–176
41. Kempthorne O (1955) The theoretical values of correlations between relatives in random mating populations. Genetics 40:153–167

42. Kimura M (1956) A model of a genetic system which leads to closer linkage by natural selection. Evolution 10:278–287
43. Kojima K (1959) Role of epistasis and overdominance in stability of equilibria with selection. Proc Natl Acad Sci USA 45:984–989
44. Lewontin RC, Kojima K (1960) The evolutionary dynamics of complex polymorphisms. Evolution 4:458–472
45. Lyubich Y (1992) ematical structures in population genetics. Springer, Berlin
46. Moran PAP (1963) Balanced polymorphisms with unlinked loci. Aust J Biol Sci 16:1–5
47. Moran PAP (1964) On the nonexistence of adaptive topographies. Ann Hum Genet 27:383–393
48. Moran PAP (1968) On the theory of selection dependent on two loci. Ann Hum Genet 32:183–190
49. Nagylaki T (1977) The evolution of one-and two-locus systems. II. Genetics 85:347–354
50. Nagylaki T (1989) The maintenance of genetic variability in two-locus models of stabilizing selection. Genetics 122:235–248
51. Nagylaki T (1993) The evolution of multilocus systems under weak selection. Genetics 134:627–647
52. Nagylaki T, Hofbauer J, Brunovsky P (1999) Convergence of multilocus systems under weak epistasis or weak selection. J Biol 38:103–133
53. Rutschman D (1994) Dynamics of the two-locus haploid model. Theor Popul Biol 45:167–176
54. Schuster P, Sigmund K, Hofbauer J, Gottlieb R, Merz P (1981) Selfregulation of behaviour in animal societies; 3. Games between two populations with selfinteraction. Biol Cybern 40:17–25
55. Schuster P, Sigmund K (1981) Coyness, philandering and stable strategies. Anim Behav 29:186–192
56. Willensdorfer M, Bürger R (2003) The two-locus model of Gaussian stabilizing selection. Theor Popul Biol 64:101–117
57. Wright S (1935) Evolution in populations in approximate equilibrium. J Genet 30:257–266
58. Wright S (1942) Statistical genetics and evolution. Bull Am Soc 48:223–246
59. Wright S (1952) The genetics of quantitative variability. In: Reeve ECR, Waddington CH (eds) Quantitative inheritance. Her Majesty's Stationary Office, London, pp 5–41
60. Zhivotovsky LA, Gavrilets S (1992) Quantitative variability and multilocus polymorphism under epistatic selection. Theor Popul Biol 42:254–283